Future-Oriented Technologies and Concepts to Increase Water Availability by Water Reuse and Desalination

In Focus – Special Book Series

Future-Oriented Technologies and Concepts to Increase Water Availability by Water Reuse and Desalination

Editor

Jörg E. Drewes

IWA PUBLISHING

Published by **IWA Publishing**
Republic – Export Building, 1st Floor
2 Clove Crescent
London E14 2BE, UK
Telephone: +44 (0)20 7654 5500
Fax: +44 (0)20 7654 5555
Email: publications@iwap.co.uk
Web: www.iwapublishing.com

First published 2021

British Library Cataloguing in Publication Data
A CIP catalogue record for this book is available from the British Library

ISBN: 9781789062496 (paperback)

Contents

Editorial: Journal relaunching as *Water Reuse*

Journal of Water Reuse and Desalination was launched in 2011 and has made great progress. It became an open access journal in 2016 as part of its mission to provide leading water knowledge to experts and communities across the globe. As an influential journal of IWA Publishing, the reputation of *Journal of Water Reuse and Desalination* has increased rapidly in recent years. This impressive growth is thanks to the valuable research, hard work and dedication of all authors, reviewers and current editors and editorial board members. At this moment, it is our great pleasure to relaunch the journal for better development and progress in the future.

Currently, the shortage of clean water and the breakthrough of treatment technologies represent challenges and opportunities for the development of water reuse. The focus on water reuse is growing rapidly worldwide, with interest expanding beyond just recycling of municipal wastewater to also include recycling of industrial and mining wastewaters for uses such as agriculture and drinking waters. Concerns and challenges for water reuse include operational, administrative, regulatory, social, and economic, and are driven by restricted availability of local water sources with which water reuse systems must be integrated. Progress is needed to address new challenges to ensure the safe and efficient use of non-conventional water resources. Indeed a broad interpretation of water reuse using non-conventional water supplies also includes desalination technologies and systems, where seawater and brackish water are considered non-conventional waters.

Water reuse systems, defined as those using non-conventional water supplies, are thus confronted with scientific and technological challenges that are quite different from those in conventional water supply and wastewater treatment systems.

With this broad concept of water reuse including all non-conventional water supplies, and with the associated significance of and great opportunities for the topic, *Journal of Water Reuse and Desalination* is re-launching with the name *Water Reuse*. This is a milestone in advancing knowledge and practices for water reuse. With the joint effort and contribution from various water academics, engineers, practitioners and all members of the water community, *Water Reuse* will grow into a great platform for presenting and exchanging ideas, works and experiences. Your interest in the journal and submissions are highly welcomed. Any suggestions or comments on how to improve our activity to deliver a better journal are very much appreciated.

Editors in Chief
Hu Hong-Ying IWA
Graduate School, School of Environment, Tsinghua University, Beijing, 100084 China

Stephen Gray IWA
Research Institutes and Centres, VU Research, Victoria University, Australia

doi: 10.2166/wrd.2020.002

Editorial: Future-oriented technologies and concepts to increase water availability by water reuse and desalination

The pressure on available freshwater resources due to urbanization and climate change impacts is increasing worldwide; so follows the need to explore unconventional resources. Water reuse and desalination are viable options for alternative water supplies, also for regions that had no need to consider these approaches in the past. However, these solutions need to be tailored to local conditions, be balanced against conventional supply options while minimizing any adverse health and environmental impacts. There is increasing interest in developing new planning tools, technological solutions and approaches in this field to foster safe and economically viable water reuse and desalination applications.

The German Federal Ministry of Education and Research (BMBF) realized these needs at an early stage and launched the major funding initiative 'WavE' (www.bmbf-wave.de/en) in 2016 entitled *'Future-oriented Technologies and Concepts to Increase Water Availability by Water Reuse and Desalination'* within the framework program 'Research for Sustainable Development (FONA)' and the research agenda on Green Economy. WavE is aimed at developing innovative technologies and management concepts to foster a sustainable increase in water availability. Three topic areas were addressed in the funding measure to provide sustainable solutions for national and international applications: (I) reuse of municipal wastewater, (II) reuse of water in industry, (III) use of saline ground- and surface water. The aim is to provide sustainable solutions for national and international applications.

In total, 13 joint research projects were funded involving more than 80 partners from academia, engineering practice, and state and federal authorities evaluating hands-on projects in Germany, Vietnam and Namibia. This special issue of *Journal of Water Reuse and Desalination* reports fully peer-reviewed research highlights from many of these comprehensive research projects addressing contemporary approaches to evaluate new technologies and concepts in non-potable water reuse applications, multi-barrier concepts for potable and non-potable reuse applications, concepts for industrial reuse and resource recovery approaches for brine streams, as well as onsite desalination technologies for developing communities.

Enjoy reading these important contributions to the field of water reuse and desalination!

Guest Editor
Jörg E. Drewes IWA
Technical University of Munich,
Munich, Germany

SPONSORED BY THE

Federal Ministry of Education and Research

doi: 10.2166/wrd.2020.000

Approaches for the evaluation of future-oriented technologies and concepts in the field of water reuse and desalination

Kristina Wencki IWA, Verena Thöne, Angela Ante, Tobias Hogen, Claudia Hohmann, Felix Tettenborn, Dagmar Pohl, Philipp Preiss and Christina Jungfer

ABSTRACT

Against the background of drastically rising global water demand and increasing pollution and overexploitation of regional water resources, the demand-driven water supply of households and industry is of central importance. Water reuse and desalination are seen as key technologies to overcome potential regional and local water shortage. In the joint projects funded by the German Federal Ministry of Education and Research (BMBF) 'Future-oriented Technologies and Concepts to Increase Water Availability by Water Reuse and Desalination (WavE)', evaluation approaches for analysing innovative technologies and concepts are being developed and assessed. All evaluation methods and criteria used were selected based on the decision situation at hand and the decision-makers preferences. Based on the analysis of six multi-criteria evaluation concepts used in selected WavE projects, this paper presents a general approach for comparative multi-criteria evaluation of water reuse systems consisting of prerequisites, minimum requirements, evaluation criteria (qualitative, semi-quantitative or quantitative) and a final aggregation of results. Exemplary sets of criteria for the application in a more industrial, municipal and/or international context are presented as an aid for the application of holistic evaluation approaches for (process) concept and technology selection in the context of water reuse and desalination.

Key words | assessment framework, desalination, multi-criteria decision analysis, water management, water reuse

HIGHLIGHTS

- The paper provides guidance on the application of different evaluation approaches for the selection of advanced treatment technologies for water reuse and desalination.
- It is based on a review of assessment methods used in six different projects funded by the German Federal Ministry of Education and Research and provides insights into challenges that were faced in the assessment processes of the individual projects.
- It presents a general approach for comparative multi-criteria evaluation of water reuse systems consisting of prerequisites, minimum requirements, evaluation criteria (qualitative, semi-quantitative or quantitative) and a final aggregation of results.

Kristina Wencki IWA (corresponding author)
Verena Thöne
IWW Rheinisch-Westfälisches Institut für Wasserforschung gemeinnützige GmbH,
Moritzstr. 26, 45476 Mülheim an der Ruhr,
Germany
E-mail: *k.wencki@iww-online.de*

Angela Ante
SMS Group,
Wiesenstraße 30, 57271 Hilchenbach,
Germany

Tobias Hogen
Environmental Process Engineering,
TU Berlin, Institute of Environmental Technology,
Straße des 17. Juni 135, 10623 Berlin,
Germany

Claudia Hohmann
Felix Tettenborn
Competence Center Sustainability and Infrastructure Systems,
Fraunhofer ISI,
Breslauer Straße 48, 76139 Karlsruhe,
Germany

Dagmar Pohl
Leibniz University Hannover, Institute of Sanitary Engineering and Waste Management,
Welfengarten 1, 30167 Hannover,
Germany

Philipp Preiss
Pforzheim University,
Tiefenbronner Straße 65, 75175 Pforzheim,
Germany

Christina Jungfer
DECHEMA e.V.,
Theodor-Heuss-Allee 25, 60486 Frankfurt am Main,
Germany

doi: 10.2166/wrd.2020.022

- Exemplary sets of criteria for the application in a more industrial, municipal and/or international context are presented as an aid for the application of holistic evaluation approaches for concept and technology selection in the context of water reuse and desalination.
- Finally, possible data sources, challenges in using them and indicators to measure data quality are presented to give guidance on data collection.

INTRODUCTION

The need for targeted and safe water reuse solutions is increasing worldwide. However, the implementation of water reuse solutions is still very limited compared to their potential due to a number of factors, e.g. low economic attractiveness and lack of public acceptance of reuse solutions, low awareness of technological advantages and poor coordination of actors from industry, authorities and the water companies. Decisions on the implementation of innovative technologies and concepts for water reuse and desalination are regularly marked by conflicting goals between economic, technical, environmental and socio-political considerations. In order to inspire decision-makers to use new technologies and to convince sceptics of the viability of innovative approaches, transparent, sound decision criteria are required.

VALUATION ISSUES IN WAVE

In the joint projects of the funding measure 'Future-oriented Technologies and Concepts to Increase Water Availability by Water Reuse and Desalination (WavE)' by the German Federal Ministry of Education and Research (BMBF), suitable evaluation approaches were developed for different user groups and local target settings, taking into account the respective decision-making situation. The internationally established applications range from the treatment of saline groundwater to the reuse of municipal and industrial wastewater for various purposes in the municipal, agricultural and industrial sectors (Table 1). Depending on the application case and object of evaluation, different evaluation methods are applied, to assist operators, customers and plant constructors in various phases of the planning process in making decisions and to provide target group-oriented results.

The evaluation in the projects WaReIp and MULTI-ReUse addresses the early conceptual design of process chains for water reuse in industry or agriculture. The focus of the project DiWaL is on the development of optimised water management concepts, including the application of Pulsed Electric Field (PEF) treatment for efficient reduction of microorganisms in process water and paints in pre-treatment and dip coating plants for automotive series painting. In the project WaKap, innovative desalination processes and raw water treatment processes for drinking water supply in areas with a poorer infrastructure are evaluated and compared to alternative water supplies. In the project WEISS, innovative treatment concepts for cooling water recycling in the steel industry are assessed.

The time horizon of the evaluation as well as the system boundaries are chosen differently in the individual projects: In the DiWaL and HighCon project, the entire life cycle of a plant is considered as in the framework of Life Cycle Assessment (LCA). In contrast, in MULTI-ReUse and WaKap, the focus is primarily on the operating phase of the systems. In WaReIp, plant construction and operation are considered.

Nevertheless, the individual evaluation procedures have some aspects in common, which result from the value system of the decision-makers and/or the respective object of evaluation. Context-specific factors, such as the region of application, can also lead to overlapping procedures.

The aim of this paper is to assist decision-makers in identifying the best possible source of water supply by providing a generic approach for comparative process assessment (section Valuation issues in wave) and providing a structured overview of assessment criteria with high relevance to the topic of water reuse and desalination (section Evaluation framework). Based on the generic flow chart and the list of criteria, which were derived from six

Table 1 | Overview of the applied evaluation approaches in selected WavE projects

Project	Subject of evaluation 'What is evaluated?'	Valuation method 'How does the evaluation work?'	User 'Who evaluates for whom?'
DiWaL	Pulsed Electric Field (PEF) treatment and other bath treatment (dip-paint coating in serial automotive painting)	Life cycle assessment/life cycle impact assessment (LCA/LCIA), total cost of ownership, stakeholder analysis	Technology developers, plant manufacturers and plant operators
HighCon	Process alternatives for the treatment of concentrates from water reuse	Material balances, energy balances, LCA	(Inter)national industrial companies
MULTI ReUse	Modular treatment plants for water reuse	Multi-criteria benefit analysis	(Inter)national industrial companies, plant operators, technology manufacturers, water and wastewater associations, water/health/environmental authorities, agricultural associations
WaKap	Modular concept for sustainable water treatment using capacitive deionisation	Multi-criteria decision analysis (MCDA) in the form of a utility analysis	Plant operators, technology manufacturers, potential funding agencies
WaReIp	Modular process chains for wastewater treatment and treatment for reuse	Parallel application and comparison of extended cost–benefit analysis, LCA and multi-criteria assessment	Industrial park and plant planners and operators
WEISS	Single and combined desalination processes (using the example of the steel industry)	Multi-criteria decision analysis with weighting based on expert interviews and LCA	Plant operators, technology manufacturers, (inter)national industrial companies, industry and plant planners

multi-criteria evaluation concepts used in selected WavE projects, it is possible to compile an evaluation methodology together with a set of criteria, tailored to a specific subject of evaluation and the decision-maker(s). If necessary, further application- or country-specific criteria can be added to the list. In section Evaluation criteria, Multi-criteria decision analysis (MCDA), cost–benefit analysis (CBA) and LCA are presented as exemplary evaluation methods for identifying preferred solutions. Finally, possible data sources and methods for addressing uncertainty are presented and discussed with regard to their applicability (section Evaluation methods and selection of preferred solutions).

EVALUATION FRAMEWORK

Despite project-specific differences, a common generic structure was identified, which applies to all six WavE projects involved. Three levels of assessment can be distinguished: prerequisites, minimum requirements and evaluation criteria (Figure 1).

Examining that prerequisites (here $n = 4$), such as fulfilling legal boundary conditions for the intended purpose of use, local availability of the raw water flows required for reuse, general openness of the decision-makers towards the solutions and a ascertained demand for water reuse, are fulfilled, represents the first step of this process. These four prerequisites lie outside the sphere of influence of the project. Failure to meet them can significantly impede the implementation and use of the investigated technologies or systems and thus leads to the exclusion of the alternative in question from the decision-making process.

Furthermore, before the actual evaluation begins, the general suitability of the alternatives is examined on the basis of various minimum requirements (here: $n = 3$). In contrast to the prerequisites, the test criteria, which are

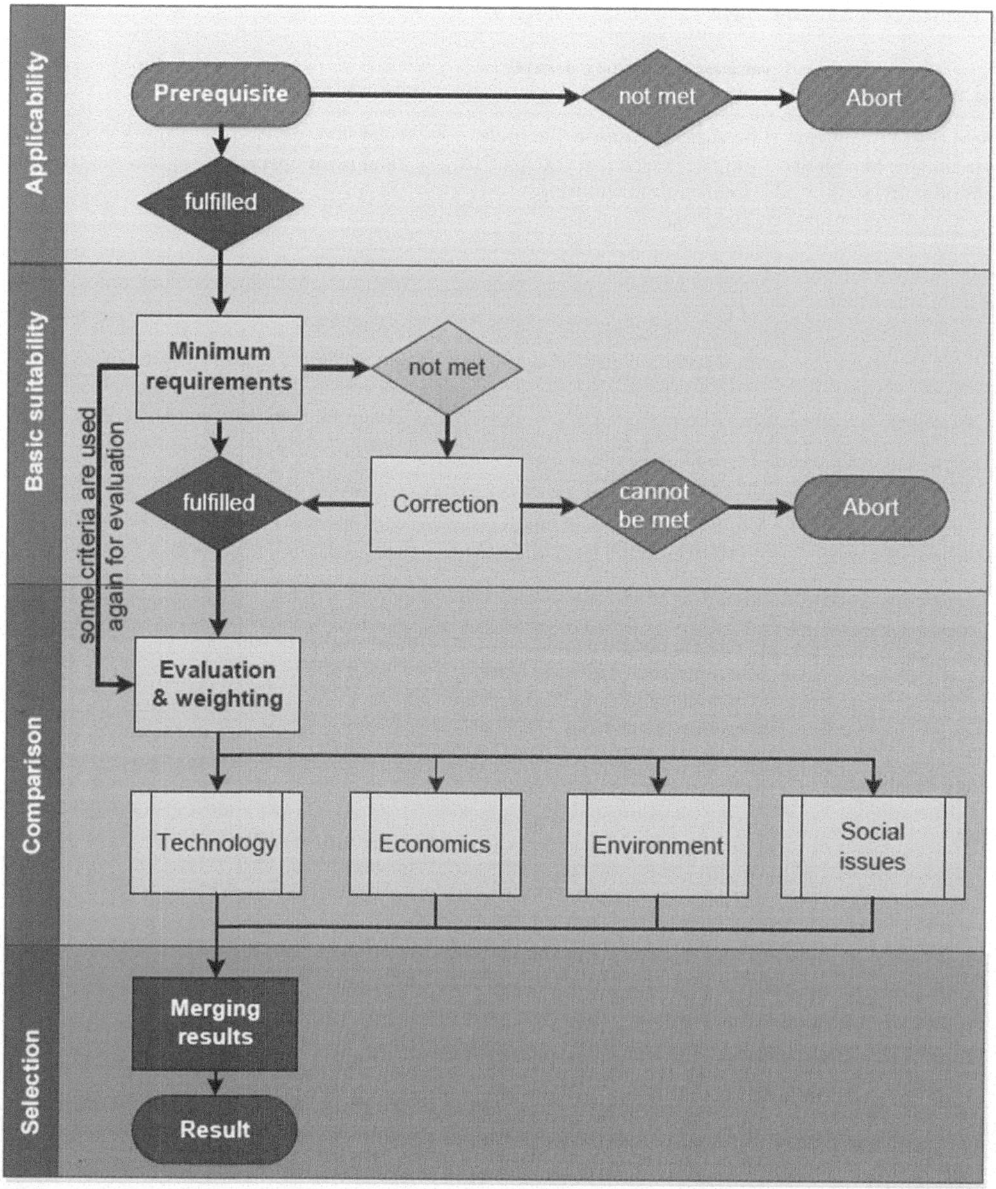

Figure 1 | Procedure for the comparative evaluation of processes, technologies or system implementations.

used to check if the minimum requirements are met, can be improved by adapting the evaluated process or system until it is generally suitable for the purpose at hand. Based on the test criteria, a technical process selection and refinement can be made. Unfulfilled minimum requirements indicate the need for action or optimisation potential of the respective solutions. Alternatives for which the minimum requirements cannot be met even after revisions are discarded as unsuitable. The minimum requirements essentially refer to the conformity of the intended technical process with operational requirements, its integrability into existing infrastructure or building and plant technology, as well as the feasibility of synchronising supply and demand of the treated water. Selected test criteria can also be included in the comparative assessment in the form of evaluation criteria: the degree to which they are (over)achieved represents the assessment basis for the characteristic concerned. An example of this is the integrability, which needs to be given in principle (test criterion) but should also be evaluated qualitatively on the basis of the effort involved (evaluation criterion).

The remaining alternatives can then be subjected to a comparative evaluation for decision-making, which can be based on a quantitative evaluation (e.g. using material or energy balances, Environmental Impact Assessment, life cycle costing) or a qualitative evaluation of individual evaluation aspects (e.g. using stakeholder analyses) or can combine elements of both evaluation methods in the form of a holistic, multi-criteria decision support approach (see section Evaluation criteria).

If the examination of some test criteria is more complex than determining the evaluation criteria, a screening for promising alternatives and/or rejection of undesirable ones can be carried out first based on the evaluation criteria, before compliance with the (remaining) test criteria is ensured.

The aim of the evaluation is to uncover strengths and weaknesses of the compared alternatives and to identify the system solution that is most suitable for a specific site. Often, however, there is no clear best solution, so that a suitable compromise has to be found. It should be noted that the evaluation criteria usually have different units or scales and are of varying importance, which makes comparison difficult.

EVALUATION CRITERIA

In order to evaluate technical alternatives within the six WavE projects under consideration, qualitative, semi-quantitative and quantitative criteria from the fields environment, social issues and economy are combined in accordance with the triple bottom line (TBL) model of sustainable development and the Sustainable Development Goals adopted by the UN. Against the background of the particularities of water supply technology (van Leeuwen & Marques 2012), the assessment dimension technology is included as a fourth evaluation dimension.

The number of evaluation criteria asked for in each project varies between 20 and 30. In total, around 50 different criteria were compiled from the participating projects covering the following aspects.

Interactions between water treatment plant and environment are recorded on the basis of resource and land use (including land use, energy and water requirements, chemical consumption), emissions into soil, water and air, and the contribution to maintaining or increasing ecosystem services (e.g. volume of additionally provided water).

The social acceptance and compatibility of a water supply alternative can be described, among other things, based on the transparency and acceptance of the decision-making process (qualitative), the acceptance of water reuse and the products manufactured (qualitative), the creation of new jobs, political legitimacy, the potential for raising awareness for sustainable resource use, nuisances caused by the operation of the system (noise, aesthetics), conflict potential due to competition over resources and land, user friendliness and affordability.

In the economic dimension, the economic profitability, local economic development potential, economic risk potential, as well as market competition, are characterised by the following criteria: possibilities of generating profits, operating and maintenance costs, specific total costs, costs for investment and commissioning, contribution to local value creation, competitive advantages through improved market positioning, technology transfer, potential for transfer to other markets, technological readiness level (TRL), risk management, as well as damage costs caused by plant failure.

In the fourth assessment dimension of technology, the TRL, minimum level of training required to operate the plant, robustness, efficiency, flexibility and level of automation of the plant, operating and maintenance costs, achievable yield, technical integrability in existing infrastructures and processes, degree of dependence on other infrastructure systems, mean time to failure, fault tolerance and occupational health and safety during operation of the plant are assessed. Based on these criteria, technical feasibility, as well as process safety, can be accounted for in the evaluation.

In general, all criteria can be divided into universal and specific criteria. Examples of universal criteria, which are used in almost all WavE projects considered, are operating costs, emissions, operating and maintenance expense and political legitimacy. In contrast, specific criteria are used in only a few projects, such as the potential for opening up

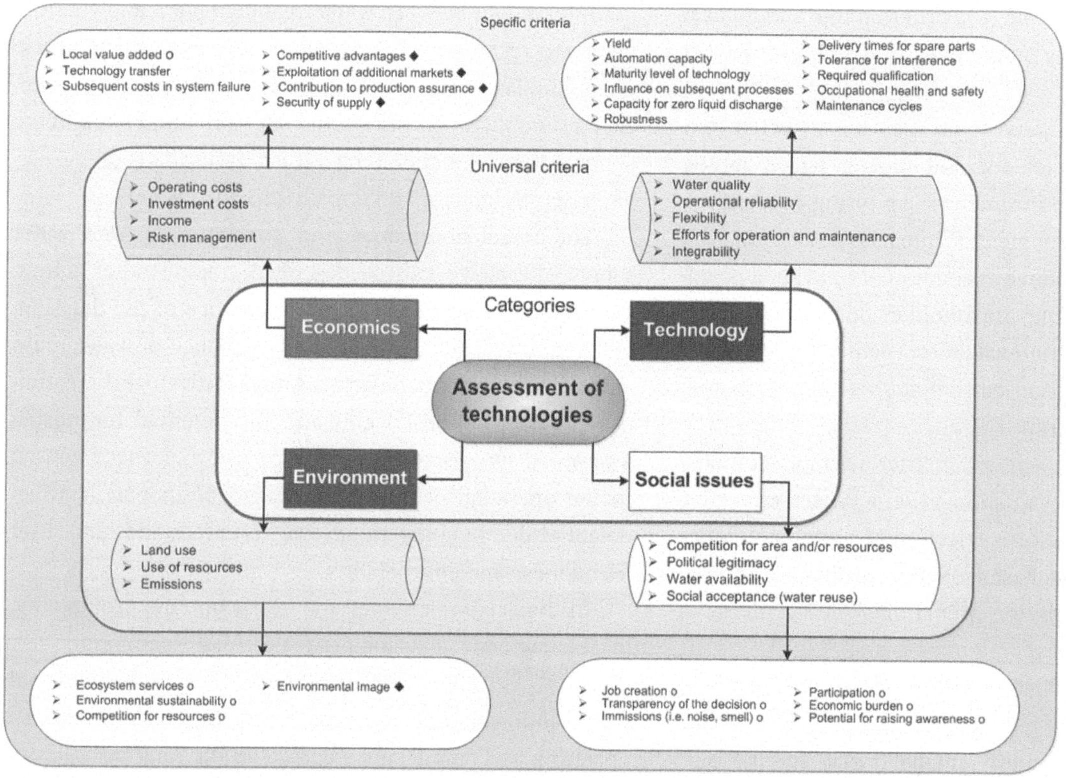

Figure 2 | Overview of evaluation criteria: relevant criteria mainly for municipal water supply: o; mainly for industrial projects: ◆.

additional markets, the potential for automation and the potential for the creation of ecosystem services or the creation of jobs. An overview of the different criteria is given in Figure 2.

Criteria that are important primarily from an economic/business perspective or from the perspective of public services of general interest are indicated by corresponding symbols.

Adopting the different criteria presented in Figure 2, it should be noted that from different perspectives, some criteria may have partly opposite implications for the evaluation. One example is the criterion 'creation of new jobs', which is seen positively from a social perspective but is viewed rather negatively from an economic perspective due to the associated personnel costs. In addition, the allocation of criteria is not always clear-cut, as criteria can have different implications in the context of different evaluation categories. This will be illustrated using the example of the criterion resource and space requirements, which can be reflected in the three categories 'environment', 'social' and 'economy':

- The use of natural resources and land is increasing constantly worldwide. It is associated with emissions and other environmental impacts throughout the life cycle of technical equipment and products. At the same time, the existing competition for land, for example, means that biological diversity, the uniqueness of landscapes, natural or archaeological monuments, fertile soils and an appropriate proportion of mineral resources must be protected (German Environment Agency 2018). Due to the limited availability of resources and land, competition situations arise in which the various types of use need to be weighed up within the 'Environment' category.
- In the context of dwindling resources and land, questions of fair distribution arise in order to avoid social inequalities and, as a result, social upheavals. The evaluation of these impacts of a project is recorded in the category 'Social'.
- From a business point of view, the acquisition of resources and land is directly linked to costs. Scarcer resources and land can be expressed in the form of

fluctuating or rising prices. Accordingly, this aspect needs to be considered in the category 'Economy'.

Depending on the decision-makers' perspective, different criteria can be used for evaluation. Two distinct positions within the WavE projects under consideration are the entrepreneurial/business perspective, focused particularly on business and marketing aspects, and the public service perspective, in which social and environmental aspects are more important. In addition, depending on the decision context, country-specific, application-specific and/or user-specific criteria may be added.

Perspective and evaluation on the part of public services of general interest

Technical infrastructure systems in urban areas ensure the supply of water, energy, heat, the disposal of wastewater and waste and the provision of traffic and green spaces. In many countries, the provision of technical infrastructure is organised as part of public services or regulated by the state (Dominguez *et al.* 2009). Water supply is a grid-bound investment-intensive infrastructure system, which ties up considerable amounts of capital. However, it has a low 'return on investment' and is accordingly described as a natural monopoly in the micro economy. Despite the great importance of public actors in providing and financing these infrastructure services, the private sector often plays an important role as well (OECD 2007). Since water resources are public goods, the complex implications of collective action should be taken into account: even if every actor benefits when action is taken, it is not worthwhile for any of the actors to invest in a solution on their own (cf. 'Prisoner's Dilemma'). Against the background of climate change, the issue of 'sustainability' for infrastructure systems is increasingly coming into focus. The public administration has a supporting, regulating, but also monitoring role in the implementation of these criteria (OECD 2007). Taking these framework conditions into account, the WavE collaborative projects have included evaluation criteria in their assessment tools that take into account important aspects of water supply in the context of public services of general interest.

The goal of the WaKap project, for example, is the development and piloting of an innovative, energy-efficient, modular combination process for the desalination of seawater and brackish or groundwater treatment. The background of the project is that the water supply in Vietnam and other Southeast Asian countries is increasingly facing challenges, such as the influence of climate change, the strong regional population growth and additional water demand due to economic development. For a permanent, sustainable use, safety-relevant and social aspects play an outstanding role in this project, in addition to ecological, technical and economic aspects.

- Transparency and acceptance in the decision-making process: The decision-making process should be communicated to the public in a transparent and comprehensible manner and should be carried out with the involvement of the relevant stakeholders. The selection of the actors considered relevant plays an important role in this process.
- Participation and co-determination in the decision-making process: In a further step, it should be examined to what extent the decision-making process is carried out and completed with the involvement of the relevant actors and to what extent equal rights for all social groups are taken into account (see here, for example, Schmitter (2000) and Grote & Gbikpi (2002)).
- Competition for resources and land: The use of natural resources and land has a variety of social impacts in addition to the consequences for the environment. Competition for land and resources with the existing environment should therefore be carefully examined, e.g. with regard to natural areas, areas for local recreation or areas for alternative uses. In this context, attention should be paid to the quality and quantity of resources and land required for a planned facility. In terms of land quality, for example, this means that natural areas with a high availability of biodiversity must be protected in particular.

Corporate perspective

The business perspective is typically focused mainly on economic and technical aspects. Apart from costs and revenues, the stable and failure-free operation of processes is often of paramount importance. This is especially true for the supply of utilities, such as water, which do not directly generate any revenue but are required for most production processes. A reliable water supply, both in terms of quantity and quality, is crucial to ensure undisturbed production and thus prevent substantial financial losses due to production downtimes and impaired product quality. Depending on the available water sources, water reuse can offer advantages but may also pose a risk in this regard. As the cost of water supply and wastewater treatment often accounts for only a fraction of the production costs and revenues, the indirect (financial) implications of water supply alternatives should be considered in the assessment.

Other criteria, e.g. the potential to improve the corporate image or to open up new markets can increase the attractiveness of a technical solution and justify possible additional expenditure. These additional aspects are usually closely related to the strategic goals of the company and may thus differ substantially in different cases. Generally, ecological aspects have gained importance in this regard in recent years. More eco-friendly technologies and processes may not only promote a green image but also facilitate compliance with future legal standards and the cooperation with local authorities. Social aspects, however, usually play a subordinate role, which is also reflected in the business-oriented assessments of the WavE projects WEISS and WaRelp. However, even though they are not as apparent in the evaluation in a corporate setting, ecological and social aspects are reflected in (external) requirements (prerequisites), such as laws and regulations that must be complied with.

Two different situations that companies may face in terms of technology selection are represented in the involved WavE projects:

- Companies that provide water reuse and desalination technologies need to decide which technologies to include in their portfolio and/or develop. This is considered in the projects WEISS and DiWaL.
- Companies that want to implement water reuse and desalination on their own site need to decide on a suitable technical solution – as in the project WaRelp.

The selected criteria and system boundaries can differ between these two situations. In the first case, the evaluation is focused mainly on the concerned technologies and assesses associated market opportunities. As this involves how well the technology satisfies the needs and preferences of potential clients, many criteria are relevant to both situations. In the second case, the question might not just be which technology to choose, but if and how to implement water reuse at all. This requires a broader view on existing water supply options (and wastewater treatment and discharge options) for the industrial site to properly compare alternatives. New market possibilities (e.g. through expansion of the production site) and competitive advantages can also be assessed, but in this case, they usually derive rather indirectly from the employed water treatment technologies.

EVALUATION METHODS AND SELECTION OF PREFERRED SOLUTIONS

In order to select a preferred solution on the basis of the various evaluation criteria, advantages and disadvantages of different alternatives have to be compared and weighed against each other. The solutions to be evaluated usually have different strengths and weaknesses, so it is not obvious which one is the best. Different scales and units of the evaluation criteria can make the comparison more difficult. In addition, resource and water management usually involves several stakeholders and decision-makers, who may have different preferences (Hajkowicz & Higgins 2008). Consequently, compromises have to be found between conflicting objectives that are advocated by different stakeholders.

There is a variety of methods that can support structured and transparent decision-making. They differ essentially in the selection of the evaluation aspects taken into account, the procedure for comparing and weighting the evaluation criteria and the determination of the best solution(s). Table 2 provides an overview of the advantages and limitations of the assessment methods used within the six projects that build the basis for this paper.

Table 2 | Experienced advantages and disadvantages of CBA, LCA and MCDA within the six WavE projects studies

Valuation method used	Experienced advantages	Experienced disadvantages
(Extended) cost–benefit analysis	• Well-suited evaluation approach to reflect economic and, in part, technical evaluation criteria • Monetary evaluation fits well into traditional, economically influenced decision-making processes and evaluation concepts	• Monetisation is ambiguous and can be a sensitive and controversial issue, especially for social aspects • Implicit or explicit weighting of different effects
Life cycle (impact) assessment	• Rigorous and comprehensive assessment of the environmental and health aspects of different alternatives using a transparent, standardised approach • Analysis can help to generate feedback loops in the early innovation phase	• Existing norms and standards do not provide support on how to identify preferred solution
Multi-criteria decision analysis	• Open method that can be adapted to specific problems • Various well-known methodological approaches allow for a use case-specific selection • Possibility of case-specific weighting of objectives and selection of evaluation criteria with different scales and units according to decision situation and decision-makers preferences • Transparent and understandable assessment process requiring little or no experience in its application	• No clear guidelines for use case-specific method selection • Detailed criteria assessment requires complementary usage of other assessment methods and tools • Objective weighting of individual criteria can only be ensured by strong stakeholder engagement • Results are mostly sensitive to users' preferences

Multi-criteria decision analysis

The aim of multi-criteria evaluation is to identify the best solution or a manageable number of equally good alternatives. A complete ranking of all alternatives is not necessarily required for this, but may be the result, depending on the evaluation method used (Figueira *et al.* 2005). Common to all multi-criteria evaluation methods is the general procedure of (i) specifying the objective and the alternatives to be considered, (ii) defining the criteria, (iii) measuring the relative importance of the criteria and (iv) aggregating the judgements (Belton & Stewart 2002). A detailed analysis of the different methods, their foundations, as well as strengths and weaknesses, can be found, for example, in Belton & Stewart (2002).

MCDA is used, for example, in the projects MULTI-ReUse and WaKap in the form of a utility value analysis. The sustainability assessment tool developed within MULTI-ReUse supports stakeholders and decision-makers in evaluating different water recycling solutions against the current system configuration in order to identify the most sustainable water supply system for the future. In this project, utility value analysis has been selected as favourable multi-criteria assessment methodology because it is comprehensive enough to deal with the diverging objectives attached to such decision cases and at the same time offers the highest level of flexibility to be applicable in various contexts. Including experts from different disciplines in the compilation and review process of criteria list, served to ensure a common understanding of terminology and consideration of contradicting stakeholders' viewpoints of water reuse implementation. The application of the decision support tool for the sustainability assessment of the two case studies provided its developers with key insights about advantages as well as limitations that were necessary to transform it into a practice-oriented tool.

Within MCDA, the weighting of the individual criteria is usually done in dialogue with the decision-makers, whereby disparate preferences of different actors can also be taken into account. It is not uncommon that persons – even in similar positions – express different opinions when asked about their preferences. In the WEISS project, for example, the importance of the category 'technology' was assessed very differently by different experts. While most experts rated it just behind cost, once it was classified as completely unimportant. With regard to environmental compatibility, the spectrum even ranged from almost unimportant to the most important criterion. In addition to the diverging preferences of the respondents, it cannot generally be ruled out that a different understanding or different ideas regarding the aspects to be classified may have contributed to the different results. Determining the evaluation criteria and their relative importance is a complex task that requires a great deal of sensitivity in the dialogue with the decision-makers. Moreover, the importance of the weights depends on the method chosen (Belton & Stewart 2002). Therefore, the influence of the choice of different weights on the resulting ranking needs to be investigated and discussed with the decision-makers (Figueira *et al.* 2005; Pajer *et al.* 2017) (see also section Evaluation methods and selection of preferred solutions).

Overall, the multi-criteria evaluation is a very open method that can be adapted to specific problems. Any evaluation criteria can be taken into account, even with different scales and units. The procedure aims to reflect as accurately as possible the preferences and value systems of the individuals and groups involved indecision-making. Thus, multi-criteria evaluation enables tailor-made, transparent decision support but also requires appropriate design in order to deliver relevant and reliable statements. The concrete formulation of preferences and discussion of the compromises to be made contributes to a common understanding among the parties involved and helps to comprehensively illuminate the advantages and disadvantages of various alternatives. Thus, multi-criteria evaluation can offer added value even beyond the sole choice of a preferred solution.

Cost–benefit analysis

As a rule, it is not only decisive what costs are caused by different decision options but also the benefits have to be seen in relation. In the so-called cost–benefit analysis (cf. Atkinson & Mourato 2015) internal costs are compared with the monetised benefits. In addition to direct costs, in particular, technical assessment criteria, such as maintenance expenditure or downtimes caused by system failure can be expressed in terms of the expected associated costs as part of the operating costs. In addition, external costs (e.g. due to greenhouse gas emissions) and benefits (e.g. increased regional water availability) can also be taken into account from an economic point of view within the framework of a so-called extended CBA. However, the approach to monetisation is ambiguous and can be a sensitive and controversial issue, especially for social aspects.

If the benefits cannot be monetised, a cost-effectiveness analysis can be carried out. This can be useful if, for example, two measures lead to the same result or if the consideration is focused on one parameter. However, if side effects are to be included in the evaluation, it is necessary to weight different effects implicitly or explicitly. For example, effects are implicitly weighted with zero if they are not included in the comparison. If alternative A (with low internal costs but higher emissions) is chosen instead of alternative B (with higher internal costs but lower emissions), the emissions are implicitly (even without LCA or other sustainability assessments) valued with the difference in internal costs, since it is implicitly assumed that the losses are lower than the difference in internal costs.

Overall, CBA is well suited to reflect economic and, in part, technical evaluation criteria. By explicitly evaluating the benefits, even solutions with different benefits can be compared without difficulty. The monetary evaluation fits well into traditional, economically influenced decision-making processes and evaluation concepts. Ecological and social criteria can only be considered in the CBA on the basis of their economic impact or by assigning a financial value. To avoid the monetisation of social and ecological criteria, the CBA can be embedded in a broader multi-criteria evaluation.

Life cycle analysis

In the context of technology assessment, LCA serves to compare the environmental performance of different technologies that produce the same product or have the same purpose (e.g. wastewater treatment or the supply of fresh water) on a global scale. Descriptions of the LCA methodology can be found, e.g. in Klöpffer & Grahl (2012), ISO 14044 (2018) or UNEP LCI (2018). The focus of LCA is on the environmental and health impacts caused directly (e.g. exhaust gases of a boiler) or indirectly (e.g. emissions caused by upstream processes such as production and transport of building materials) during different life cycle phases of a product or a service. Various impact categories, such as climate change, land occupation, acidification or use of non-renewable resources, are considered in order to provide a comprehensive picture of the environmental impacts caused per functional unit. Within each impact category, the effects of all processes considered in the LCA are quantified and summed up by means of an impact category indicator (e.g. kg CO_2-eq per functional unit). Effects are related to a so-called functional unit. When comparing different alternative technologies, products or services, a comparison is possible. When mapping water recycling within closed system boundaries, the choice of functional unit may well differ from that in an open system. Since in the project DiWaL, two technologies for efficient reduction of microorganism in the pre-treatment for dip painting of the car body are compared, the functional unit is 'one car body without complaints'. In the WavE project HighCon, for example, the wastewater volume to be treated in m^3 was defined as a functional unit. The treated water flows back into the process and thus reduces the fresh water withdrawal and the effort for further treatment for production. In this case, the material and energy required to treat the fixed volume of wastewater is thus offset by savings on the fresh water side, reduced waste flows and recovered inorganic salts as raw materials. Similarly, in projects for water reuse, the m^3 of water provided can also be selected as a functional unit, which facilitates a comparison of potential water resources. The associated implications for wastewater treatment and discharge need to be considered in a closed system.

ISO 14044 (2018) does not specify how to identify the preferred solution based on the environmental and health impacts identified. It does, however, provide a very rigorous and comprehensive assessment of the environmental and health aspects of different alternatives using a transparent, standardised approach. As a basis for decision-making, the impact categories identified in the LCA can be used as evaluation indicators, supplemented by other relevant evaluation criteria, in a multi-criteria evaluation.

To apply ISO 14044 (2018) is especially important for cross-study comparisons, e.g. for similar products or 'functional units' from different companies.

Car body painting plants have a high production volume where a great deal of water is consumed. In the DiWaL project, the environmental performance of PEF treatment is compared with the surge dosing of biocides which is applied to reduce the microbial contamination of paint and process water. The environmental analysis conducted within the DiWaL project combines elements of the LCA with a parametrised energy and material flow model and further input from stakeholder consultation via different dialogs and workshops. This approach is called Integrated Innovation and Sustainability Analysis (Gasde *et al.* 2019, 2020). Applying simplified approaches similar to LCA (e.g. using expert guess and stakeholder interviews for data assumptions), scenario and hotspot analysis, allowed an evaluation of the potential environmental impact of the PEF treatment early in the development phase of the innovation. As far as possible, quantitative results regarding environmental and economic sustainability have been derived. However, more importantly, the analysis helped to generate feedback loops in the early innovation phase. By addressing these, the relevant sources of negative impacts can be indicated. Consequently, the design of the corresponding product was adjusted. There are different types of so-called prospective or ex-ante LCAs which try to take aspects into account which will change in the future (Cucurachi *et al.* 2018). The PEF treatment will start to operate only in the near future (ca. 5 years). Moreover, once such a device is installed, it will continue to operate for maybe 15 years and more. Further devices will go into operation even later. Hence, in order to make a fair comparison of this innovation with its current alternatives, the LCI data were adjusted to future conditions. For example, the underlying electricity generation mix is very relevant for the evaluation of the PEF treatment. To select a reasonable time horizon is one of the conventions which have to be decided in the frame of the definition of the goal and scope of the analysis.

DATA BASIS, IMPACT ASSESSMENT (CRITERIA CHARACTERISTICS) AND SENSITIVITY ANALYSIS

The data required for the evaluation depend on the project and on the evaluation criteria, dimensions and methods chosen. The following types of data are found in several WavE projects:

1. System- or plant-specific operating data (energy/chemical consumption, cleaning performance, personnel and maintenance requirements, etc.)

2. Geometry, material properties, mass and volume of materials used (e.g. for building a device)
3. Technical boundary conditions: inlet quality, water quantities and requirements, information on existing systems
4. Costs, tariffs
5. Characterisation factors for the assessment of environmental impacts (impact factors for LCA), e.g. ReCiPe (Goedkoop *et al.* 2013)
6. Legal and organisational frameworks
7. Decision-maker and stakeholder goals and preferences

Not all the data required are specific to the project or the system under evaluation. Therefore, these data may be available elsewhere and do not have to be collected in the course of the project. Depending on the data source, different challenges arise for its use (Table 3).

Empirical values, experience (e.g. from projects already implemented) and literature data are usually readily available for established technologies and processes. However, it should be examined whether the transferability of such data to innovative approaches is guaranteed. Increasingly, literature data on studies and experiments that have been carried out are also being questioned as to their reproducibility.

Data from laboratory tests and pilot plants carried out in the project offer a very high reliability, provided that sufficient attention is paid to careful data preparation. The scalability to industrial scale (up-scaling) should already be taken into account in the design of experiments.

Modelling enjoys unbroken popularity as a supplement or alternative to laboratory tests or piloting. Here, the conflict of objectives between the level of detail and the practicability represents the main challenge. The quality of a model prediction is always related to the quality of the input data.

One tool for obtaining qualitative or semi-quantitative data is stakeholder and/or expert consultation. Here the aim is to bring together experiences and complementary knowledge from different perspectives. In order for this method to be successful, a common understanding of the facts to be evaluated and a common terminology is crucial.

Statistical data are used, for example, to estimate population figures, water demands or price indices – often also to be able to forecast future developments. In addition to the challenge typical for statistical data of using a suitable sample size and distribution as a basis, extrapolations of past developments into the future need to be understood and should be presented as hypotheses.

As a basis for LCA, very large amounts of data on the environmental impacts of production, transport and disposal processes are required, which are usually not available or cannot be collected within a project. Proprietary databases such as the ecoinvent data base offer extensive information on these topics. If the data are to be extended by proprietary data sets, there is a risk that these will not be consistent with the existing data.

Laws and regulations supply information on effluent or discharge values and other emission limits which need to be

Table 3 | Possible data sources, challenges in using them and indicators to measure data quality

Data source	Challenge	Measures to ensure data quality
Literature	Transferability, reproducibility	Thorough literature review based on case-specific criteria, consideration of grey literature as complementary source for local data
Laboratory tests and piloting	Data processing, up-scaling	Long test periods under stable conditions with continuous data recording
Modelling	Model quality, quality of the input data	Model calibration and validation based on historical data
Experience	Transferability	Exchanging with practitioners and other scientists
Stakeholder interviews	Different, ideally complementary knowledge and experience, lack of common understanding and common terminology	Defining a common language and shared understanding of the problem and targeted solution
Statistical data	Future development, sample size and quality	Making use of methods for statistical quality control
Proprietary databases	Transparency, costs, expandability	Favoring public data bases
Laws and regulations	Uniqueness, physical consistency	Periodic review (local/regional) laws and regulations

complied with. These can vary greatly from region to region. In addition to the transferability from one site to another, which needs to be checked, this also means that the data are often not physically consistent. This can lead to the fact that an alternative that is considered to be sensible on the basis of the state of the art cannot be implemented or that legal requirements are physically contradictory.

Since the evaluation for decision support is carried out in the planning/design phase, the required data can usually not be collected on the real system. Therefore, the data used for the evaluation are subject to uncertainties that may also vary for different alternatives. How reliably prerequisites and minimum requirements can be fulfilled and evaluation attributes estimated, both for the expected use of the technology and with regard to future developments or unexpected boundary conditions, can therefore play an important role in the decision-making process. This is also reflected in some of the evaluation criteria, such as the TRL or the robustness and flexibility.

Further uncertainties or ambiguities arise in the selection of preferred solutions. For example, the choice of weighting factors in classical MCDA methods is highly subjective and not necessarily clear to decision-makers (see also section Evaluation framework) (Figueira *et al.* 2005).

A simple and intuitive way to consider uncertainties of the input data and/or the evaluation is to consider different predefined scenarios (i.e. different sets of input parameters or data sets). This method is often used to describe possible future developments, but it can also be used to map different plausible, mutually exclusive conditions for planning and evaluation (e.g. good/bad biodegradability of organic substances in the treated water, low/high population growth). In the MULTI-ReUse project, for example, scenarios with different weightings of the evaluation criteria are compared in order to investigate the influence on the resulting ranking of the alternatives.

The definition of the scenarios significantly influences how robust and comprehensive the additional knowledge gained is. To keep the uncertainty analysis manageable, only a limited number of scenarios can be considered. This limits the possible combinations of uncertain factors that can be investigated. However, a limitation of the considered contingencies can be quite useful if not all combinations are likely or possible. An advantage of the scenario analysis is that it is easy to implement and the results are straightforward to understand and interpret.

Alternatively or in addition, a systematic investigation of the influence of uncertainties in the underlying data or the assessment steps on the evaluation result can be carried out by means of a comprehensive (global) uncertainty and sensitivity analysis (see Saltelli *et al.* (1999)). Depending on the question posed and the evaluation framework, the possible intervals (or probability distributions) of the input data are examined to see how they affect the criteria specification, aggregated evaluation indicators or the ranking of the alternatives. The sensitivity analysis also allows identification of the most relevant uncertainties in the input data as well as their interactions, so that a targeted refinement and expansion of the data basis can be carried out. Possibilities for integrating uncertainty and sensitivity analysis into the model-based planning of water reuse concepts in industrial parks are being investigated in the WaReIp project.

The prerequisite for a broad sensitivity analysis is a formalised and, as far as possible, automated calculation of the target variables, since many calculation runs are usually required. In addition, the uncertainty or possible bandwidths of the input data should be estimated as realistically as possible in order to obtain meaningful results. The implementation, evaluation and interpretation of the results tend to be more complex than when considering individual parameter sets/alternative data sets and may not be entirely trivial. On the other hand, global sensitivity analysis provides a more complete picture of the possible effects by simultaneously considering the existing uncertainties and allows a well-founded prioritisation of the influencing factors.

CONCLUSION AND OUTLOOK

Within the framework of the WavE joint research projects, a variety of evaluation approaches have been adopted and developed to assess the advantages of innovative process technologies over conventional supply structures in the context of water reuse and desalination.

The working group on evaluation methods in WavE has succeeded in developing a generic approach for the comparative evaluation of procedures, technologies and system solutions. Via checking prerequisites and test criteria, compliance of the proposed solutions with the existing legal framework and other prerequisites can be achieved by optimisation of the proposed solutions in an iterative process. Furthermore, a structured overview of evaluation criteria was developed, which, in addition to the three evaluation dimensions environmental, social and economic of the TBL model of sustainable development, also includes the field of technology. Considering all these evaluation criteria in the assessment enables the users to find the most preferable of the feasible solutions for a specific decision case. Together, the evaluation procedure and the evaluation criteria grouped in the form of individual clusters offer intuitively usable assistance for preparing individualised, multi-criteria evaluations. The results of the working group thus show a high relevance with regard to manifold questions of water supply and reuse in practice.

The choice of criteria and methods for the evaluation of sustainable technologies and concepts is always application-specific and should be oriented towards the overall objectives, the actors involved in the decision (e. g. perspective of public services or business perspective) and the decision-making framework (scope, implications, duration of the decision-making process) in order to identify the most suitable solution for the individual decision-making situation. The differences between the evaluation approaches developed in the WavE projects underline this need for adaptation to the specific situation. However, many similarities were identified both in the general approach and methodology as well as the chosen criteria. Nevertheless, it should be noted that each evaluation method goes hand in hand with limitations that have to be considered in the selection process.

MCDA is a family of decision-making methodologies that may include any criteria defined by the stakeholders and can be shaped very specifically to any project. However, no out-of-the-box application is possible. LCA allows very rigorous assessment of global ecological impact but does not include other technical, economic or social criteria. In contrast, CBA can encompass aspects of all four dimensions as long as they are monetised, i.e. expressed in terms of financial revenues or costs. LCA and CBA do not directly point out the most preferable solution but can be valuable complements in an MCDA framework.

Practical applicability of an evaluation methodology is strongly dependent on the availability of reliable and robust data. Assessing technical, economic, ecological and social aspects of water reuse and desalination solutions requires data from many different sources. Especially in planning, data availability and reliability is a crucial issue for prognosis as the required information concerns concepts and technologies that are not yet implemented. As the data form the basis on which decisions are taken, data quality deserves close attention. Considering the uncertainty of the underlying data and assumptions can improve the robustness of the developed alternatives and the taken decisions, thus minimising risks (or abating negative consequences).

A thorough consideration of technical, economic, ecological as well as social criteria are recommended to be considered in any decision case. In both public and corporate projects, ecological aspects have become increasingly important, although the motivation for including respective criteria in the decision-making process may not always be the same. In public projects, well-being and preservation of nature and resources seem to be more focused, whereas in corporate projects, green image and compliance are the environmental aspects that highest priority is assigned to. In general, social and ecological aspects are less important for the assessment in corporate projects but are reflected in prerequisites that stem from laws and regulations.

A multi-criteria evaluation, which follows the generic approach of the WavE projects presented, helps to implement a clearly structured decision-making process. A detailed examination of the objectives and requirements, exchange of viewpoints and preferences, as well as the development of a common understanding among the project participants during the decision-making process, can promote the development of sustainably satisfactory solutions, especially when closely linked to process development. Through appropriate, accompanying documentation of the selected procedure and the methodology applied, the evaluation approach offers a high degree of transparency for all those involved. The evaluation results support the decision-makers in communicating the decisions made internally and externally.

Multi-criteria evaluation methods cannot liberate the actors of the final decision, because none of the presented methods can fully reflect all aspects to be considered.

Therefore, it is important to facilitate a profound understanding of the pros, cons, potentials and risks of the available alternatives. To this end, the presented evaluation methods are a helpful communication and decision support tool.

ACKNOWLEDGEMENTS

The research leading to these results has received funding from the German Federal Ministry of Education and Research as part of the funding measure 'Future-oriented Technologies and Concepts to Increase Water Availability by Water Reuse and Desalination (WavE)'.

DATA AVAILABILITY STATEMENT

All relevant data are included in the paper or its Supplementary Information.

REFERENCES

Atkinson, G. & Mourato, S. 2015 Cost-Benefit Analysis and the Environment. OECD Environment Working Papers, No. 97, OECD Publishing, Paris, https://doi.org/10.1787/5jrp6w76tstg-en.

Belton, V. & Stewart, T. J. 2002 *Multiple Criteria Decision Analysis. An Integrated Approach.* Springer US, Boston, MA. https://doi.org/10.1007/978-1-4615-1495-4.

Cucurachi, S., van der Giesen, C. & Guinée, J. 2018 Ex-ante LCA of emerging technologies. *Procedia CIRP* **69**, 463–468. https://doi.org/10.1016/j.procir.2017.11.005.

Dominguez, D., Worch, H., Markard, J., Truffer, B. & Gujer, W. 2009 Closing the capability gap: strategic planning for the infrastructure sector. *California Management Review* **51** (2), 30–50. doi:10.2307/41166479.

Figueira, J., Ehrgott, M. & Greco, S. 2005 *Multiple Criteria Decision Analysis. State of the Art Surveys.* Springer Science + Business Media Inc., New York, NY. Available from: https://link.springer.com/book/10.1007%2Fb100605.

Gasde, J., Preiss, P. & Lang-Koetz, C. 2019 Integrated innovation and sustainability analysis in collaborative R&D projects. In *The ISPIM Innovation Conference – Celebrating Innovation: 500 Years Since DaVinci, June 16–19, 2019*, Florence.

Gasde, J., Preiss, P. & Lang-Koetz, C. 2020 Integrated innovation and sustainability analysis for new technologies: an approach for collaborative R&D projects. *TIM Review* **10** (2), S.37–S. 50. doi:10.22215/timreview/1328.

German Environment Agency 2018 Flächennutzungen: Dessau-Roßlau. https://www.umweltbundesamt.de/daten/flaeche-boden-land-oekosysteme/flaeche/folgen-der-flaechennutzung#textpart-1 (accessed 28 February 2020).

Goedkoop, M., Heijungs, R., Huijbregts, M., De Schryver, A., Struijs, J. & van Zelm, R. 2013 *ReCiPe 2008 – A Life Cycle Impact Assessment Method Which Comprises Harmonised Category Indicators at the Midpoint and the Endpoint Level*, 1st edn (version 1.08) Report I: Characterisation.

Grote, G. & Gbikpi, B. 2002 *Participatory Governance. Political and Societal Implications.* Springer VS, Wiesbaden.

Hajkowicz, S. & Higgins, A. 2008 A comparison of multiple criteria analysis techniques for water resource management. *European Journal of Operational Research* **184** (1), 255–265. doi:10.1016/j.ejor.2006.10.045.

ISO 14044 2018 Environmental management – Life cycle assessment – Requirements and guidelines (ISO 14044:2006 + Amd 1:2017).

Klöpffer, W. & Grahl, B. 2012 *Ökobilanz (LCA). Ein Leitfaden für Ausbildung und Beruf (Life Cycle Assessment (LCA). A Guide to Training and Work).* Wiley-VCH Verlag, Weinheim, Germany.

Organisation for Economic Co-operation and Development (OECD) 2007 *Infrastructure to 2030. Vol. 2: Mapping Policy for Electricity, Water and Transport. OECD Futures Project 'Global Infrastructure Needs: Prospects and Implications for Public and Private Actors'.* OECD Publishing, Paris.

Pajer, S., Streit, M., Torsney-Weir, T., Spechtenhauser, F., Muller, T. & Piringer, H. 2017 Weightlifter: visual weight space exploration for multi-Criteria decision making. *IEEE Transactions on Visualization and Computer Graphics* **23** (1), 611–620. doi:10.1109/TVCG.2016.2598589.

Saltelli, A., Tarantola, S. & Chan, K. 1999 A role for sensitivity analysis in presenting the results from MCDA studies to decision makers. *Journal of Multi-Criteria Decision Analysis* **8** (3), 139–145. doi:10.1002/(SICI)1099-1360(199905)8:3 < 139::AID-MCDA239 > 3.3.CO;2-3.

Schmitter, P. C. 2000 *How To Democratize the European Union. And Why Bother?* Rowman & Littlefield Publishers, Lanham.

UNEP LCI 2018 *Lifecycle Initiative: Life Cycle Assessment – A Product-oriented Method for Sustainability Analysis – UNEP LCA Training Kit Module e – Impact Assessment.* Available from: https://www.lifecycleinitiative.org/wp-content/uploads/2013/06/Module-e-Impact-assessment.pdf (accessed 28 February 2020).

van Leeuwen, K. & Marques, R. 2012 *Current State of Sustainability of Urban Water Cycle Services. Part I – Towards a Baseline Assessment of the Sustainability of Urban Water Cycle Services. Trust Deliverable D11.1.* Available from: https://pdfs.semanticscholar.org/c7d6/0b99552cc2850685f2216b2963e5df76b4ca.pdf (accessed 28 February 2020).

First received 30 March 2020; accepted in revised form 21 June 2020. Available online 31 August 2020

Advanced processed wastewater for different uses: constellations favouring future implementation of a multimodal water reuse concept

Engelbert Schramm, Dennis Becker and Michaela Fischer

ABSTRACT

Advanced wastewater treatment is often used to produce one single water quality. In recent years, technologies have been developed that allow the production of different qualities that are fit for their purpose. These technology bundles are still not being marketed, and market requirements are unclear. Two constellations in West Basin, California and in Oman were analysed to shed light on the different constellations of actors, resource situations and institutions. The first led to the industrial reuse of several water qualities, while the second produced an application in holiday resorts, leisure and food production. A hypothetical solution was contrasted with an historical case. The analysis of the constellations showed that multi-sectoral investments and dependencies require strong cooperation arrangements and long-term agreements. Governmental institutions were revealed to be suitable for coordinating the process, especially during the initial phase, but also in view of security of supply. A (comparative) examination allows an initial, still provisional systematic overview of other constellations that favour systems with recycled water of different qualities. Further research is required to understand the welfare and distribution effects of multimodal water pricing policies and the feasibility of co-financing of agricultural irrigation and opportunities for more sustainable water reuse.

Key words | advanced treatment, agriculture, industry, tourism, water reuse

Engelbert Schramm (corresponding author)
Michaela Fischer
ISOE – Institute for Social-Ecological Research, Hamburger Allee 45, 60486 Frankfurt am Main, Germany
E-mail: *schramm@isoe.de*

Dennis Becker
DECHEMA Gesellschaft für Chemische Technik und Biotechnologie e.V,
Theodor-Heuss-Allee 25, 60486 Frankfurt am Main, Germany

HIGHLIGHTS

- Application of a constellation analysis to find out the conditions for advanced treatment and supply with two or more water qualities.
- Introduction of the concept of multimodality.

INTRODUCTION

Water reuse can support 9 of the 17 Sustainable Development Goals: zero hunger, good health and well-being, clean water and sanitation, decent work and economic growth, industrial innovation and infrastructure, sustainable cities and communities, responsible consumption and production, life below water and partnerships for the goals.

Advanced treatment was initially developed for a small number of high-value water reuse applications, in particular groundwater recharge and drinking water production (Rizzo *et al.* 2019). For most applications, simpler treatment processes were initially considered sufficient. In the past, treatment was very simple, especially for agricultural applications; in many cases, secondary wastewater treatment with additional hygienisation is still considered sufficient (see European Union 2020). However, examples from many countries now show that such simple wastewater

doi: 10.2166/wrd.2020.043

treatment can lead to environmental damage such as groundwater pollution, for example through the emission of environmental chemicals and anthropogenic micropollutants, especially contaminants of emerging concern (CEC, e.g. endocrine disruptors; Seis *et al.* 2016). Such long-term consequences are considered unacceptable in sustainability assessments (Kennedy & Tsuchihashi 2005). Accordingly, efforts are being made, such as advanced water treatment that includes further process stages (Kennedy & Tsuchihashi 2005; Drewes *et al.* 2019). These innovations potentially increase the range of sustainable applications for water reuse. In addition to agricultural irrigation, it can also be used for livestock breeding, for irrigating tourist facilities and public green areas, and for a variety of industrial, commercial and domestic purposes (Drewes *et al.* 2019).

In many cases, advanced treatment has so far only been applied to produce a single water quality. In view of the numerous possible applications, however, it is also conceivable that the wastewater treatment plant (WWTP) effluent is treated to produce different qualities for various uses. The provision and supply of different water qualities, which are used for different purposes, are abbreviated here as multimodal. Multimodal supply is the (usually simultaneous) supply of water of varying quality, which is differentiated and usually carried out using different transport carriers (usually pipelines). Accordingly, multimodal water use refers to the use of several separately transported water qualities (usually) for different purposes.

To date, systematic investigations of constellations that favour treatment and distribution systems with recycled water (in different qualities for various purposes) have not been carried out. The aim of the research was to identify promising examples of multimodal reuse of water in order to draw conclusions about the possibility of targeted dissemination of technologies to support sustainable water reuse.

The application of constellation analysis (CA) allows the identification of methods for introducing the advanced and multimodal treatment technologies tested in the MULTI-ReUse project, which can reduce the risks both to human health and the environment. The research questions were:

- What are the (existing) constellations for water reuse of different water qualities where advanced treatment is appropriate?
- Which aspects promote their realisation and sustainable operation?

The identified cases were examined using CA to assess whether they lead to a sustainable solution.

METHODS AND MATERIALS

The CA concept

The method used in this paper is based on CA, as explained by Schön *et al.* (2004). CA was developed to analyse and solve complex problems involving social, technical and natural aspects that are closely interlinked. Furthermore, CA claims to bridge prevailing problems in interdisciplinary settings and to integrate practical and scientific knowledge of various disciplines and societal groups (Ohlhorst & Schön 2015). Elements of the constellation are heterogeneous – individual or collective actors as well as technical norms and laws, technical objects and natural resources. Each of these element types is treated as equally important in the CA; the elements can either be the components of a specific problem situation or parts of a solution. It is possible to identify promoting and hindering factors of innovations, and a special feature of CA is the visualisation of the constellation, which helps to identify networks, dynamics and the relationships between the elements. Conventional water reuse is analysed in Nölting & Mann (2018). The process can be summarised in the following steps:

- identification of the elements of the constellation and description of their relations,
- analysis of the functional principles and characteristics of the constellation under consideration, including the identification of destabilising and stabilising elements,
- examination of the dynamics in the constellation and its development and
- evaluation of the constellation and strategy making (optional).

The CA was applied in an interdisciplinary research process in which competencies from ecology, engineering (from water management, agriculture and chemical engineering, landscape planning) and social sciences were used. Thus, the knowledge considering the elements and their relationships could be drawn from various scientific fields and combined with practical knowledge.

Broad expertise and the graphically based design procedure of CA allow a quasi-experimental optimisation and stabilisation of constellations in its development process. Thus, it is possible to simplify or improve the constellation, e.g. by omitting or replacing actors or by changing relationships and incorporating technical innovations. Inhibiting factors (e.g. acceptance problems of traditional farmers) can also be changed by focusing on new products and sales markets.

Concept application

In this paper, constellations of two different cases are considered and evaluated. The cases describe examples of water reuse in different sectors, which are applied in practice in the fields of industry, agriculture, tourism and landscape irrigation. The first case has its starting point in the constellation that led to the reuse of different water qualities by industry in California's West Basin district in 1994. The second case deals with the water reuse in the Arab peninsula and analyses so far hypothetical applications in aquaculture, tourism and landscape.

Based on an initial market analysis for the MULTI-ReUse methods (Becker *et al.* 2017), typical problem situations were identified and described. In an interdisciplinary group process, several cases were discussed with the aim of identifying the benefits of providing different water qualities and highlighting bottlenecks for sustainable development. For different cases (an industrial application and an agricultural and domestic application), a situation was selected that is as conducive as possible in order to present possibilities that favour advanced water reuse and where the financial feasibility of the supply of different water qualities can be assumed.

THE MULTI-REUSE CONCEPT

The MULTI-ReUse process produces three different, precisely defined water qualities from a base product of municipal wastewater already being conventionally treated. The water is processed further either at the same treatment plant or at its site of use (cf. Nahrstedt *et al.* 2020). Thus, no raw wastewater is used, but rather the effluent from the WWTP. Due to its orientation towards user requirements, it can be seen as a fit-for-purpose approach. In the production process, which is quality-assured using specific monitoring procedures, certain modules of an advanced water treatment system are used according to the defined quality. The MULTI-ReUse process also includes observation of the microbiological quality of the treated water, which is close to real-time control. The process has been tested since 2018 in a trial and demonstration plant (pilot plant) at the Nordenham WWTP by the MULTI-ReUse consortium. There, the following qualities were produced, including side by side if required (cf. Nahrstedt *et al.* 2020).

- Process water 1 is practically free of undissolved substances and pathogenic bacteria, but still contains nutrient salts and CEC. The water can be used industrially for washing processes (e.g. for street cleaning) or for cooling processes with low requirements (e.g. for the concentration of dissolved salts). It can also be used for domestic applications (e.g. toilet flushing and washing machine) and for irrigation of urban greenery and energy and industrial plants. However, the nutrient content may affect water storage and distribution (which needs to be considered during the operation of the plants).
- Process water 2 is practically free of undissolved ingredients and pathogenic bacteria and viruses. The concentration of nutrients is considerably reduced compared with process water 1 (this increases the microbiological stability of the water, i.e. less microbial regrowth during storage and distribution). It also contains significantly fewer CEC of anthropogenic origin. The water can be used for industrial washing or cooling processes with higher quality requirements, as well as for high-quality agricultural applications (e.g. underground irrigation of fruit and vegetables for human consumption and supply of livestock) or for groundwater recharge.
- Reverse osmosis (RO) water (process water 3) is free of particles and pathogenic germs, and ions and macromolecules (e.g. CEC) dissolved in the water are removed as far as possible by RO in further process steps. This highly treated water can, therefore, be used in a wide

range of applications as process water, e.g. as boiler feed water, for the production of ultrapure water or as mixed water for dilution purposes. However, the low ion concentration and free carbonic acid cause the water to have a corrosive effect on certain metallic materials, which should be compensated for by suitable materials during water storage and distribution (alternatively, the water can be adjusted by buffering and pH correction). In comparison with the other two water qualities, a maximum of microbiological stability is also achieved (cf. Nocker *et al.* 2020). RO water is used in industrial processes and agriculture, for example, especially in greenhouse horticulture.

MATERIALS USED FOR THE CA

Oman and California were selected because of the advanced water treatment and water reuse in both countries. Furthermore, there is good availability of information in English in both cases. Different sources of information were used to understand developments and the current situation with regard to water reuse in California and Oman.

In the case of California, droughts and early institutional developments (West Basin Municipal Water District established in 1947) created an early case in advanced water reuse, which is well documented. The CA was based on scientific and online information. The most detailed and comprehensive information is presented in the website of the West Basin Municipal Water District (https://www.westbasin.org). One of the main scientific sources was Lazarova *et al.* (2012).

With regards to water reuse in Oman, there is an abundance of scientific publications, especially on technologies for different water reuse applications. Furthermore, information was supplied by institutions in the water sector in Oman. Haya Water in particular provides detailed information on their website about the technology currently used and the challenges of water reuse. Moreover, policy documents and interviews in professional magazines and newspapers were used to understand historical developments and current challenges. An analysis of the water sector in general, and of trends and investments can also be found on websites such as German Trade and Investment. Furthermore, scientific and online information on agriculture and tourism was used for the analysis.

RESULTS

CA: supplying industries with different water qualities

The United States was chosen because it is one of the world's pioneers in water reuse and there is good data availability. It is an example of an industrialised country where industrial water reuse plays a remarkable role. Experiences from Northern America can be transferred to similar environments on a global scale, such as Europe, Mexico, China, India and the Middle East (particularly, the member states of the Gulf Cooperation Council) (Becker *et al.* 2017).

Preconditions and framework analysis

The United States, with an area of approximately 9.06 million km^2, is the third largest country in the world. Due to the enormous size of the country, the population density is relatively low at 33 inhabitants per km^2 (BMWi 2018), despite a population of more than 321 million (as of 2016). In some states, the increased salinity in aquifers and surface waters poses a serious water quality problem. The steady increase in salinity is a result of excessive irrigation and increasing urban water use. In California, the biggest salinity problems are in the aquifers in the southern Central Valley and the Salton Sea. There is currently no sign of a slowdown in the trend (BMWi 2018). Further saline inputs to groundwater are via saltwater intrusions in coastal areas.

Economy and legal framework

The US water sector generates annual revenues of more than US$ 157 billion (as of 2015) and, with average growth of about 4%, is one of the promising future markets for global investors and new water-related technologies. The market is influenced by various external factors such as regulations or the financing instruments offered, resulting in a state of constant change in the water sector (Water Technology 2016). Water abstraction law in the USA is regulated very heterogeneously due to the size of the country. In the east and mid-west of the country, most states apply the Riparian Law, which links water abstraction to land ownership. The person who owns the land with the water source may use the water within reasonable limits. What is meant by reasonable use, however, remains a matter of interpretation and, in case of doubt, will be decided by the competent courts. In some countries, water abstraction also requires a permit from the competent environmental authority. In the western states, on the other hand, the Prior Appropriation Doctrine is predominantly represented. According to this legal text, the use of water for the benefit of the general public (Beneficial Use) enjoys priority. Water abstraction is, therefore, independent of land ownership. In California and the other remaining states, aspects of both systems are combined into a hybrid system (BMWi 2018). At the federal level, there are also numerous laws and executive orders that specify the practical implementation of environmental laws. The 'Clean Water Act' and the 'Safe Drinking Water Act' are the two most important laws in the area of water quality and wastewater treatment. The Office of Water (BMWi 2018) is responsible for implementing these two laws. The Clean Water Act regulates all wastewater discharged into US waters and sets pollutant limits and quality standards for surface waters. The Safe Drinking Water Act exists to protect drinking water and applies equally to groundwater and surface water used for drinking water production. Plant operators have to comply with it. However, the introduction and enforcement of these laws is the responsibility of the individual states and heavily dependent on environmental and industrial influences (BMWi 2018). In addition, there are separate guidelines and standards at the state level for the use of recycled wastewater. The pioneer in this case was once again the state of California. California's legal regulations on this are more detailed than all previously enacted laws and regulations, and serve as the basis for a large number of water reuse projects (Becker *et al.* 2017). California's Regulations related to Recycled Water, Title 22 of the Code of regulations, Title 22, Division 4, Chapter 3, include four water types for water reuse (Becker *et al.* 2017).

Water reuse as a challenge

The West Basin Municipal Water District is a water agency that provides imported drinking water to 17 cities (and unincorporated areas) in Los Angeles County. The agency was created by a vote of the people in 1947 to reduce local groundwater over-exploitation and to make local water supplies more reliable (e.g. by groundwater recharge). In 1948, West Basin became a member agency of the Metropolitan Water District of Southern California and began wholesaling imported water from the Colorado River, and later also from Northern California (Lazarova *et al.* 2012).

West Basin is the sixth largest water district in the State of California and serves a population of more than 900,000 people. The West Basin service area consumes 271 million km^3 of water annually. The managed aquifers are near the Pacific Ocean. Due to the abstractions, saltwater intrusion was occurring here as early as the 1940s.

By 1947, it had become apparent that concentrating water extraction on (desalinated) seawater (and supporting the groundwater by infiltration) was too expensive compared with importing water from other basins. It was only the differentiation by type of use that made it possible to identify industry as a priority customer for reclaimed water. During the 1976/1977 drought, a water conservation programme was implemented; regional industry was very active in reducing water use, and the West Basin's Board of Directors discussed the possibility of using secondary treated wastewater from the Hyperion Treatment Plant in Playa del Rey for industrial use. The nearest industrial user to this plant was the Chevron refinery in El Segundo. Initial consideration was given to water reuse for Chevron's cooling towers in 1978, and a pilot programme was launched (Joham 1978).

Description of the constellation

In the major drought of the late 1980s and early 1990s, West Basin's Board of Directors led the agency in developing alternative local water supplies that included water recycling for industrial use and landscape management. Industrial applications were seen as the main applications; in terms of volume, applications with seawater intrusion barriers and landscape irrigation (which were actually the agency's objective) were secondary (Lazarova *et al.* 2012). An overview of the constellations of actors, resources, legal constraints and technical solutions in the California West Basin is given in Figure 1.

The West Basin Municipal Water District is the key actor in this constellation, along with the facility manager and operator (SUEZ) of the Edward C. Little Water Recycling Facility (ECLWRF), and the following three satellite treatment facilities:

- Carson Regional Water Recycling Facility (CRWRF),
- Chevron Nitrification Facility (CNF) and
- ExxonMobil Water Recycling Facility (EMWRF).

The City of Los Angeles' Hyperion Wastewater Treatment Plant (HWWTP) is currently the sole source of supply for West Basin's treatment facilities and recycled water systems. ECLWRF is the only treatment facility that receives secondary treated wastewater effluent as supply from the HWWTP. The satellite facilities rely on water from ECLWRF as a supply source, which produces tertiary water (Title 22) for them as well as stabilised RO water for the West Coast Barrier Blending Station for the barrier for seawater intrusion (Lazarova *et al.* 2012). CNF further treats Title 22 recycled water received from the ECLWRF through a biological nitrification process to remove ammonia for cooling water tower application. EMWRF (XOM) further treats Title 22 recycled water through biological nitrification and microfiltration (MF)/RO processes for use at the ExxonMobil Refinery in Torrance. The water produced is used for cooling tower and boiler feed applications. CRWRF further treats Title 22 recycled water through MF/RO and biological nitrification, before it is blended and sent to the BP Refinery in Carson for industrial use.

ECLWRF and the three satellite facilities produce five different qualities of recycled water (Lazarova *et al.* 2012):

- tertiary water (Title 22) for a wide variety of industrial and irrigation uses,

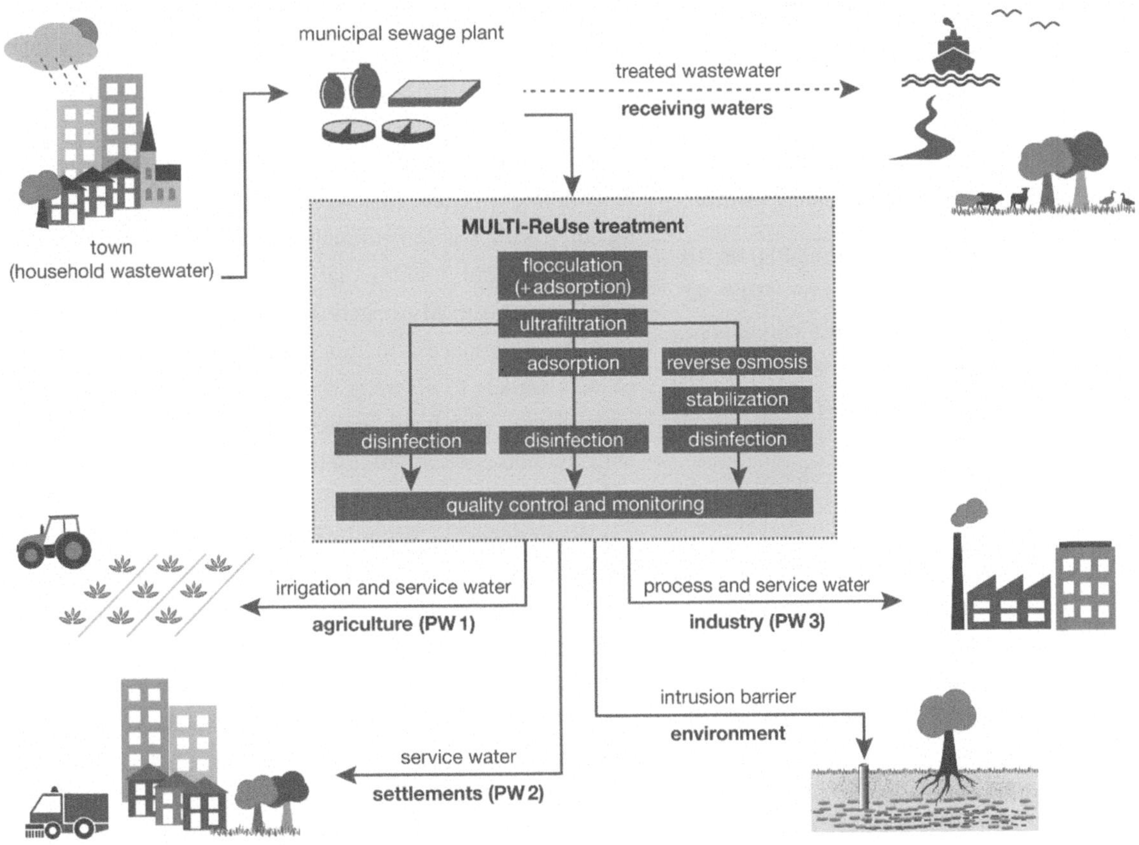

Figure 1 | Process diagram of the MULTI-ReUse-treatment technologies and applications (adapted from Becker *et al.* (2017)).

- nitrified water for industrial cooling towers,
- pure RO water for refinery low-pressure boiler feed water,
- ultrapure RO water for refinery high-pressure boiler feed water and
- barrier water or stabilised RO water.

Regarding the use of the recycled water from ECLWRF, the majority (53%) goes to industrial supply, while the seawater intrusion barrier uses 36%, landscape irrigation 10% and mixed use 1%.

Key milestones of the constellation are as follows (Lazarova *et al.* 2012):

- West Basin launched its water recycling programme in the mid-1990s.
- Contractual obligations with the industry to be supplied and the operation of the recycling facilities (by SUEZ) led to a public/private partnership.
- West Basin has invested more than $500 million dollars in this programme.
- Daily production capacity is 150,000 m^3/d (SUEZ group 2020) and in 25 years has produced 750 million m^3 of recycled water – every cubic metre recycled is a water volume not imported from Northern California or the Colorado River.
- The water is used for various purposes such as refinery cooling towers and boilers, street cleaning and irrigation.

The California Department of Public Health (DPH) issued the California Wastewater Reclamation Criteria (Title 22 regulations) in 1978. The department controls compliance and has not been a fundamental veto player against water reuse.

An example is given below of a typical communication and decision process among industrial customers in their switch from imported water to the use of recycled water at ECLWRF:

- contact made between West Basin and customer,
- basic information is exchanged – potential use and amount of recycled water/role of West Basin in the process,
- determination of the feasibility of West Basin serving the site,
- determination of interest by customer,
- development of drawings of retrofit based on customer plans,
- determination of feasibility for customer conversion,
- submission of plans to the California DPH for review,
- site inspection prior to construction by DPH,
- retrofit construction with ongoing inspection by DPH and West Basin, completion of retrofit construction and final inspection by DPH including cross connection testing and
- after approval by DPH, completion of conversion by connecting recycled water.

Although customers have unique needs, it has been possible to meet them with only a few classes of water produced, and tertiary water (Title 22) representing the majority. The satellite facilities treat the tertiary water (Title 22) further for the specific purposes of industrial partners. The so-called barrier water or stabilised RO water produced by ECLWRF, a secondary treated wastewater purified by MF and followed by RO and disinfection, is used as a seawater intrusion barrier. For most applications, MF is coupled with RO. The key advantages of this double membrane system are considered to be highly efficient in CEC removal and desalination (Lazarova *et al.* 2012).

ECLWRF in El Segundo, California, is nowadays one of the largest and a well-known example of water recycling facilities with advanced treatment in the United States. Its five different types of water, which are conditioned together with the three satellite facilities, meet the needs of the industrial, commercial and municipal customers of the West Basin Municipal Water District. One of the main keys to success in this example was the willingness of local leaders to diversify water resources in partnerships with federal and local authorities and operate the plant with public–private partnership integration (Lazarova *et al.* 2012).

When departing from the historical course, the question can be addressed in the CA as to which stage of development the Californian constellation favoured multimodal water reuse. The technical possibilities of the MULTI-ReUse procedure were used for this investigation. An overview of a hypothetical favouring constellation of actors, resources, legal constraints and technical solutions in the California West Basin is given in Figure 2.

With the West Basin Municipal Water District, a very strong actor was already created in the 1950s that was

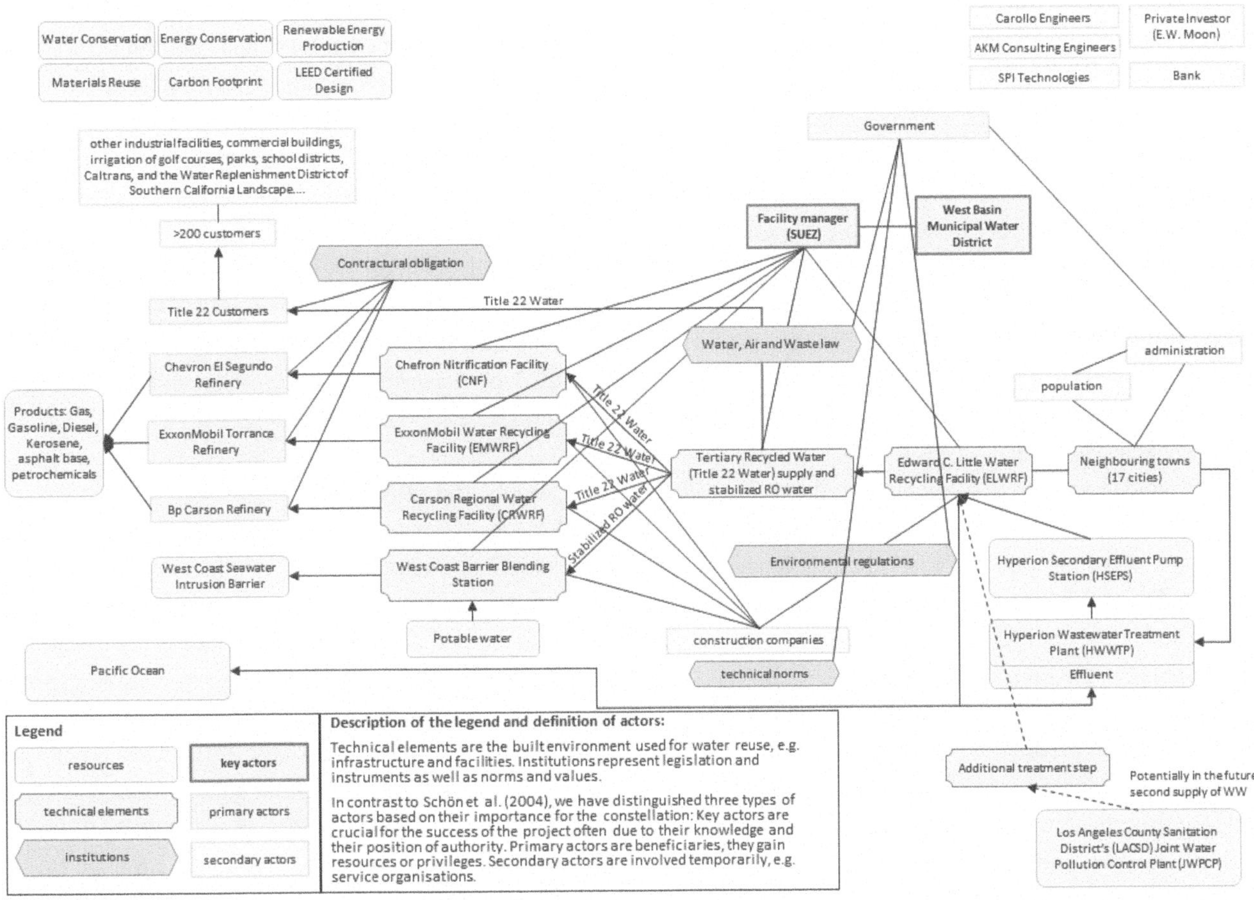

Figure 2 | Constellation of actors in the Californian West Basin 1994, illustration is based on information from https://www.westbasin.org and Lazarova *et al.* (2012).

able to promote water reuse. At the end of the 1970s, it became clear that there were potentially different groups of customers and uses, including industry, agriculture and groundwater protection. Different water qualities were needed side by side to supply them. Recycling capacities could only be established after the demand was contractually guaranteed. Due to the distances to the treatment plant, one option could be to initially treat the water only to PW 1 quality and to process it to higher qualities only in the vicinity of the customers. If several qualities are transported in parallel in different pipelines, these satellite facilities could only be used for quality control and disinfection. By mixing water from the PW 2 and PW 3 treatment lines, other wishes of industrial customers can also be satisfied. By mixing water qualities PW 2 and PW 3, the intrusion barrier is no longer dependent on the addition of tap water, which means less water has to be brought into the area from outside.

Assuming sufficient quantities of wastewater, it is technically possible to supply all interested parties with water of a higher quality than tertiary treatment effluent. For applications where this water enters the soil (and possibly also contributes to groundwater recharge), this water is preferable to Title 22 water. Its use can also be recommended for healthcare reasons. Due to methodological limitations, it is not possible to investigate in the CA whether all customers would agree to a corresponding payment for better quality. Should a large number of these users decide to use conventionally treated wastewater, it is nevertheless possible to opt for multimodal water use due to the industrial customers (Factories 1–3) and the importance of intrusion prevention.

CA: MULTIMODEL SUPPLY OF TOURIST AREAS, AGRICULTURE/AQUACULTURE AND LANDSCAPE

Oman was selected as a potential country for multimodal purpose water reuse for several reasons, explained below. Beneficial for the analysis was good data availability and the possible transferability of the results to other countries in the Arabian Peninsula.

Preconditions and framework analysis

The Sultanate of Oman covers the south-eastern corner of the Arabian Peninsula. With the exception of the humid Dhofar region and some mountains, the climate is arid or semi-arid. Potential evaporation varies between 2,200 mm/year in the interior and 1,660 mm/year on the southern Salalah plain (FAO 2008). Rainfall in Oman can exceed 350 mm/a in the mountains and is less than 50 mm/a in the desert, in the foreland the average is 100 mm/a (Lehane 2015).

Oman is a high-income country and its economy is heavily depending on oil production; other important sectors in terms of GDP are tourism, shipping, mining, manufacturing and gas-based industries (The Heritage Foundation 2019). Agriculture, forestry and especially fishery are traditional activities, but account for only 1.3% of the GDP (FAO 2017). Economic growth in Oman is estimated to be slowing down, and diversification investments to reduce dependencies will lead to growth in the medium term (World Bank 2019). Oman has a population of 4.4 million people, with half of the population living in densely populated areas in the north of the country, mainly in al-Batinah and Muscat (FAO 2008). Due to a rural-urban shift, only 21.5% of the population lives in rural areas (FAO 2017), with 16% of the population employed in the agricultural sector (Kotagama & Al-Farsi 2019).

Most of the farmers run small farms with low technology, while 0.2% of the farmers account for 21% of the land and have modern technologies and access to markets (FAO 2017). Only 7% of the total area (22,000 km^2) is suitable for farming; an additional 14,000 km^2 are suitable, but less productive (McDonnell 2016). Currently, 680 km^2 are used for agriculture (McDonnell 2016), which is located mainly in the northern part of the country. Agriculture is based on irrigation: as well as dates, bananas and limes, vegetables (eggplants, carrots and tomatoes) and cereals are also produced. With the number of greenhouses in Oman increasing steadily in recent years, supported by the Ministry of Agriculture and subsidies as a climate change adaptation strategy (Choudri *et al.* 2013), further research has been done on cooling technologies, water supply and energy consumption to increase the sustainability of greenhouse production (Mendoza 2019). Oman's fishery is the largest producer in the region, with fish farming, currently rather small, supported by the government, e.g. by Fish Farming Centres and Fisheries Training Centres. Furthermore, there is potential for increasing the value of the products from agriculture and fisheries (FAO 2017).

Water supply and sanitation

Water supply in Oman is mainly provided by groundwater (83%), with most of it used for agricultural production (Haya Water n.d.). Over-extraction leads to declining groundwater levels and saline intrusion in coastal aquifers and on many farms. The area of extreme degraded land almost doubled between 2005 and 2010 (from 34,385 to 65,595 ha) (Al-Balooshi & Charabi 2012). Water reuse has been supported by national policy since 2000 (McDonnell 2016), and according to the Oman Salinity Strategy (MAF & ICBA 2012), the use of treated wastewater for irrigation is suitable to protect the land from degradation. Al-Batinah, the 250-km-long narrow coastline on the Gulf of Oman between Muscat and Suhar, is a fertile, frequently irrigated plain whose groundwater is fed by the gorge-like wadis from the Oman Mountains, which only occasionally carry water. The region, which traditionally exported dates and dried lemons, has developed a great deal of water-intensive horticulture along the coast, which has reached its limits. Meanwhile, the demand for groundwater is so great that plans are being made to ensure irrigation by seawater desalination.

Haya Water is the public wastewater service company, which started in 2005 in Muscat and by 2014 had taken over WWTPs and infrastructure of the entire country from the Ministry of Regional Municipalities & Water Resources. Currently, Haya Water is 'rebranding as a water reuse company' (Freyberg 2018) and has prepared a master plan for the reuse of treated water in Muscat in order to increase water reuse in the different sectors (Haya Water n.d.).

Investments in sewage infrastructure, such as WWTPs, have been focusing on urbanised areas and around 50% of Muscat's inhabitants are not currently connected to the sewer system. For Muscat, the aim is to connect 80% of the households (National Water Sector Master Plan), and this is expected to be accomplished by 2026 (Freyberg 2018).

Haya Water has started to build a network for the distribution of treated wastewater in the north of the country in the Muscat and al-Batinah region. As of 2018, 230 km of the distribution network for treated wastewater was completed (Espey 2018). As well as palm trees, parks and district cooling in Muscat (Freyberg 2018), four golf courses, sport stadiums and five farms are also supplied with the irrigation water. The Ministry of Agriculture supports these farmers by providing modern irrigation systems, financial support for water fees and advice on crop cultivation.

Currently, two different qualities of irrigation water are produced in Oman (Table 1):

1. irrigation water for fodder, cereal and seed crops, vegetable consumed after cooking, etc. as well as pastures (Standard B)
2. irrigation water for vegetables and fruits likely to be eaten raw, as well as public parks, lawns and recreational areas (Standard A).

Table 1 | Quality standards for vegetables and fruits likely to be eaten raw, etc.

Parameter Max. allowable limits	Standards A (mg/l)	B (mg/l)
Biochemical oxygen demand (BOD)	15	20
Chemical oxygen demand (COD)	150	200
Suspended solids (SS)	15	30
Total dissolved solids (TDS)	1,500	2,000
Electrical conductivity (EC) (μs/cm)	2,000	2,700
pH (within range)	6–9	6–9
Chloride ion	650	650
Ammoniacal nitrogen	5	10
Nitrate	50	50
Organic nitrogen (Kjeldahl)	5	10
Phosphorus	30	30
Sodium	200	300
Sulphate	400	400
Faecal coliform bacteria (per 100 ml)	200	1,000

Note: (A) and other crops, etc., (B) according to the Regulation for wastewater reuse and discharge, MD 145/1993 (MRMWR 1993).

Current problems in Oman for water reuse

Market for water reuse: One of the main challenges facing Haya Water is finding customers to utilise the irrigation water. Farmers do not want to pay for it since traditionally the groundwater was free of charge (Freyberg 2018) and electricity is subsidised (McDonnell 2016). Currently, an excess amount of treated wastewater is discharged into the marine environment, as there is no optimal water reuse but only 52% demand.

Health risks: Technology and skilled labour are required to increase the safety of water reuse (Haya Water n.d.). Little research has been done so far on the health and environmental risks connected to water reuse (Al-Riyami *et al.* 2018). Sampling of treated wastewater in Muscat shows compliance with Omani standards (COD, BOD, pH, EC, turbidity and total suspended solids (TSS)), however 'such a reuse application of wastewater treatment needs to comply with public health and environmental requirements, both locally and internationally' (Baawain *et al.* 2019). Studies by Al-Farsi *et al.* (2018) indicate that in Oman, frequently consumed pharmaceuticals remain in tertiary treated wastewater, which can accumulate in the soil and plants if the water is reused for irrigation.

Public acceptance: The public has general understanding of water treatment technologies and supports the idea of water reuse; however, there are concerns with regards to health and environmental risks. A survey of 115 people showed that Omanis have a general idea of wastewater treatment technologies and are very positive about irrigation of non-food crops and different green spaces or urban uses such as cooling, fire hydrants and toilet flushing (Baawain *et al.* 2019). Irrigation of edible crops was not preferred by 49%, somewhat preferred by 23% and well preferred by 28%. In addition, the discharge of the treated effluent into the marine environment was not preferred by 50% and only well preferred by 9%. According to Baawain *et al.* (2019), this reflects environment and health concerns and the wish to protect the environment.

Financing: The need to increase the cost efficiency of operations and find investments for technology includes the search for international partners for investments, but also for possible partnerships (Freyberg 2018).

Description of the constellation

For Oman, water reuse has been under discussion for some time; there is a high level of acceptance, however, for the irrigation of edible crops, but there are concerns among the public. Where there are sewer systems in Oman, 98% of the wastewater is already reusable in terms of treatment technology. So far, corresponding efforts have been made for the capital Muscat in particular.

Just a few years ago, Oman was a country for adventurous individual travellers, but in the meantime, state-sponsored tourism has developed that meets international standards. In the last ten years, some government plans have been postponed to develop large areas for high-quality tourism, in particular to build extensive hotel complexes in the northern coastal region (e.g. Al Madina A'Zarqa).

Nevertheless, demand in the area of superior individual tourism (diving, snorkelling and fishing) is growing strongly. It can, therefore, be assumed that there will be an increasing number of private initiatives. These initiatives offer windows of opportunity to construct sewerage and wastewater treatment plants at the same time, and to protect the sensitive coral reefs and fishing grounds. An overview of a hypothetical favouring constellation of actors, resources, legal constraints and technical solutions is given in Figure 3.

If the water is treated according to the MULTI-ReUse process (cf. Figure 4), it can be reused for various purposes. In hotels, this water can be applied for garden irrigation and toilet flushing (PW 1), but also to produce fruit, vegetables and fish for tourism and local needs (PW 2). PW 3 is the highest quality, as after ultrafiltration, RO is also applied. This water is suitable for hydroponic greenhouses, where

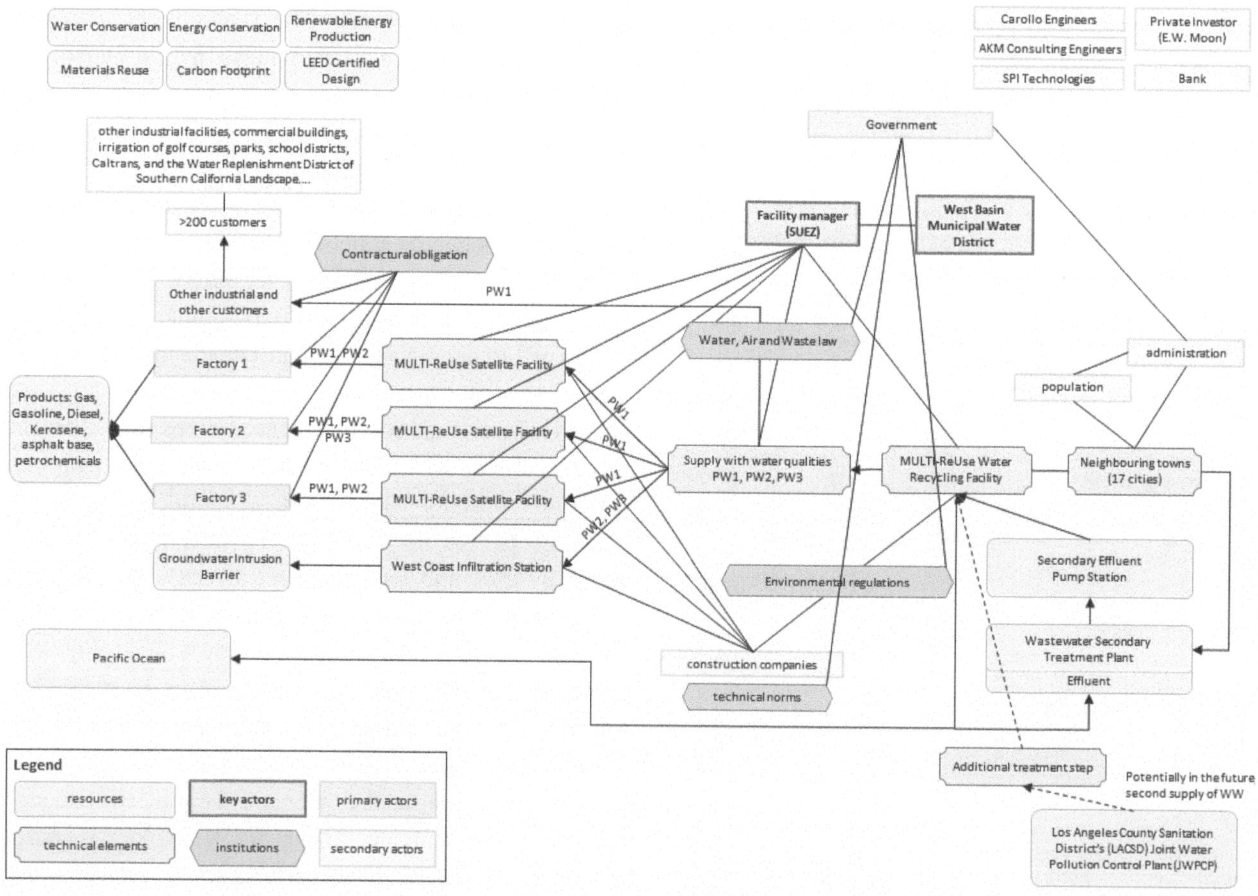

Figure 3 | CA of hypothetical industrial and other water reuse in California (for definitions of legend and actors cf. Figure 1).

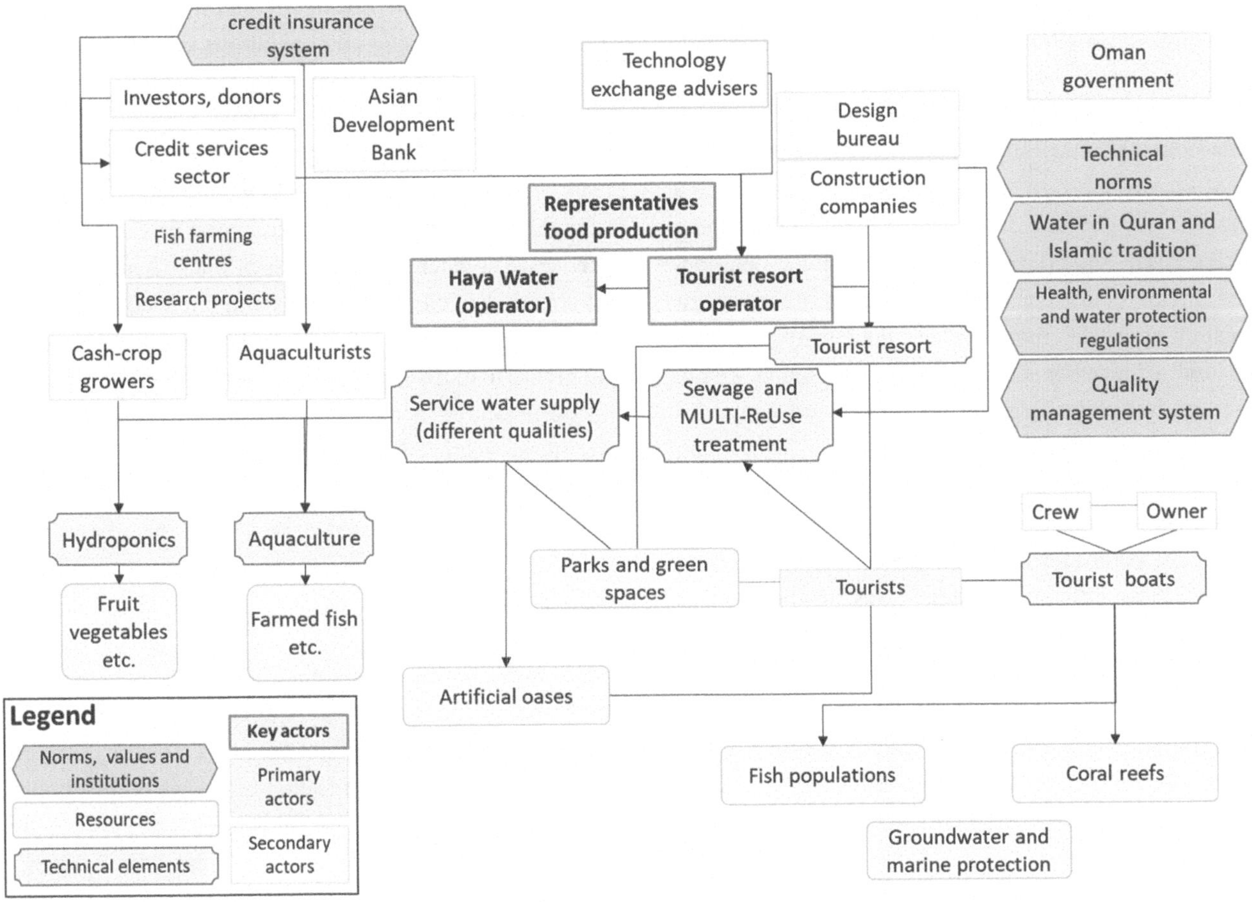

Figure 4 | CA of hypothetical rural water reuse with tourist resorts and food production in Oman (for definitions of legend and actors, cf. Figure 1).

salinity in the irrigation water can be reduced by mixing it with PW 3. The occasional use of RO water for the dilution of secondary treated wastewater for agricultural irrigation is also suggested by Bunani *et al.* (2015). Moreover, water for the pool can be produced with the help of PW 3. Multimodal water reuse can, therefore, be developed here.

When planning new tourist spots and green areas in water-scarce regions, the water infrastructure should be the starting point. Since different water qualities are supplied to different water users, the location of buildings and facilities should be based on treatment and supply of water, otherwise construction, maintenance and pumping might increase costs tremendously.

Wastewater must be treated and discharged in such a way that the sea is not polluted. After all, this is the main attraction for scuba divers and snorkelers in the coral reefs and for fishing tourists. At the same time, the hotel sector also requires high-quality fruit, vegetables and meat, preferably from locally controlled production. Given the scarcity of water, it is also necessary to irrigate oases.

For this, a certain constellation of actors must be available and willing to cooperate under a shared vision. As Haya Water is currently '… rebranding as a water reuse company…' (Freyberg 2018), they would be key actor and most suitable for guiding the coordination process between the different actors (see CA (Figure 3) for an overview). Next to Haya Water, the operator of the tourist resort and representatives of aquaculture and horticulture are also key actors. An institution that has been formed in Oman to support aquaculture is fish farming centres. Moreover, research and support for agricultural production such as horticulture comes from universities and government institutions. Aquaculturists and horticulturists with interest in cash crops might be suitable partners representing primary actors, alongside technology exchange advisers, design bureaus and construction companies who are secondary actors in the constellation.

One possible option to start and succeed in this process of knowledge exchange and negotiation during the initial phase of the project is to set up a board, e.g. chaired by a member of Haya Water, since the different competences for water reuse should be merged here. Joint decision-making on the board can include planning processes for the location of the different buildings, the quality assurance system and arrangements necessary for the supply chain. Tourist and leisure facilities include hotels and numerous (artificially created) palm oases, for example, which can also be used as lying or picnic areas, as well as a golf course and a leisure park with water-based facilities, e.g. rowing areas, coral and fishing grounds. It is conceivable to develop these different business plans on the basis of long-term cooperation. This can be based on long-term contracts, e.g. for the supply of fish, fruit and vegetables for the tourist gastronomy.

The constellation must remain stable in the long run in order to encourage cross-sectoral investments. Long-term contracts can bring about the desired stability if the resort is well accepted by tourism. A good prerequisite for this is a clean coastal environment, which depends on the success of water reuse in the Muscat region. Whether the drinking water is also produced by another, spatially isolated line of the plant using seawater desalination and mineralisation, or is obtained from the outside via a district water pipeline, depends on the local conditions.

The WWTP can either be financed and built by the state or with private funds. A private finance initiative is conceivable here. Veto players in the decision process can prevent or block water reuse. No veto players have been identified in the CA, showing a supportive situation.

The commission board is an instrument important for the cross-sectoral coordination in the initial and concept phase. In addition, the coordination later on during the operation can be arranged in different ways. The management of the tourism, agriculture and landscape park can be executed by a public–private partnership, but also by a public authority or a private company on its own. Other forms of operation are based on self-administration, such as a cooperative or an association.

For the constellation as analysed for Oman, and given the fact that also during the operations complex processes have to be coordinated, including quantitative adjustments and qualitative monitoring, a public authority such as Haya Water, which is also the operator of the WWTP, might be suitable to take over the operational coordination of the different facilities. Oman has good preconditions for water reuse in different sectors: Haya Water is an institution that already has the competences and knowledge for water reuse in agriculture. If Haya Water becomes a key actor in water reuse in tourism and food production (e.g. aquaculture and horticulture), it might be possible to cover the costs since these sectors have higher profit margins than traditional agriculture.

DISCUSSION

Multimodality means that several (e.g. 'dual') supply networks must be set up and operated on a permanent basis. Crucial to the cost of water is the length of the pipes. To save costs and reduce problems due to reinfection of the water, it is important that the advanced treatment takes place close to the water user. The WWTP, however, could be at some distance since one pipe is sufficient to supply treated wastewater for advanced treatment. The Californian example highlights this: the central facility is located near the WWTP, but has satellites in direct proximity to the customers. If only small quantities of one quality are required (e.g. RO water for a hotel's dishwasher), transport, e.g. by trucks, can be considered instead of a cost-intensive network of pipes.

The results of this CA confirm and specify a preliminary market analysis for multimodal water reuse (Becker *et al.* 2017). Fairly similar constellations to that in California can be assumed in comparable situations, especially in more industrialised countries such as in Europe or Mexico, China, India, South Africa and countries in the Middle East. As a result, multimodal water reuse could be implemented there. Comparable constellations to that in Oman can be expected in tourism destinations in Europe, the USA, the Middle East and North Africa, but also in countries in the Far East, Mexico, the Caribbean, New Zealand, Australia, South Africa and Kenya, so that opportunities for multimodal water use might also be realised there.

In order to prevent water shortages, it is necessary to involve intermediaries for the water management of a region with water stress. Intermediaries can facilitate processes and carry out a comprehensive problem analysis. They need to be independent of the 'free play of forces' otherwise a short-term analysis of the initial constellation alone may lead to the adoption of the wrong strategy. The Californian example shows that a high level of regional suffering (several supply crises caused by drought) leads to a rethink and an initially rejected task is pursued after all. In both cases, the CA clarifies that a large number of cooperation arrangements must be mastered in order to implement the discussed innovation successfully in the long term. Cooperation management can support this (Kerber *et al.* 2017).

Both constellations are characterised by joint use of new water resources. The coordination for planning and operation can either be carried out by a utility or water agency as shown in the CA, but also by private entities such as the users themselves (e.g. in a cooperative). A private initiative of this kind can start by means of integrated business planning for supply chain design and collaborative planning. For the conversion of land into tourism, leisure facilities and food production systems area development are required. This involves the construction of a new WWTP, including advanced treatment and distribution of various (guaranteed) qualities of process water. In the initial phase, as well as the coordination of planners, water technology advisers, aquaculture and horticulture advisers, the financing of initial investments and financial architecture for the different businesses is required. In this real estate development scenario, a key actor might be a private enterprise experienced in area development and infrastructure projects.

For constellations with industry partners as well as for those with tourism partners, there was an initial investigation (not documented here) about whether such private partners could also play a key role as a 'central node' and act as a catalyst. Public actors would seem to be more capable of taking all the different interests involved in water reuse into account.

In the two constellations examined, state (or intermunicipal) intermediaries play a decisive role. A utility or an intermediary has a professional perspective on the users' water quality requirements. Therefore, it is possible to find a suitable typology of process water, pooling the different user requirements in order to increase cost efficiency. These classes are also related to the WWTP effluent and the available technology.

In one of the two cases, a constellation in which the national wastewater company is the key actor coordinating the important customers has proven to be stable. As a counterweight in this constellation, it will be necessary for the tourist resort and other users to participate responsibly on a board. However, state (or inter-municipal) independence is not required for the implementation and actual operating phase of the multimodal water reuse. Once the planning phase is finalised, the intermediary can also assign an operator from the private sector (e.g. SUEZ).

CONCLUSIONS

Reclaimed wastewater can be used for many different purposes. In practice, when developing a market for advanced water reuse, it is problematic to limit this to a single use and a single user group since an increase in water stress will affect all user groups. In this respect, a fit-for-purpose approach can be pursued. Based on further constellation analyses and supplementary investigations, it will be possible to introduce multimodal water reuse systematically.

Multimodal water reuse is a technically complex and cost-intensive process. There are constellations in which advanced water reuse is favoured. The CA makes it plausible that advanced treatment for reuse will be implemented more frequently in future due to a reduction in the associated risk to human health and the environment. Furthermore, if constellations can be established by forward-looking governmental agencies, private actors and civil society, such cases will increasingly replace simpler forms of water reuse. The target visions linked to water reuse in the United Nations' Sustainable Development Goals can then be realised sustainably.

The CA revealed a greater complexity regarding cooperation between actors compared with traditional water reuse. Hence, the venture needs to be supported by reliable (perhaps even strong) institutions as essential network nodes, supported by steering committees and participative planning processes for example.

Another drawback of multimodal water reuse is the cost of advanced treatment. In countries where the price of (irrigation) water is not subsidised, the supply of sufficiently treated water will only be available to target groups with sufficient capital. If water treatment can be covered by cross-financing, it might be possible to provide farmers with irrigation water at prices that are adapted to their income and the market prices of their products (e.g. by cross-subsidising the water tariff of irrigation water with the tariff of RO water). In this way, local food production, small-scale agriculture and landscape quality will be preserved.

Based on specific local preconditions, other opportunities for sustainable water supply always need to be considered. Moreover, further research is required to understand the welfare and distribution effects of multimodal water pricing policies and the feasibility of co-financing agricultural irrigation.

The following favourable constellations can be noted:

1. Multimodal water reuse can be established either when sewage transport and (semi-) centralised wastewater treatment is already in place, or in the case of new development.
2. A multimodal approach is possible in areas in which water users carry out activities that require a wide range of different water qualities.
3. Multimodal water reuse is appropriate in constellations in which cooperation between heterogeneous actors (from operators of WWTPs to monitoring institutions and customers) can be initiated and maintained in the long term.
4. In principle, multimodal water reuse is suitable for countries that:

    - offer a high degree of legal security and contractual reliability (customers have to take the water for an agreed period of time),
    - are comparatively wealthy and have the possibility to engage qualified personnel (advanced wastewater treatment depends on energy and material input as well as on knowledge and skills) and
    - have different water users capable and willing to pay the cost (in a way that ultimately covers costs).

The stated conditions allow countries to be identified in which constellations can be formed beneficially for multimodal water reuse. Industrialised countries and countries in transition are eligible. These include BRIICS, as well as Thailand, South Korea, OPEC countries and possibly some other MENA countries as well.

ACKNOWLEDGEMENTS

The authors acknowledge financial support of the project 02WAV1403 by the German Federal Ministry of Education and Research.

DATA AVAILABILITY STATEMENT

All relevant data are included in the paper or its Supplementary Information.

REFERENCES

Al-Balooshi, A. & Charabi, Y. 2012 *Assessment of Land Degradation in Al-Bathina Region*. Oman. Available from: https://www.squ.edu.om/Portals/97/Activities/Workshop/WWD2017/HayaWater.pdf (accessed 27 March 2020).

Al-Farsi, R., Ahmed, M., Al-Busaidi, A. & Choudri, B. S. 2018 Assessing the presence of pharmaceuticals in soil and plants irrigated with treated wastewater in Oman. *International Journal of Recycling of Organic Waste in Agriculture* **7** (2), 165–172. http://dx.doi.org/10.1007/s40093-018-0202-1.

Al-Riyami, I. M., Ahmed, M., Al-Busaidi, A. & Choudri, B. S. 2018 Antibiotics in wastewaters: a review with focus on Oman. *Applied Water Science* **8** (7), 2046. http://dx.doi.org/10.1007/s13201-018-0846-z.

Baawain, M., Sana, A., Al-Yahyai, R. & Amoatey, P. 2019 Sustainable reuse options of treated effluents: a case study from Muscat, Oman. *Journal of Environmental Engineering and Science* 1–9. http://dx.doi.org/10.1680/jenes.19.00019.

Becker, D., Frey, A., Jungfer, C., Krömer, K., Kulse, P., Maaßen, S., Schramm, E., Wencki, K., Zimmermann, B. & Zimmermann, M. 2017 *Marktpotentiale der Wasserwiederverwendung: Anforderungen und Kriterien in unterschiedlichen Sektoren und mögliche Zielmärkte für das MULTI-ReUse-Verfahren (Market Potential of Water Reuse: Requirements and Criteria in Different Sectors and Possible Target Markets for the MULTI-ReUse-Process)*. ISOE Materialien Soziale Ökologie 49.

BMWi – Federal Ministery for Economic Affairs and Energy (ed.) 2018 *Wasserwirtschaft Kalifornien (Water Management in California): Zielmarktanalyse 2018 mit Profilen der*

Marktakteure (Analysis of Target Market 2018 Including Profiles of Market Players). Berlin.

Bunani, S., Yörükoğlu, E., Yüksel, Ü., Kabay, N., Yüksel, M. & Sert, G. 2015 Application of reverse osmosis for reuse of secondary treated urban wastewater in agricultural irrigation. *Desalination* **364**, 68–74. http://dx.doi.org/10.1016/j.desal.2014.07.030.

Choudri, B. S., Al-Busaidi, A. & Ahmed, M. 2013 Climate change, vulnerability and adaptation experiences of farmers in Al-Suwayq Wilayat, Sultanate of Oman. *International Journal of Climate Change Strategies and Management* **5** (4), 445–454. http://dx.doi.org/10.1108/IJCCSM-11-2012-0061.

Drewes, J. E., Schramm, E., Ebert, B., Mohr, M., Beckett, M., Krömer, K. & Jungfer, C. 2019 Potentiale und Strategien zur Überwindung von Hemmnissen für die Implementierung von Wasserwiederverwendungsansätzen in Deutschland (Potentials and strategies to overcome barriers of the implementation of ideas for water reuse in Germany). *KA Korrespondenz Abwasser, Abfall* **66** (12), 995–1003.

Espey, R. 2018 *Oman Investiert in Umwelttechnik: Ausbau des Abwassernetzes/Moderne Müllentsorgung im Aufbau*. Dubai/Maskat. Available from: https://www.gtai.de/GTAI/Navigation/DE/Trade/Maerkte/suche,t=oman-investiert-in-umwelttechnik,did=1902744.html (accessed 10 October 2019).

European Union 2020 *Regulation (EU) 2020/741 of the European Parliament and of the Council of 25 May 2020 on Minimum Requirements for Water Reuse*. Official Journal of the European Union L 177/32.

FAO – Food and Agriculture Organization of the United Nations (ed.) 2008 *Country Profile – Oman: Version 2008*. Rome.

FAO – Food and Agriculture Organization of the United Nations (ed.) 2017 *Country Programming Framework for the Sultanate of Oman*.

Freyberg, T. 2018 *Taking Water Reuse Haya Across Oman*. Available from: https://www.waterworld.com/municipal/technologies/article/16201270/taking-water-reuse-haya-across-oman (accessed 10 October 2019).

Haya Water (ed.) n.d *Haya Water: Presentation by Buthaina Mahfoud Al Whaibi*. Available from: https://www.squ.edu.om/Portals/97/Activities/Workshop/WWD2017/HayaWater.pdf.

Joham, J. 1978 West Basin's plans for water reuse. In: *Industrial Water Reuse: Conference Proceedings*, Culver City, pp. 45–48.

Kennedy, L. & Tsuchihashi, R. 2005 Is water reuse sustainable? Factors affecting its sustainability. *Arabian Journal for Science and Engineering* **30**. Available from: https://www.researchgate.net/publication/228379299_Is_Water_Reuse_Sustainable_Factors_Affecting_Its_Sustainability (accessed 25 March 2020).

Kerber, H., Kunkis, M. & Schramm, E. 2017 Kooperationsmanagement: Ein Instrument zur Differenzierung der Wasserinfrastruktur (Cooperation management: an instrument for differentiation of water infrastructure). In: *Wasserinfrastruktur: Den Wandel gestalten: Technische Varianten, räumliche Potentiale, institutionelle Spielräume. Edition Difu – Stadt Forschung Praxis, 16 (Water Infrastructure: Shape the Change: Technical Variants, Potentials of Area, Institutional Scopes)* (M. Winker & J. H. Trapp, eds). Difu, Berlin, pp. 219–235.

Kotagama, H. B. & Al-Farsi, H. 2019 Impact of the domestic labor market on sustainability of agriculture in Oman. *Journal of Agricultural and Marine Sciences [JAMS]* **23** (1), 24–24. http://dx.doi.org/10.24200/jams.vol23iss1pp24-28.

Lazarova, V., Oelker, G. & Oelker, W. 2012 Production of custom-made recycled water for various reuse purposes: lessons learned from one of the world's largest recycling facilities. *Water Practice and Technology* **7** (3). http://dx.doi.org/10.2166/wpt.2012.052.

Lehane, S. 2015 *The Sultanate of Oman: Food and Water Security to 2025*. (accessed 9 October 2019).

MAF – Ministery of Agriculture and Fisheries & ICBA – Sultanate of Oman International Center for Biosaline Agriculture Dubai (ed) 2012 *Oman Salinity Strategy: Annex 4. Governance, Legal/Regulatory Frameworks and Policies in Agriculture and Water*.

McDonnell, R. 2016 *Groundwater use and Policies in Oman*. IWMI report 14

Mendoza, J. 2019 First Sustainable Greenhouse Research Facility Launched in Oman. *Oman Daily Observer*, (16.03.2019). Available from: https://www.omanobserver.om/first-sustainable-greenhouse-research-facility-launched-in-oman/ (accessed 21 January 2020).

MRMWR – Ministry of Regional Municipalities and Water Resources – Oman (ed.) 1993 *Ministerial Decision 145/93 Dated 13 June 1993: Regulations for Wastewater Re-use and Discharge*.

Nahrstedt, A., Gaba, A., Zimmermann, B., Jentzsch, T., Krömer, K., Tiemann, Y., Harsanyi, L., Buchta, P., Dölchow, U., Lipnizki, J., Mende, K., Koch, T. & Rohn, A. 2020 Reuse of municipal wastewater for different purposes based on a modular treatment concept. *Journal of Water Reuse & Desalination* (this issue). https://doi.org/10.2166/wrd.2020.040.

Nocker, A., Schulte-Illingheim, L., Müller, H., Rohn, A., Zimmermann, B., Gaba, A., Nahrstedt, A., Mohammadi, H., Tiemann, Y. & Krömer, K. 2020 Microbiological changes along a modular wastewater reuse treatment process with a special focus on bacterial regrowth. *Journal of Water Reuse and Desalination* (this issue). https://doi.org/10.2166/wrd.2020.012.

Nölting, B. & Mann, C. 2018 Governance strategy for sustainable land management and water reuse: challenges for transdisciplinary research. *Sustainable Development* **26** (6), 691–700.

Ohlhorst, D. & Schön, S. 2015 Constellation analysis as a means of interdisciplinary innovation research – theory formation from the bottom up. *Historical Social Research/Historische Sozialforschung* **40** (3), 258–278.

Rizzo, L., Malato, S., Antakyali, D., Beretsou, V. G., Đolić, M. B., Gernjak, W., Heath, E., Ivancev-Tumbas, I., Karaolia, P.,

Lado Ribeiro, A. R., Mascolo, G., McArdell, C. S., Schaar, H., Silva, A. M. T. & Fatta-Kassinos, D. 2019 Consolidated vs new advanced treatment methods for the removal of contaminants of emerging concern from urban wastewater. *The Science of the Total Environment* **655**, 986–1008. http://dx.doi.org/10.1016/j.scitotenv.2018.11.265.

Schön, S., Nölting, B. & Meister, M. 2004 Konstellationsanalyse: Ein interdisziplinäres Brückenkonzept für die Technik, Nachhaltigkeits- und Innovationsforschung (Constellation analysis: an interdisciplinary concept for technic, sustainability and innovation research). Discussion paper Nr. 12/04.

Seis, W., Lesjean, B., Maaßen, S., Balla, D., Hochstrat, R. & Düppenbecker, B. 2016 *Rahmenbedingungen für die umweltgerechte Nutzung von behandeltem Abwasser zur landwirtschaftlichen Bewässerung (Frame Conditions for an Environmental Friendly Usage of Treated Sewage Water for Agricultural Irrigation).* UBA-Texte 34/2016. Dessau-Roßlau.

SUEZ group (ed.) 2020 *SUEZ Renews the Contract to Operate One of the Largest Wastewater Recycling Plants in the United States.* Available from: https://www.suez.com/en/news/press-releases/suez-renews-contract-operate-one-of-the-largest-wastewater-recycling-plants-united-states (accessed 25 March 2020).

The Heritage Foundation (ed.) 2019 *Index of Economic Freedom: OMAN.*

Water Technology (ed.) 2016 *U.S. Water Industry Revenues Rise to $160 Billion: Water Instruments and Information Segment Leads Growth.* Available from: https://www.watertechonline.com/wastewater/article/16205061/us-water-industry-revenues-rise-to-160-billion (accessed 23 March 2020).

World Bank (ed.) 2019 *OMAN.* Available from: http://documents.worldbank.org/curated/en/984551556570728787/pdf/Macro-Poverty-Outlook-for-Oman.pdf (accessed 21 January 2020).

First received 1 May 2020; accepted in revised form 1 September 2020. Available online 25 November 2020

Reuse of municipal wastewater for different purposes based on a modular treatment concept

Andreas Nahrstedt, Anil Gaba, Barbara Zimmermann, Timo Jentzsch, Kerstin Kroemer, Yannick Tiemann, Lajos Harsanyi IWA, Patrick Buchta IWA, Uli Doelchow, Jens Lipnizki, Katharina Mende, Thomas Koch and Anja Rohn

ABSTRACT

Due to water scarcity and water pollution, the importance of water reuse is increasing more and more. As part of a German research programme on water reuse, the effluent of a wastewater treatment plant in the coastal region of northern Germany was used to investigate within the project MULTI-ReUse the direct treatment of tertiary effluent for usage in different applications in industry or agriculture. A modular constructed pilot system has been operated to optimize different treatment chains producing different water qualities simultaneously. The technological focus was put on membrane technologies, namely ultrafiltration (UF) and reverse osmosis (RO), and also biofiltration, adsorption and disinfection were part of the piloting. Beside the development of monitoring strategies for ensuring biological and chemical safe water qualities, the operational stability and the safe transport of water to the consumers were examined. The direct treatment of wastewater is a demanding task due to the lack of dilution and hydraulic retention time in the receiving water (environmental buffer). However, the multiple barrier approach guaranteed constant secure water. Fine adjustments of individual processes were particularly important. A stable operation of the UF could be realized in particular by using more or less intermittent inline coagulation as coating. The RO performance could be improved significantly by using monochloramine as disinfectant to minimize biofouling.

Key words | adsorption, organic micropollutants, pathogen removal, reverse osmosis, ultrafiltration, water reuse

Andreas Nahrstedt (corresponding author)
Anil Gaba
Barbara Zimmermann
Timo Jentzsch
Anja Rohn
IWW Water Centre gGmbH,
Moritzstr. 26, 45476 Mülheim an der Ruhr,
Germany
E-mail: *a.nahrstedt@iww-online.de*

Kerstin Kroemer
Yannick Tiemann
Oldenburgisch-Ostfriesischer Wasserverband,
Brake (Unterweser),
Germany

Lajos Harsanyi IWA
Patrick Buchta IWA
inge water technologies GmbH,
Greifenberg,
Germany

Uli Doelchow
Jens Lipnizki
LANXESS – IAB Ionenaustauscher GmbH Bitterfeld,
Bitterfeld-Wolfen,
Germany

Katharina Mende
SUEZ Water Technologies & Solutions,
Ratingen,
Germany

Thomas Koch
De.EnCon GmbH,
Oldenburg,
Germany

HIGHLIGHTS

- The first project in Germany on direct reuse of municipal wastewater for industrial purposes.
- Optimization of UF operation by initial hydroxide coating followed by low levels of coagulant dosing.
- Implementation of an ultra-low-pressure reverse osmosis membrane (ULPRO).

INTRODUCTION

Water reuse can be of interest for various reasons. On the one hand, more frequent and severe drought conditions due to climate change (Solomon *et al.* 2007), excessive water consumption and competing environmental, industrial and agricultural needs for water can stress the availability of conventional freshwater resources (Drewes & Horstmeyer 2015). On the other hand, the low quality of regionally available fresh water resources gives reason to

doi: 10.2166/wrd.2020.040

focus on the use of alternative resources. In this context, collected wastewater has been explored as an additional water resource since decades (Asano *et al.* 2007). Up to date, numerous wastewater reclamation facilities have been successfully implemented for different water reuse purposes (Curl *et al.* 2019). Even the direct potable reuse is becoming more and more accepted as state-of-the-art technologies have proved to provide the desired water quality with high reliability.

Despite the successes that water reclamation projects have achieved internationally, in Germany this topic is still being discussed very critically and defensively by many public actors (UBA 2017). However, according to climate change predictions, more warm and dry summers are to be expected in central Europe. The effect on the groundwater level and surface water reservoirs could be noticed in several areas in Germany during the years 2018 and 2019 (Hellwig *et al.* 2020). The expected simultaneous rise in sea levels also leads to a disturbance in the balance between groundwater and the seawater interface, which results in increasing salinization of the groundwater body. This makes the North Sea coast of Europe a good example for the problem of saltwater intrusion (Auken *et al.* 2012).

Against this background, a regional water board in Lower Saxony, Germany, decided to explore the possibilities of municipal wastewater reuse for industrial purposes in a local case study. Nordenham is a mid-sized German town with water-intensive industries and trades. It is located in the Wesermarsch, a coastal region in the northwest of Germany, which is affected by saltwater intrusion. The entire district does not have its own drinking water supply. In the surrounding districts, drinking water obtained from groundwater resources is treated and pumped over long distances to the industrialized area. Here, the water demand for industrial processes with very low water quality requirements is covered as much as possible with water from the high salinated river Weser. However, for the majority of applications, drinking water is used (Kroemer *et al.* 2019). To lower or to completely avoid the demand for drinking water in cooperating industrial companies could lead to a significant relief of the used drinking water aquifers. This case study was a main part of the MULTI-ReUse project of a German multicentre research consortium. Within the project, a treatment system for water reuse has been tested at the site of the municipal wastewater treatment plant (WWTP) of Nordenham with the aim of producing process water in three different qualities.

The technological studies focused on the membrane processes ultrafiltration (UF) and reverse osmosis (RO) supplemented by further treatment steps. Depending on the intended use, the processes can be combined into different treatment chains (Fit-for-Purpose) (EPA/600/R-12/618 2012). By choosing these state-of-the-art technologies, focus was on the demonstration of technologies under site-specific conditions. Additionally, the development of new RO membranes and microbial monitoring methods was supported. This article describes the set-up and the main results of the pilot plant operation, beside the aspects of waste disposal and water transport are discussed for the specific case study.

MATERIALS AND METHODS

Pilot plant

The MULTI-ReUse pilot plant consists of modular units for three treatment lines (Figure 1):

1. *Line 1 for ReUse Water 1*: pre-filtration (sieve), optional adsorption with powdered activated carbon (PAC), flocculation, UF and UV disinfection; water for rinsing and cooling purposes.
2. *Line 2 for ReUse Water 2*: pre-filtration, flocculation, UF, UV disinfection, sand filtration and granulated activated carbon (GAC) filtration; water for irrigation purposes.
3. *Line 3 for ReUse Water 3*: pre-filtration, optional PAC and flocculation, UF, chemical disinfection, RO and UV; water for industrial purposes.

Two UF lines were operated in parallel. The treatment lines were protected by a sieve filter (AMIAD TAF-750) with a mesh size of 200 μm. The sieve filter was rinsed automatically. Coagulation was used prior to UF to reduce fouling by decreasing the cake resistance, limiting pore blockage and increasing the backwash efficiency (Barbot *et al.* 2008). Coagulation also increases the removal of organic matter, phosphate and disinfection by-product (DBP) precursor (Guigui *et al.* 2002). Aluminium

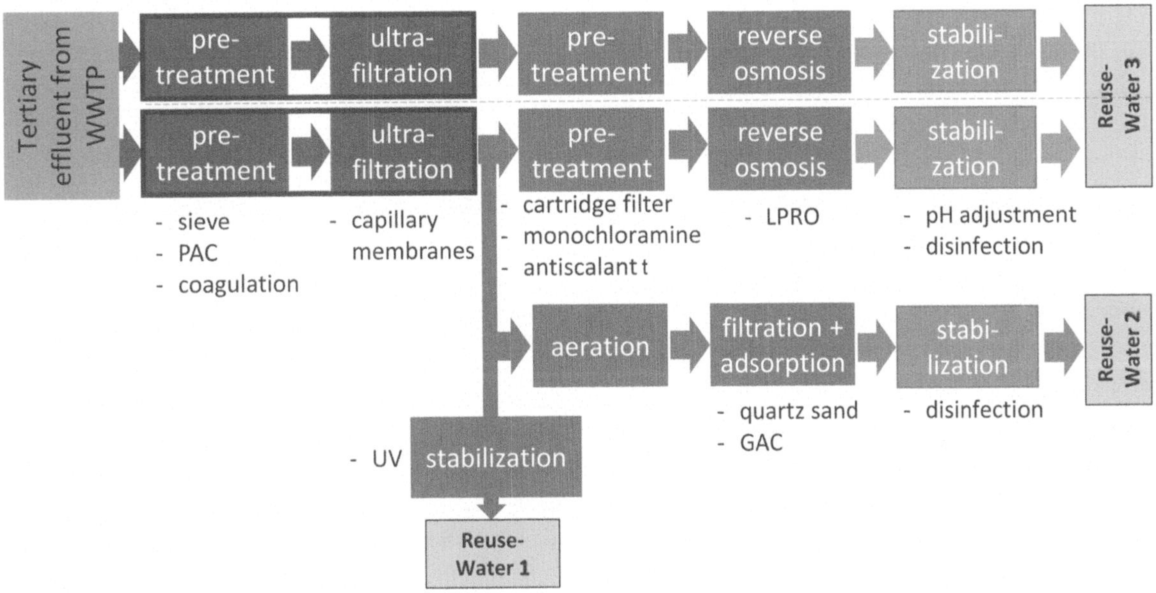

Figure 1 | Flow chart of the pilot plant.

hydroxychloride (feralco PLUSPAC FD ACH)- and ferric-based coagulants ($FeClSO_4$ or $FeCl_3$, CG CHEMIKALIEN) were used with a dosage of $[Al^{3+}] = 2\text{–}3$ mg/L and $[Fe^{3+}] = 6\text{–}7$ mg/L. The average hydraulic retention time for microfloc formation was optimized to 40 s.

Hard coal-based PAC (Chemviron PULSORB® WP260-90 S) with a mean size of 0.3 μm (number concentration, Beckman Coulter LS 13,320) was used for the removal of organic micropollutants and DBP precursor (Lin *et al.* 1999; Ivancev-Tumbas *et al.* 2008; Stoquart *et al.* 2012).

After pre-treatment, the water was filtrated with two large-scale UF modules (dizzer XL 0.9 MB 80 WT, INGE; polyethersulphone, capillary inner diameter 0.9 mm; pore diameter 0.02 μm and membrane surface area 80 m^2). The flux varied from 60 to 70 L/(m^2·h), and the filtration time was set to 45 min. Chemical enhanced backwash (CEB) with NaOH (pH ~12.2) and H_2SO_4 (pH ~2.1) was done one up to two times per day. Cleaning-in-place (CIP) intervals were dependent on the amount of fouling but generally occurred one up to two times per year with 300 mg Cl_{free}/L (pH ~12.2) by dosing NaClO and 4 g of oxalic acid/L (pH ~1.8). For ReUse Water 1, the filtrate was finally disinfected by UV radiation (sterilAir® AQD64-4 K).

For ReUse Water 2, the water was further treated successively by inline aeration, biologically active deep bed filtration (bed height 2 m in two serial columns; quartz sand, grain diameter 1–2 mm; filtration rate 5.5 m/h, empty bed contact time (EBCT) 20 min) and GAC filtration (bed height 2 m in two serial columns; Hydraffin AR, 8 × 30 mesh, Donau Carbon; filtration rate 5.5 m/h, EBCT 20 min). Deep bed filtration is required for catalytic and biological oxidation of dissolved Mn(II). GAC filtration was operated for removing micropollutants and DOC. Both filtration steps also serve for biodegradation of BDOC (biodegradable dissolved organic carbon), nitrification (microbiological stabilization) and the growth of autochthonous bacteria to lower the regrowth potential of pathogens (seed experiments).

To prevent massive biofouling on RO membranes and their pre-filters, NaClO and NH_4Cl solutions were added consecutively on line 1 directly into the UF filtrate stream in the stoichiometric ratio of 2:1. Both react *in situ* to 1 mg/L of monochloramine:

$$NH_4Cl + NaClO \rightarrow NH_2Cl + H_2O + NaCl \quad (1)$$

The hydraulic retention time of monochloramine was, due to the 1.8 m^3 UF filtrate tank, approximately 23 min. Free chlorine concentrations up to 0.1 mg/L were tolerated according to the RO manufacturer's requirements. Therefore, the dosage of sodium bisulphite as a scavenger for free chlorine was not necessary. Free and bound chlorine concentrations in the RO feed, concentrate and permeate were checked with the photometric DPD (N, N-diethyl-p-phenylenediamine) method. Dosing times of 22 h down to 4 h were tested, and total and intact cell counts (ICCs) were monitored online with flow cytometric analysis (OBA, Metanor AG). For preventing scale formation on the RO membrane, 3.7 mg/L of the scale inhibitor (Free-Flow 4, Dr Naehring) was added.

Non-woven fabric cartridge filters (WFMB melt-blow cartridge, polypropylene, nominal pore size 1 μm) protected the RO membrane against coarse particles. The load with TSS (total suspended solids) and biofilms of the filter was monitored by the differential pressure measurement. Two series of RO pressure vessels were operated in parallel, each fed by its own UF module. In each series, three 4-inch (101.6 mm) RO elements were housed in three RO pressure vessels. With this configuration, LPRO (low-pressure RO) elements from the Lewabrane® product series were tested in parallel (LANXESS Lewabrane® B085 ULP 4040, B085 FR 4040, B085 LE 4040; membrane area 7.9 m^2 per element). Finally, the RO permeate was disinfected with UV-light (sterilAir® AQD 64C-4 K).

Analytical methods and calculations

The chemical analytics of the water samples was done by the accredited laboratories of EUROFINS, IWW and OOWV. The analysis of the microbiological parameters was carried out by the laboratory of the Public Health Department of Aurich. The applied methods are listed in Table 3 (Appendix).

Several parameters have also been analysed onsite, such as the SDI (Silt Density Index, Portable SDI/MFI-Analyzer, Convergence), free chlorine and bound chlorine by the DPD method (chemicals: Huebers GmbH; photometer: ALLDOS 310-055-1000), pH, O_2 and conductivity (WTW Multi 3630 IDS) and turbidity (Nephla, HACH-Lange).

Since water temperatures can have a significant influence on the measured membrane permeability, it was normalized to the reference temperature of 20 °C for the purpose of monitoring the membrane resistance, including fouling effects (fouling layer and pore blockage). The following equation cited by Mallevialle *et al.* (1996) is used to normalize the UF permeability $P_{\vartheta,\mathrm{UF}}$ to 20 °C in which $\mathrm{TCF}_{20,\mathrm{UF}}$ is the temperature correction factor and ϑ the actual water temperature in °C.

$$\mathrm{TCF}_{20,\mathrm{UF}} = \frac{P_{\vartheta,\mathrm{UF}}}{P_{20,\mathrm{UF}}} = \exp\left[-0.0239\left(\frac{\vartheta}{[^\circ\mathrm{C}]} - 20\right)\right] \quad (2)$$

To assess the fouling of the RO membrane, the normalized permeate flow was calculated according to the international standards (ASTM D4516 2010).

Sampling scheme

Throughout the pilot phase, water samples were taken monthly as random samples along the entire process chain. Sampling for the investigation of organic micropollutants was carried out as 24-h mixed samples at selected times in order to characterize the micropollutant retention of the RO and the GAC. The device for taking the mixed samples collected a distinct volume into a sampling vessel automatically every 30 min.

RESULTS AND DISCUSSION

Operation conditions

The pilot study started in August 2017. Up to December 2018, each treatment step was commissioned and optimized individually. In the period from January to December 2019, all treatment steps were operated in combination and further optimized to demonstrate the stability of the processes and the water quality.

In the WWTP of Nordenham, physical, chemical and biological processes are combined for the treatment of the municipal wastewater, which is collected from households, small and medium enterprises, and a hospital. The last treatment step of the WWTP is a clarifier. The sewer system consists of both separate and combined rainwater and sanitary sewers. In dry weather periods, the effluent volume flow is ~4.000 m^3/d. During rain events, it increases by the factor of three. The effluent shows the seasonal variations of high ammonium and nitrite peaks during periods with low temperature and can be characterized as a typical municipal WWTP effluent unaffected by big industry. One special characteristic of the raw and treated wastewater is the high content of dissolved manganese, which enters the sewer system with the groundwater. The results of the major physical–chemical parameters that have been monitored in a monthly routine are listed in Table 2.

As hygienic safety of customers and operators is a prevalent goal in water reclamation, a regular monitoring of traditional microbial hygiene indicator bacteria was performed. Figure 2 shows the results of this monitoring for the WWTP effluent.

Although the purpose for water reuse in the case study at Nordenham was not irrigation but industrial use, the elimination of micropollutants by RO and GAC has been investigated at the pilot plant, because water reuse for agricultural or landscape purposes is of interest in some areas in Germany. Table 1 shows a selection of micropollutants that were found in the wastewater effluent of the WWTP of Nordenham.

Produced water qualities: hygienic aspects

During the entire pilot phase, the UF proved to be an efficient microbiological barrier: the filtrates were – apart from the commissioning phase – free of *Escherichia coli* and coliform bacteria (<1 MPN/100 mL) as well as intestinal enterococci, *Clostridium perfringens* and *Legionella* sp. (<1 CFU/100 mL). If the maximum values in the feed (usually coliform bacteria $>1\times10^5$ CFU/100 mL and enterococci up to 2.2×10^6 CFU/100 mL, see Figure 2) are taken into account, permanent bacterial retention of more than 5–6 log units was achieved by the UF. Beside the traditional microbiological hygiene indicator bacteria, the total cell counts have been monitored by means of flow cytometry. The results of online flow cytometric measurements (sampling in cycles of 15 min) presented 10^6–10^7 intact cells/mL in the feed and 10^2–10^3 intact cells/mL in the filtrate of the UF. Thus, this real-time monitoring method indicated only 3–4 log units of bacterial retention. The different retention results can be explained by the differences in the investigation methods. As with the flow cytometry all floating cells are detected, also autochthonous microbial flora from the non-sterile filtrate containing parts (e.g. cells which are released from biofilms inside the inner wall of the filtrate pipes) are counted, which actually should not be taken into account for the evaluation of the UF retention. This means, by looking at the total cell counts in the non-sterile system, that the UF retention is likely to be underestimated. As

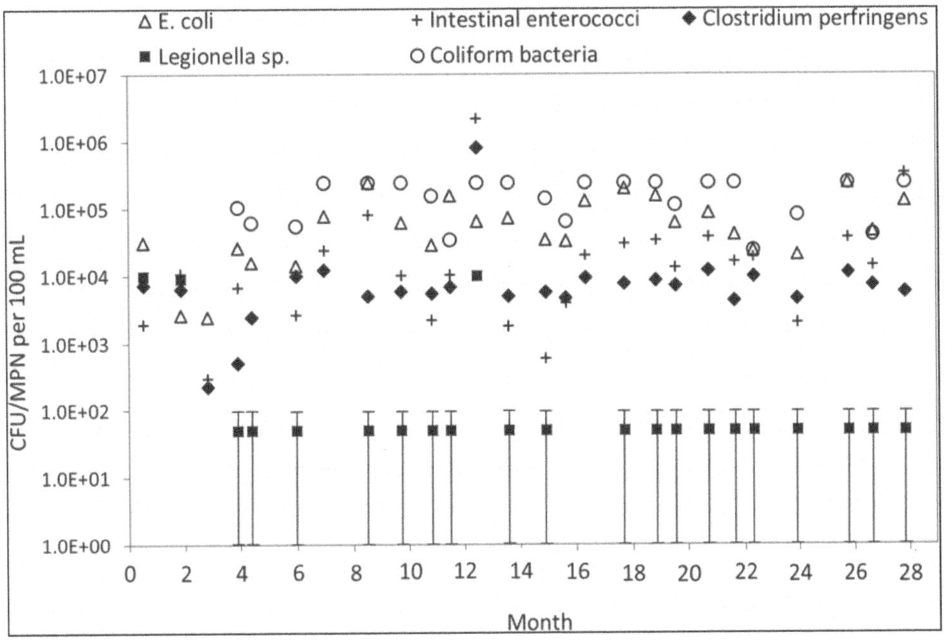

Figure 2 | Microbiological findings of hygiene indicator bacteria in monthly random samples of the WWTP effluent after pre-filtration (sieve, 200 μm) during the entire period of the pilot plant operation.

Table 1 | Selection of detected organic micropollutants in the effluent of the WWTP (24-h mixed samples)

		Minimum	Average	Median	Maximum	*n*
1H-Benzotriazole	µg/L	4.3	6.0	6.4	7.7	8
4-Methyl-1H-benzotriazole	µg/L	6.0	10.2	10.5	15.4	8
5-Methyl-1H-benzotriazole	µg/L	0.3	0.6	0.6	0.9	8
Acesulfame	µg/L	1.0	2.0	1.3	3.4	7
AMPA	µg/L	1.4	2.1	1.8	3.1	8
Carbamazepine	µg/L	0.3	0.7	0.7	1.0	9
Clarithromycin	µg/L	0.1	0.2	0.2	0.4	9
Diclofenac	µg/L	1.4	2.6	2.5	3.6	9
EDTA	µg/L	0.1	12.6	15.1	27.8	7
Glyphosate	µg/L	0.3	0.4	0.4	0.7	8
Iomeprol	µg/L	0.1	1.3	1.2	2.5	8
Iopromide	µg/L	2.5	8.8	8.5	16.4	8
Metoprolol	µg/L	0.7	1.7	1.4	3.5	9
Sulfamethoxazol	µg/L	0.0	0.0	0.2	0.3	9
TMDD	µg/L	0.2	1.2	1.0	3.5	8

microbiological monitoring was a main focus within the MULTI-ReUse project, a separate publication on this topic has been submitted by Nocker *et al.* (2020). For further detailed information on methods and results, this publication is recommended. Virus retention was not examined during the MULTI-ReUse project. However, for the chosen UF membrane type, a retention of >4 log units was detected in a preliminary project in the laboratory experiments using MS2 bacteriophages (Lipp *et al.* 2017) with a diameter of approximately 24–26 nm (Wick & McCubbin 1999).

The above-mentioned hygiene indicators could not be detected in the filtrates of the sand and GAC filters and in the RO permeates as well. These results are certainly supported by the bacteria retention of the UF, which protects the whole subsequent treatment steps from microbial contaminations of the WWTP effluent. The total cell counts in the filtrates of the sand and GAC filters were significantly higher (10^5–10^7 cells/mL) than in the UF filtrate. This result was expected as the filters were operated to enable biological activities inside the filter bed. The RO permeates showed total cell counts on the same low level as the UF filtrates. However, a bacterial regrowth in the UF filtrate and the RO permeate was detected after storing the samples for 7 days at 22 °C, if no chemical disinfectants were applied. This proves that the water samples still contained assimilable nutrients and an autochthonous bacterial flora which is able to multiply under these conditions. For further information, refer Nocker *et al.* (2020).

Produced water qualities: physical–chemical parameters

Results on the chemical composition of the water generated and the WWTP effluent are shown in Table 2.

The ReUse Water 1 is free of particles and pathogens. Flocculation before the UF reduced the DOC concentration on average to 10 mg/L (−19%) and the concentration of P_{total} to 0.061 mg/L. The concentrations of the metals such as aluminium and iron contained in the wastewater effluent were also significantly reduced by 61 and 93%, whereas dissolved salts pass through the UF membrane.

Further treatment of ReUse Water 1 by filtering it through a biological active quartz sand filter and a GAC filter leads to ReUse Water 2 quality. ReUse Water 2 is additionally low in manganese (−68%), ammonium (−70%) and organic micropollutants after the GAC passage. Figure 3 shows the removal efficiency of the GAC filter for several micropollutants at increasing operating times,

Table 2 | Average values and standard deviation of the analysis results of the WWTP effluent and three produced ReUse Water types

Parameter	Unit	WWTP effluent Average St. dev.	n	ReUse Water 1 Average St. dev.	n	ReUse Water 2 Average St. dev.	n	ReUse Water 3 Average St. dev.	n
pH	–	6.9 ± 0.1	25	6.8 ± 0.1	24	6.8 ± 0.1	18	5.3 ± 0.2	23
Conductivity (25 °C)	µS/cm	1.314 ± 375	26	1.329 ± 376	25	1.401 ± 394	18	26.6 ± 12.0	24
Turbidity	NTU	2.26 ± 1.9	24	0.14 ± 0.11	24	0.24 ± 0.18	18	0.19 ± 0.22	21
N total	mg/L	8.3 ± 2.9	25	7.6 ± 2.7	26	8.4 ± 2.4	16	0.39 ± 0.19	23
P total	mg/L	0.37 ± 0.14	26	0.06 ± 0.06	26	0.09 ± 0.08	17	0.05 ± 0.06	23
COD	mg/L	39 ± 9	27	27 ± 11	27	21 ± 5	15	<5	24
TOC	mg/L	12.4 ± 2.5	30	10.0 ± 3.4	27	8.7 ± 2.5	17	0.39 ± 0.43	24
SAC_{254}	1/m	31.1 ± 17.4	21	23.9 ± 8.5	24	22.6 ± 12.5	17	0.21 ± 0.17	22
SAC_{436}	1/m	2.97 ± 1.69	6	1.12 ± 0.4	24	0.9 ± 0.3	18	0.05 ± 0.06	24
AOX	mg/L	0.05 ± 0.07	22	0.03 ± 0.02	21	0.02 ± 0.01	18	<0.01	18
Al total	mg/L	0.07 ± 0.03	33	0.03 ± 0.03	32	0.02 ± 0.01	18	0.009 ± 0.009	25
Fe total	mg/L	0.60 ± 0.19	35	0.04 ± 0.03	31	0.02 ± 0.02	18	0.007 ± 0.007	25
Mn total	mg/L	0.41 ± 0.12	36	0.41 ± 0.10	33	0.13 ± 0.21	16	0.003 ± 0.002	25
Cl^-	mg/L	215 ± 102	24	215 ± 102	28	245 ± 113	18	2.9 ± 2.0	24
SO^{4-}	mg/L	95.7 ± 25	23	91.8 ± 26	28	101 ± 25	18	0.6 ± 0.4	24
NH^{4+}	mg/L	1.75 ± 2.47	27	1.86 ± 2.52	26	0.39 ± 0.66	18	0.10 ± 0.09	22
NO_2^-	mg/L	1.10 ± 1.94	27	1.12 ± 1.94	27	0.24 ± 0.41	18	0.06 ± 0.05	24
O_2 (aq)	mg/L	3.53 ± 1.37	25	6.31 ± 0.69	25	2.17 ± 1.82	18	5.73 ± 1.76	24
Ca^{2+}	mg/L	78.8 ± 3.2	29	78.8 ± 12.0	29	77.5 ± 13.8	17	0.1 ± 0.1	25
Silicate total	mg/L	27.9 ± 5.0	20	27.6 ± 4.9	26	27.2 ± 5.0	18	0.25 ± 0.1	25
Base-neutralizing capacity	mmol/L	0.9 ± 0.4	27	1.0 ± 0.4	27	0.8 ± 0.3	16	0.7 ± 0.4	22
Alkalinity	mmol/L	4.5 ± 0.8	28	4.4 ± 1.3	27	3.9 ± 0.8	18	0.2 ± 0.2	22

characterized by the treated 'bed volume' (m^3 filtrate/m^3 GAC filter bed). From Figure 3, the progress of the breakthrough of some micropollutants can be observed. As one extreme, the benzotriazoles can be mentioned, which were removed at nearly 100% over the complete operation time. On the other side, at the end of the pilot phase, a certain mass of already adsorbed sulfamethoxazole had been replaced by competing organics with a desorption effect: the concentration sulfamethoxazole was higher in the filtrate than in the raw water (negative removal). Due to the biological degradation of the DOC in both filters and adsorption on the GAC, the concentration of DOC is reduced by 10% additionally to the removal caused by flocculation (total retention 30%).

ReUse Water 3 is the water with the highest purity. The average concentration of TOC is 0.3 mg/L (−97%). By using the RO membrane filtration, strong attention was paid to the elimination of salts and hardness formers, as these are critical in many industrial process steps. However, it was not the aim to receive a water free of salinity, but to find a good compromise between the low TDS content of the permeate on the one hand and the low energy demand of the pumps on the other hand.

The retention of organic micropollutants was constantly high. The lowest retention was detected for all benzotriazoles (83% on average). For the purpose of groundwater recharge, depending on the retention of the RO membrane and the non-polar micropollutants contained in the RO permeate, a final GAC filtration step could be operated optionally. This depends on the properties of the micropollutants remaining in the permeate and also on the need for a redundant barrier.

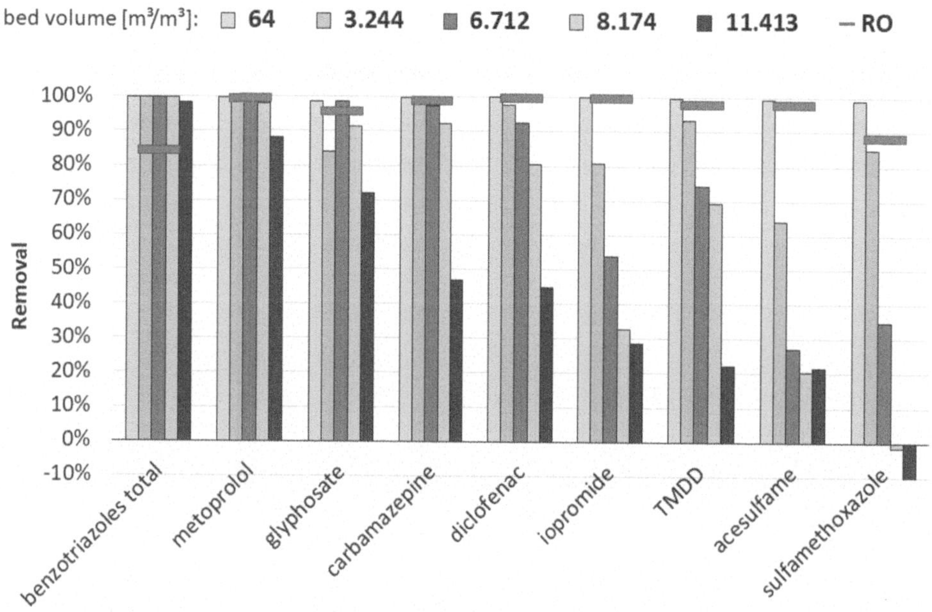

Figure 3 | Retention of micropollutants in the GAC filter according to treated bed volumes (m³ filtrate/m³ GAC filter bed) and the averaged retention by the RO membrane over the entire operation time.

Process stability UF

To find optimal operating conditions, coagulants based on both aluminium ($Al_2(OH)_5Cl\cdot$ 2–3H_2O) and iron ($FeClSO_4$ and $FeCl_3$) were tested for UF. After commissioning, both UF lines were operated in parallel with a flux of 60 L/($m^2\cdot$h) and 3 mg Al^{3+}/L. The contact time of the coagulant was 20 s. The results show that the UF performed well with high stability. Low increases in transmembrane pressure (TMP) with constant flux resulted in high permeability of around 300 L/($m^2\cdot$h·bar). The operational behaviour for a period of 2 weeks (week 0–2) is shown in Figure 4.

After this period, UF2 was operated with $FeClSO_4$ at a concentration of 5 mg Fe^{3+}/L, while for UF1 the Al-dose was kept constant. This resulted in an unstable filtration process and permeability dropped steeply in UF2. An increase of the CEB frequency from 1 to 2 CEBs/d could not stabilize the filtration process. Increasing the iron concentration up to 10 mg Fe^{3+}/L improved the performance and verified the suitability of all tested iron- and aluminium-based coagulants for the use in this application.

The different filtration behaviour can be explained by the different characteristics of the coagulants. Al-based coagulants are supposed to build up denser cake layers than Fe-based coagulants even at low dosages, which are more efficient in retaining foulants (Ma *et al.* 2014). Nevertheless, the Fe-based coagulants were chosen for further optimizations because quite high silica contents are present in the raw water and the danger of irreversible aluminosilicate scaling on the RO membranes was supposed to be avoided.

Over the course of the pilot study, the type of dosage of the coagulant was changed several times, particularly when using $FeCl_3$. With continuous coagulant metering, frequently sharp TMP increases at the beginning of filtration intervals were observed (see Figure 5). Presumably, the dosed amount of coagulant (6 mg Fe^{3+}/L) was insufficient to include or adsorb the organic matter into flushable hydroxide flocks. This resulted in pore blockage especially at the beginning of the filtration intervals, when the layer of coagulant on the membrane is still thin. CEBs were able to reduce the TMP, but were not efficient enough to restore the initial level. To counteract an initial increase in TMP, a higher coagulant dosage was used at the start of each filtration process (12 mg Fe^{3+}/L). After a certain time (e.g. a third of the filtration time), the dosage rate was significantly reduced (3 mg Fe^{3+}/L). This method is a combination of the so-called initial hydroxide coating (Buchta *et al.* 2017) on the membrane and continuous dosing. When using this

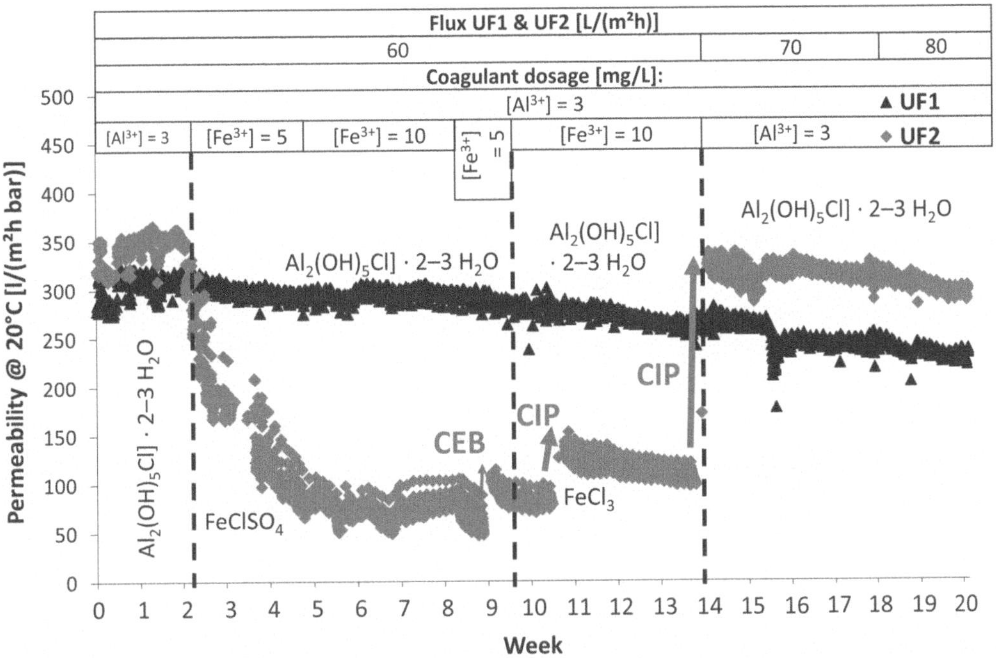

Figure 4 | Performance of UF1 and UF2 by treating the WWTP effluent with several amounts of Fe(III)- and Al(III)-based coagulants.

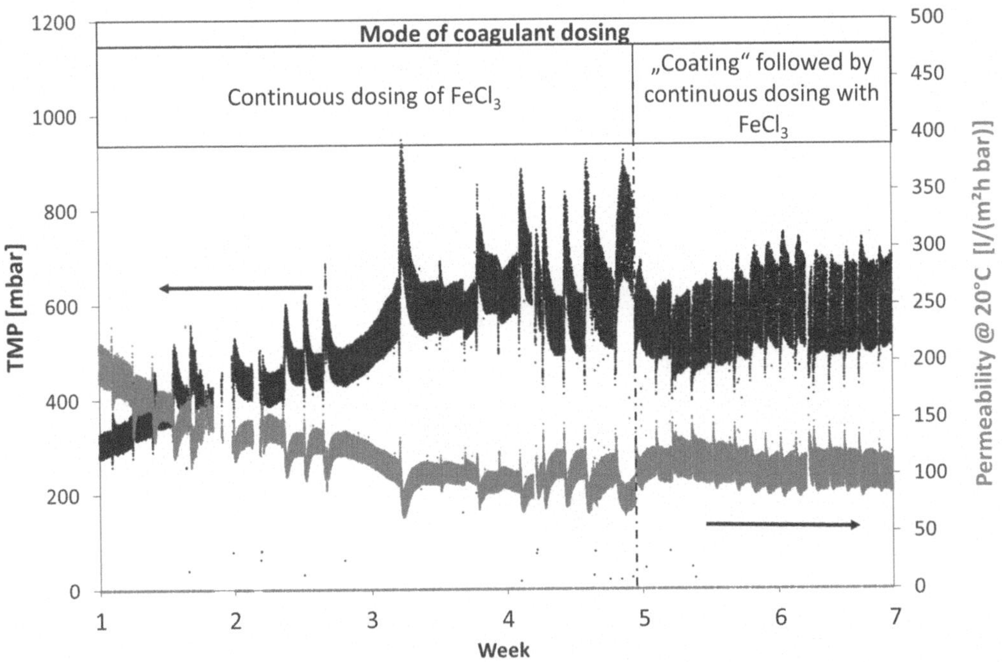

Figure 5 | Improving the performance of the UF by changing the mode of coagulant dosing.

method, a filter cake is built up on the membrane surface rapidly during the high-dosage phase directly at the beginning of a filtration cycle. This represents an additional barrier to dissolved and colloidal matter which protects the membrane from foulants and can be removed easily by the hydraulic backwash (Ma *et al.* 2014). The different

TMP courses are shown in Figure 5. It is evident that the stable continuous dosing led to stronger TMP peaks compared with the combined dosing strategy. The operation became much more stable despite adding the same amount of 6 mg Fe^{3+}/L coagulant on average for the overall filtration cycle.

Figure 6 shows that the UF capillary membrane is very sensitive to malfunctions such as the failure of a metering pump (e. g. caused by air bubbles) for the addition of H_2SO_4 during a CEB. In this case, the operation stabilized automatically after the problem was resolved. In the event of a failure of the coagulant metering pump, however, it can be assumed that the membrane will not recover on its own, but will be needed to be subjected to intensive cleaning. Operating problems of the WWTP, which were confirmed after consultation with the operator, also showed high stress levels for the UF process. A deterioration in the quality of the WWTP effluent can disturb immensely. In some cases, the increase of the coagulant dosage rate and also increase of the CEB frequency (2 CEBs/d) were able to stop the increase of the TMP. It was finally necessary to switch off the pilot plant and to clean the UF with NaOH and NaClO (pH 12.2 and 300 mg Cl_{free}/L).

Process stability RO

During the first year of the pilot phase, the focus for the RO process was lying on the membrane development and the design of the spiral-wound elements. A new membrane type for the filtration of low to moderate saline waters has been successfully established. The aim of keeping a high retention of ~98% for TDS, 97% for TOC and >90% for the majority of organic micropollutants, while reducing the operational pressure by about 30% compared with a standard product, could be reached. Further details of these experiments are described elsewhere (Ogier & Lipnizki 2019)

Biofouling and scale formation can reduce the performance of the RO. To ensure operational stability for the production of ReUse water 3, monochloramine (NH_2Cl) was used in addition to antiscalant for RO1 to avoid biofouling. RO2 served as a reference line without monochloramine disinfection. The dosage of monochloramine was noticeable, both in the differential pressure of the cartridge filter and also in the permeability of the filtration process. Especially in the summer months due to high water temperatures of above 20 °C and the highly

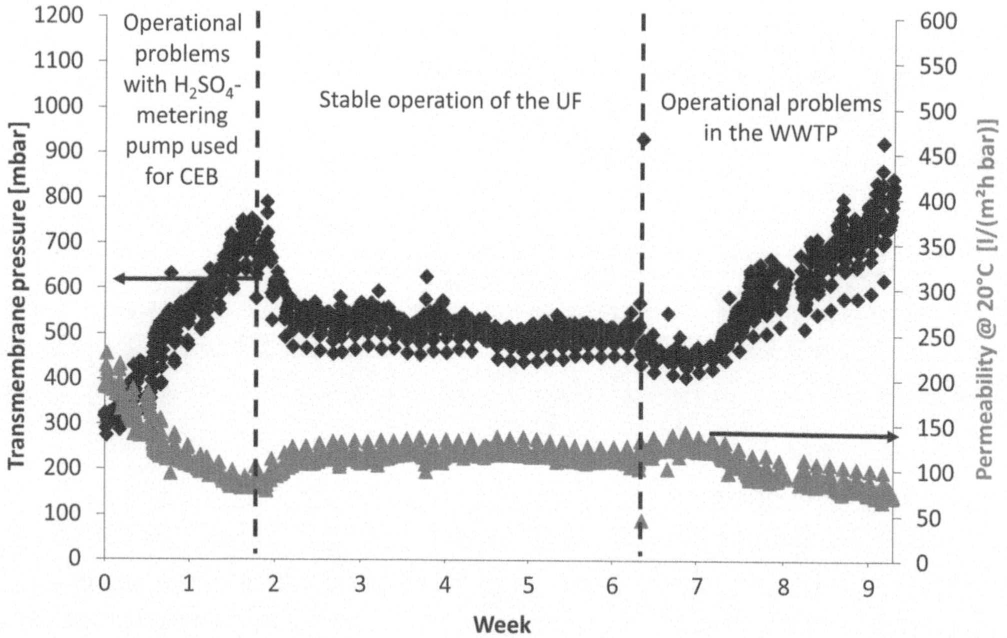

Figure 6 | Stable operation of the UF and effects in case of operational disturbances both at the treatment plant and at the pilot plant.

concentrated inlet water (lack of rain water in the mixed water sewer system), the biological activities were clearly visible on the cartridge filters. At times with peak water temperatures, the cartridge filters had to be replaced after 2 weeks at the latest. By using monochloramine, this period could be extended up to 2 months. Furthermore, the cleaning of the UF filtrate tank (storage for backwash water) in 2-week intervals was no longer necessary. Figure 7 shows the effect of 1 mg NH_2Cl/L for a dosing time of 22 h/d on the cartridge filter upstream to RO1 in comparison to that of RO2 over a period of 4 weeks (see Figure 8) in the hot summer of 2019. The deposition on these cartridge filters was examined by preparing a distinct piece of each cartridge with aqua regia solution and analysing the extract by ICP-OES. The summarized mass of all analytes in relation to the initial sample weight was significantly lower at RO1 (9.7 g/kg) compared with RO2 (23 g/kg). The chemical elements with the highest share among all elements identified by this method were iron (RO1: 2.8 g/kg; RO2: 11 g/kg) and calcium (RO1: 3.5 g/kg; RO2: 6.2 g/kg). The fact that the Fe deposition on the RO2 cartridge was about four times higher than on the RO1 cartrige and the amount of phosphorous was also significantly higher (RO1: 0.16 g/kg; RO2: 1.0 g/kg) confirms the assumption that the biological activity was effectively controlled by the disinfectant dosage.

The concentration of chloramine (i.e. bound chlorine) was regularly checked, using the DPD method, before the water entered the RO pressure vessel to ensure that the membrane would be in contact with the disinfectant. The average concentration of bound chlorine was 0.57 ± 0.21 mg/L depending on the quality of the feed water.

To reduce the dosing amount of NH_4Cl and NaClO to the bare minimum, tests were carried out with different dosing times between 22 and 4 h/d. Flow cytometry was used to assess the effectiveness of the disinfection. An optimal dosing time of 13 h/d could be determined without increasing the ICCs. Detailed results from these studies are presented by Nocker *et al.* (2020). The calculated salt retention based on the conductivity was similar in both RO lines. This observation and the results of a membrane autopsy after this operating phase of 6 months indicated that the membranes have not been deteriorated by the disinfection. Due to the membrane passage of chloramine, bound chlorine could be detected with concentrations of 0.39 ± 0.24 mg Cl_{bond}/L in the permeate.

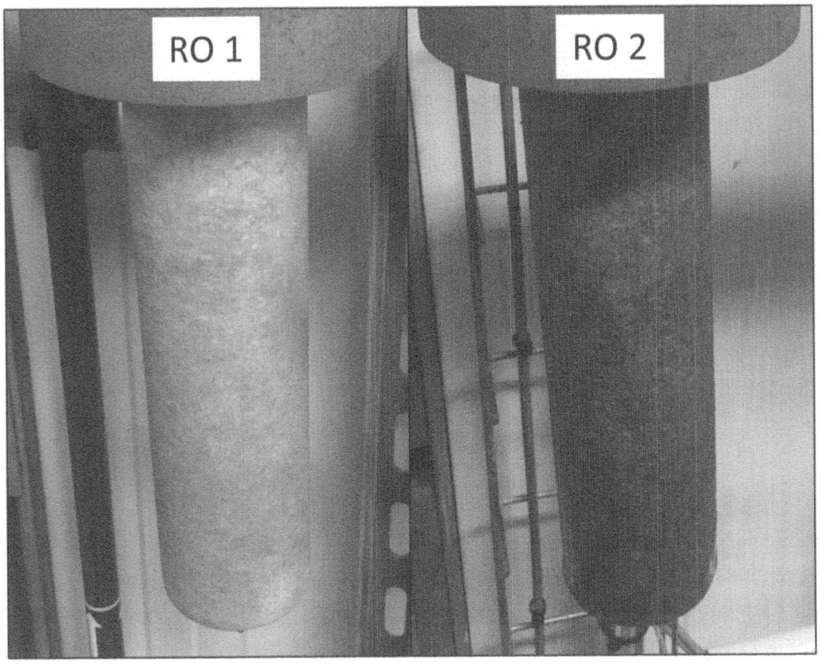

Figure 7 | Cartridge filters of RO1 (with NH_2Cl) and RO2 (without NH_2Cl) in comparison after 4 weeks of operation (26 June 2019–23 July 2019).

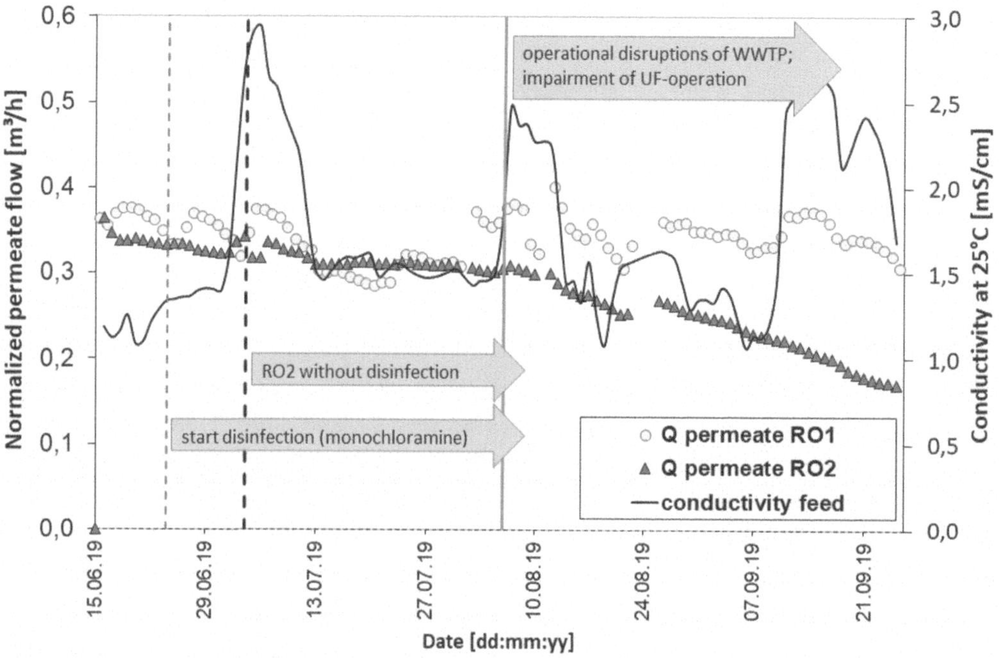

Figure 8 | Permeabilities of RO1 (with NH_2Cl) and RO2 (without NH_2Cl) and conductivity in the feed.

The efficiency of the filtration operation via the RO was monitored via the process data measured online. Figure 8 shows the normalized permeate flow for RO1 (with NH_2Cl dosage) and RO2 (without NH_2Cl dosage) during the summer of 2019. During this operational period, conductivity peaks of up to 3,000 μS/cm in the RO feed were detected. The cause could not be determined precisely. Despite these conditions, the RO lines showed a stable operation over several months. A malfunction of the WWTP operation at the beginning of August 2019 led to an impaired operation of the upstream UF lines (Figure 6, week 6). Consecutively, the high stress load on the RO membranes caused a continuous decline of the normalized permeate flow in RO2. By using monochloramine, however, the operation of the RO1 was stabilized and the operating time until the next CIP cleaning could be extended even under feedwater conditions.

Waste residues

Membrane processes offer many advantages. But one important disadvantage is the volume of concentrates and residues produced during the treatment process and their correct post-treatment or disposal. When operating UF and RO membranes, chemically contaminated water is generated, which must be disposed of properly. This includes the sludge-containing backwash water (9.1%, i.e. recovery 99.1%) from UF processes and chemical cleanings of UF (CEB and CIP) and RO (CIP), as well as the concentrate of RO with scale inhibitors. In the MULTI-ReUse project, the wastewater from the pilot plant was returned to the WWTP inlet. In the case of a large-scale reclamation plant, the relation between raw wastewater and residues from the reclamation plant could approximately reach a ratio of 4:1. The high salinity of the RO concentrate in the complete return of all waste residues could result in reduced performances of the biological treatment in the WWTP, as well as for the reclamation plant itself. Disposing of the RO concentrate therefore requires a separate discharge or further treatment.

One option, which was still being examined at the time this publication was written, is a discharge of RO concentrate into the River Weser. This seems to be the easiest way of disposal; however, a discharge permit is required. The Weser River is a navigable waterway and the river section at the case study site belongs to the estuary (brackish water). The costs for the discharge permit, which is dependent on the quality and volume, should be considered.

Backwash water of UF is largely unaffected by dissolved water additives. The hydroxides of the coagulation agent and PAC, if use is required, are the only water treatment substances added into the treatment process prior to the UF. The particulate matter, including metal-containing flocs (Fe^{3+} or Al^{3+}) and PAC, will be eliminated during the conventional WWT process. In addition, PAC stabilizes the biological treatment of the WWTP and its sludge treatment (dewatering) (Menzel 1997). Its loading capacity can be widely exploited because on the one hand the adsorption in the PAC-UF process does not reach equilibrium conditions and on the other hand the PAC will be applied (counter-current) in two treatment steps with a high concentration level during the biological treatment and a lowered level during the UF step (DWA 2019).

The chemically contaminated backwash water of the CEB (NaOH: pH ~12; H_2SO_4: pH ~2) and cleaning solutions from CIP cleanings for UF and RO, containing NaOH (pH ~12); oxalic acid/HCl (pH ~1.8) and small amounts of EDTA (only for RO cleaning), should first be neutralized in storage tanks. If no other potentially harmful cleaning chemicals (such as hypochlorite) are used, this waste stream can then be fed into the WWTP inlet.

Sludge-containing backwash water from the deep bed or GAC filtration as they were used in the treatment chain of ReUse Water 2 are free of additional chemicals. The contained particulate matter can easily be eliminated by the conventional WWT process; therefore, a disposal into the WWTP inlet should be suitable (DWA 2019).

Selection of material for the transport piping system

For transporting the ReUse waters to the consumers (point of use), an additional distribution system has to be implemented. The following materials for the piping should be considered:

- metallic materials (metal): cast iron, unalloyed or low alloyed steels,
- metal with internal cement mortar lining (metal/CM) and
- polymeric materials (polymer): most likely polyethylene (PE).

Metallic pipes without CM only have to be considered if an old, existing drinking water distribution system is intended to be used for the distribution of the ReUse waters. The chemical properties of the ReUse waters (according to the values in Table 2) and the corrosion behaviour of the respective materials have to be correlated.

ReUse Water 1 can be described by a high neutral salt content (chloride, sulphate and nitrate). Additionally, the calcite solubility, which is associated with pH, calcium concentration and other parameters of the carbonate system of the water, is relatively high so that the formation of protective surface layers inside metallic pipes may be hindered. This can also be a threat for cement-based materials as the lime within the cement can be dissolved. Beside the weakening of the structure of the cement, this can also lead to the release of particles into the water. Also manganese in ReUse Water 1 can precipitate. These particles and precipitates can cause blocking of valves, depending on the amount of water to be transported (freight). The content of ammonium can cause problems with copper alloys (brass), which might be important when considering the materials of the operating networks of the consumers (e.g. valves and fittings).

The characteristics of ReUse Water 2 are quite similar to those of ReUse Water 1. The manganese and ammonium content could be lowered significantly by additional aerobic biological filtration.

In ReUse Water 3, the concentrations of many water constituents have been lowered significantly, including neutral salts. On the other hand, the alkalinity is almost completely eliminated. As this parameter acts as a buffer within the corrosion system, the water can be characterized as poorly buffered from the corrosion chemical point of view. The pH is relatively low and the carbonate system of this water is far away from equilibrium, leading to very high calcite solubility.

The most important characteristics and the consequences for the selection of materials for distribution networks are summarized by a traffic light colour scheme in Figure 9.

Polymeric materials (e.g. PE pipes) are appropriate for all three ReUse Waters. In the case of ReUse Water 1, manganese has to be considered as described above. Metallic- or cement-based materials require further stabilization of the ReUse Waters (e.g. pH rising and buffering). Due to the high neutral salt content in ReUse Waters 1 and 2, metallic

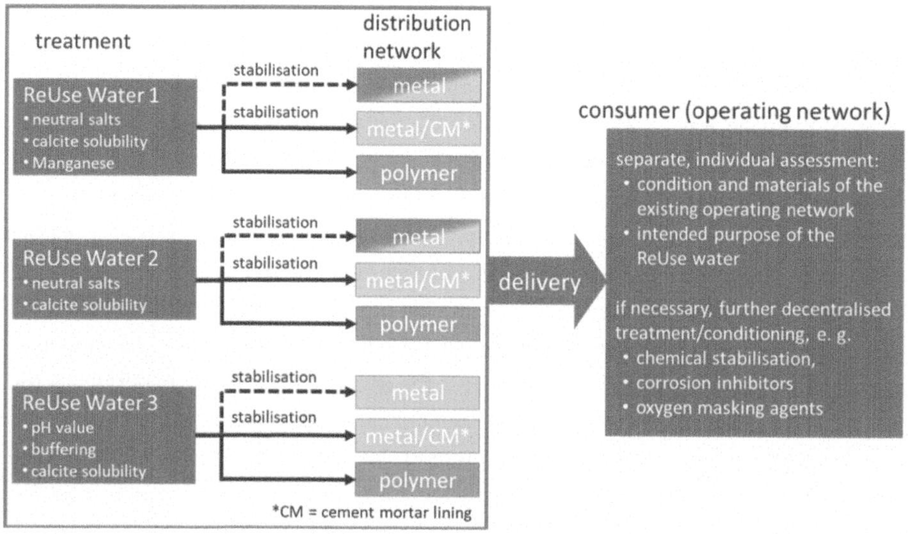

Figure 9 | Decision tree for the selection of materials for distributing the ReUse Waters.

materials can only be used under certain conditions regarding the network structure and the service conditions and might require the addition of corrosion inhibitors. If old, existing drinking water networks made of metallic materials are to be used for the waters 1 or 2, a detailed and individual assessment of the condition of the existing network is necessary in advance.

Regarding the service conditions, it has to be considered that they may differ between drinking water and reuse water. In contrast to a drinking water network, the conditions in a reuse water network – especially for industrial purposes – can be characterized by a constant, plannable consumption.

The materials and conditions of the respective operating networks of the consumers of reuse waters also have to be taken into account, as well as the intended purpose of usage. Due to the great variety between the consumers, this can only be done by individual assessments and will not be further discussed here.

CONCLUSION

In the MULTI-ReUse project, a modular treatment concept for the reuse of WWTP effluent was developed. To ensure consistent stable water quality, reliable, i.e. robust treatment technologies were used on the one hand, and extensive online monitoring in combination with laboratory analyses on the other.

The importance of considering the treatment efficiencies of the WWTP and the reclamation plant as one unit has to be emphasized. There is still a lack of data for the evaluation of the influence of operational problems in the WWTP (e.g. in drought or heavy rain periods) on the membrane processes in the reclamation plant. It is also suspected that certain chemical water constituents occasionally found in the WWTP effluent can disrupt the in/out UF process. The experiments showed that under normal circumstances at least 6 mg Fe^{3+}/L should be used for a stable filtration process when using $FeCl_3$ and at least 2 mg Al^{3+}/L when using $Al_2(OH)_5Cl{\cdot}2\text{–}3\ H_2O$ as a coagulant. A combined coagulant coating strategy was successfully used to stabilize the UF process.

For protecting the RO membrane against biofouling and keeping the performance, high *in situ* formation of monochloramine prior to the RO membrane was highly effective. Furthermore, it suppresses the regrowth of bacteria in the RO permeate by passing the membrane and stabilizing the water. Especially in hot summers and drought periods, the potential of biofouling formation is quite high. Consequentially, energy and maintenance costs for intensive cleaning of membrane pipes can be reduced and the treatment be more efficient.

The evaluation of possible materials for a distribution network has shown that in most cases individual assessments are necessary regarding the operating conditions and the purpose of usage.

ACKNOWLEDGEMENTS

The Project MULTI-ReUse was funded by the BMBF Federal Ministry of Education and Research of Germany within the research cluster WAVE (grant agreement no. 02WAV1403A-I). The authors are responsible for the content of this publication.

DATA AVAILABILITY STATEMENT

All relevant data are included in the paper or its Supplementary Information.

REFERENCES

Asano, T., Burton, F. L., Leverenz, H. L., Tsuchihashi, R. & Tchobanoglous, G. 2007 *Water Reuse. Issues, Technologies, and Applications*. Metcalf & Eddy, Inc., Boston, MA, USA.

ASTM D4516 2010 *Standard Practice for Standardizing Reverse Osmosis Performance Data*. ASTM International, West Conshohocken, PA, USA.

Auken, E., Bosch, A., Courtens, C., Elderhorst, W., Euwe, M., Gunnink, J., Hinsby, K., Jansen, J., Johnsen, R., Kok, A., Lebbe, L., Louw, P. G. B., Noorlandt, R., Oude Essink, G., Pedersen, J., Rasmussen, P., Scheer, W., Siemon, B., Sonnenborg, T. & Wiederhold, H. 2012 *Groundwater in A Future Climate - The CLIWAT Handbook*. Central Denmark Region, Aarhus.

Barbot, E., Moustier, S., Bottero, J. Y. & Moulin, P. 2008 Coagulation and ultrafiltration: understanding of the key parameters of the hybrid process. *Journal of Membrane Science*. doi:10.1016/j.memsci.2008.07.054.

Buchta, P., Kripahle, A., Vial, D., Winkler, R. & Berg, P. 2017 In-out ultrafiltration in tertiary wastewater applications – comparison of different operational strategies. *Desalination and Water Treatment* **73**, 145–154. doi:10.5004/dwt.2017.20461.

Curl, J., Brown, A. R., Wait, S., Dai, N. & Vorheis, J. 2019 Solving future water challenges: trends in water reuse. *AWWA* **111** (8), 40–45.

Drewes, J. E. & Horstmeyer, N. 2015 Recent developments in potable water reuse. In: *Advanced Treatment Technologies for Urban Wastewater Reuse, The Handbook of Environmental Chemistry* (D. Fatta-Kassinos, D. Dionysiou & K. Kümmerer, eds). Series Volume 45, Springer International Publishing, Switzerland, pp. 269–290.

DWA 2019 Aktivkohleeinsatz auf kommunalen Kläranlagen zur Spurenstoffentfernung - Verfahrensvarianten, Reinigungsleistung und betriebliche Aspekte. *DWA-Themen* T1/2019, DWA, Hennef.

EPA/600/R-12/618 2012 *Guidlines for Water Reuse*. Environmental Protection Agency, Washington, DC, p. 643.

Guigui, C., Rouch, J. C., Durand-Bourlier, L., Bonnelye, V. & Aptel, P. 2002 Impact of coagulation conditions on the in-line coagulation/UF process for drinking water production. *Desalination* **147** (1–3), 95–100. doi:10.1016/s0011-9164(02)00582-9.

Hellwig, J., de Graaf, I. E. M., Weiler, M. & Stahl, K. 2020 Large-scale assessment of delayed groundwater responses to drought. *Water Resources Research* **56** (2). doi:10.1029/2019WR025441.

Ivancev-Tumbas, I., Hobby, R., Kuchle, B., Panglisch, S. & Gimbel, R. 2008 p-Nitrophenol removal by combination of powdered activated carbon adsorption and ultra- filtration – comparison of different operational modes. *Water Research* **42**, 4117–4124. doi:10.1016/j.watres.2008.07.009.

Kroemer, K., Koch, T., Nahrstedt, A., Patrick, B., Doelchow, U. & Glaenzer, U. 2019 MULTI-ReUse: Wasserwiederverwendung zur Brauchwasserversorgung von Industrie und Gewerbe. *Korrespondenz Wasser-Abwasser* **66** (6), 456–463.

Lin, C. F., Huang, Y. J. & Hao, I. J. 1999 Ultrafiltration processes for removing humic substances: effect of molecular weight fraction sand PAC treatment. *Water Research* **33**, 1252–1264. doi:10.1016/s0043-1354(98)00322-4.

Lipp, P., Hambsch, B., Boesl, M., Nahrstedt, A. & Herzog, R. 2017 Standardisiertes Testverfahren zur Beurteilung des Virenrückhalts von Ultrafiltrationsmembranen. *Energie Wasser-Praxis* **9**, 34–41.

Ma, B., Yu, W., Liu, H. & Qu, J. 2014 Comparison of iron (III) and alum salt on ultrafiltration membrane fouling by alginate. *Desalination* **354**, 153–159.

Mallevialle, J., Odendaal, P. E. & Wiesner, M. R. 1996 *Water Treatment Membrane Processes*. McGraw-Hill, New York.

Menzel, U. 1997 *Optimierter Einsatz von Pulveraktivkohle zur Elimination Organischer Reststoffe aus Klaranlagenablaufen. Stuttgarter Berichte zur Siedlungswasserwirtschaft. Band 143*. Oldenbourg, Munich.

Nocker, A., Schulte-Illingheim, L., Mueller, H., Rohn, A., Zimmermann, B., Gaba, A., Nahrstedt, A., Mohammadi, H., Meckenstock, R., Tiemann, Y. & Kroemer, K. 2020 Microbiological changes along a modular wastewater reuse treatment process with a special focus on bacterial regrowth. *Journal of Water Reuse and Desalination* **10** (4), 380–393.

Ogier, J. & Lipnizki, J. 2019 Prozessoptimierung einer Umkehrosmose als vierte Reinigungsstufe. *Gwf-Wasser Abwasser* **11**, 78–82.

Solomon, S., Qin, D., Manning, M., Chen, Z., Marquis, M., Averyt, K. B., Tignor, M. & Miller, H. L. 2007 *Climate Change 2007:*

The Physical Science Basis. Cambridge University Press, Cambridge.

Stoquart, C., Servais, P., Bérubé, P. R. & Barbeau, B. 2012 Hybrid membrane processes using activated carbon treatment for drinking water: a review. *Journal of Membrane Science* **411–412**, 1–12. doi:10.1016/j.memsci.2012.04.012.

UBA 2017 *Recommendations for Deriving EU Minimum Quality Requirements for Water Reuse*. Scientific Opinion Paper. Umwelt Bundesamt, Dessau-Rosslau.

Wick, C. H. & McCubbin, P. E. 1999 Passage of MS2 bacteriophage through various molecular weight filters. *Toxicology Mechanisms and Methods* **9** (4), 265–273.

First received 14 April 2020; accepted in revised form 29 July 2020. Available online 1 September 2020

Framework conditions to design sustainable business models for decentralised water treatment technologies in Viet Nam for international technology providers

Mona Fritz, Claudia Hohmann and Felix Tettenborn

ABSTRACT

The expansion of water-intensive industrial activities and the impacts of climate change are jeopardising the sufficiency of safe drinking water in several Southeast Asian countries. One is Viet Nam, where geogenic arsenic contamination further limits the availability of freshwater resources with a simultaneous increase in water demand. Innovative and sustainable water treatment technologies are required to meet these challenges. Equally, we assume that the provision of safe drinking water requires tailored business models (BMs). In this study, we focus on the key stakeholders and framework conditions to design tailored BMs providing safe drinking water to the low-income and middle-income population in Viet Nam. We consider decentralised technologies to be suitable due to their lower investment costs for implementation and the avoidance of strong path dependencies. We therefore conducted a literature review and interviews with international experts in the domain of decentralised water treatment technologies. Our results show that relevant aspects include a lack of financial resources, specific characteristics associated with Vietnamese culture, e.g. the importance of relationships and trust in the business domain, lack of education and vocational training, market saturation suggesting co-operation with existing water suppliers, lack of suitable partners, and deficiencies in the institutional environment.

Key words | base of the pyramid markets, decentralised water treatment technology, emerging markets, sustainable business models, Vietnamese water supply market

Mona Fritz
Claudia Hohmann (corresponding author)
Felix Tettenborn
Fraunhofer Institute for Systems and Innovation Research ISI,
Karlsruhe,
Germany
E-mail: *claudia.hohmann@isi.fraunhofer.de*

Mona Fritz
Rhine-Waal University of Applied Sciences,
Kleve,
Germany

Claudia Hohmann
Faculty of Geosciences,
University of Utrecht,
The Netherlands

HIGHLIGHTS

- Application of emerging market and base of the pyramid concepts to the case of decentralised water treatment technology in Viet Nam.
- Using sustainable business models as a lens to uncover framework conditions that are relevant to ensure access to safe drinking water across income groups in an emerging country.
- Offering a European perspective on technology supply for decentralised water treatment technologies in Viet Nam.
- Improving the implementation of innovative decentralised water technologies with promise of sustainability using sustainable business models.

doi: 10.2166/wrd.2020.016

INTRODUCTION

In recent decades, there has been progress made in the access to improved water sources and to basic water services in East and Southeast Asia (WHO & UNICEF 2000, 2017). Improved (drinking) water sources are 'those which by nature of their design and construction have the potential to deliver safe water' (WHO & UNICEF 2017, p. 8). Basic water services refer to drinking water from an improved source, provided that the collection time is not more than 30 min for a round trip (WHO & UNICEF 2017). According to WHO & UNICEF (2017), 95% of the population in Eastern Asia and South-Eastern Asia enjoyed access to improved water sources in 2015. In Viet Nam (following the UN, we use the spellings 'Viet Nam' and 'Vietnamese' in this paper), 94% of the population had access to improved water sources and 91% could access them within 30 min, including queueing. However, regional differences exist as well as differences depending on income levels. In 2015, the regional population enjoying access to basic water services ranged from 84 to 99% (WHO & UNICEF 2017). Inequalities between population groups exist: 100% of the wealthiest population quintile and 80% of the poorest population quintile had access to basic water services (WHO & UNICEF 2017).

Despite the promising progress in access rates, the current availability and quality of water in Viet Nam are jeopardised by increased industrial activity, climate change impacts such as a rising sea level contributing to saltwater intrusion in available freshwater sources (AHK Vietnam 2012; Niederste-Hollenberg *et al.* 2012; Luong *et al.* 2018), and (agricultural) pollutants (Wegner 2015; Reichenbach *et al.* 2018). These developments are endangering the availability of safe drinking water (AHK Vietnam 2012). Regionally, groundwater resources are overexploited, which results in decreasing water tables, land subsidence and saltwater intrusion (Niederste-Hollenberg *et al.* 2012). With the utilisation of groundwater set to increase in the future, these problems are expected to become more severe (Niederste-Hollenberg *et al.* 2012).

The complex challenges of the water sector call for innovation towards more sustainable technologies (Sousa-Zomer & Cauchick Miguel 2018). Solutions are needed that are innovative and tailored to local conditions (Sousa-Zomer & Cauchick Miguel 2018) as well as affordable for low-income groups (Gebauer & Saul 2014). The importance of affordable and safe drinking water as well as the participation of local communities in water management has even been recognised in one of the current UN development goals: Sustainable Development Goal (SDG) 6 of the Agenda 2030 (UN n.d.). In this context, the research project 'Modular concept for sustainable treatment of arsenic affected and saline groundwater in Vietnam – WaKap' (funded by the German Federal Ministry of Education and Research) has developed and piloted a modular, decentralised and innovative process for removing salt and arsenic from ground-, sea- and brackish water (Cañas Kurz *et al.* 2018; Luong *et al.* 2018). The hybrid system consists of modules for decentralised water treatment, which can be combined with renewable energy production modules including storage to form customised system solutions. An important goal was to improve the energy supply of the hybrid process by integrating regenerative energy sources (photovoltaic and wind), in order to enable a self-sufficient and decentralised operation of the plants (without grid connection). In general, decentralised technologies have several advantages: they can be tailored to local conditions and avoid the path dependencies typical for centralised infrastructure, allowing adaptation to changing conditions in the surrounding settlements or the environment (Sartorius 2007). They allow the reduction of environmental and social impacts, enhanced public participation and cost recovery (Domènech 2011).

As technology is not a 'panacea for sustainability' (Silvestre & Silva Neto 2014), we consider business models (BMs) an important part of these solutions. BMs are 'a simplified and aggregated representation of the relevant activities of a company' (Wirtz *et al.* 2016, p. 41). In this article, the notion of a sustainable business model (SBM) is used. A SBM is defined here as an organisation's proposition, creation, delivery, capture, and exchange of value in a sustainable way, realised 'for, and in collaboration with, a broad range of stakeholders' (Geissdoerfer *et al.* 2016, p. 1219). SBMs are understood as contributing to society and/or ecology by striving for sustainable social, ecological, and economical values and considering (negative) externalities. However, the water sector lacks sufficient knowledge of innovative BMs, and profit-oriented organisations are often suspected of exploiting the poor by selling them the 'human right' to water (Gebauer & Saul 2014).

This article aims to identify the framework conditions for SBMs that ensure access to safe drinking water for the low-income and middle-income population in Viet Nam using decentralised water treatment technologies. We determine the key stakeholders for such BMs as well as the framework conditions in the low- and middle-income segments of the Vietnamese water supply market. In this paper, we focus on the perspective of European actors and draw upon three concepts for our research design: the base of the pyramid (BoP) concept and the emerging market theory are used to determine the framework conditions specific to the Vietnamese water supply market and the low-income population there. The literature on SBMs (innovation) is used to identify existing innovations in the sector as well as potential BM responses to the determined framework conditions.

Low-income populations in emerging countries are increasingly recognised as a large and attractive market (Ablaza *et al.* 2007). Emerging countries or markets are only loosely defined in the literature. Here, emerging markets are understood as low- to middle-income countries with prospects for rapid economic growth and integration into global markets. They are not yet 'developed countries', but are striving for this category and might share certain features. Emerging markets and low-income markets are characterised by specific features and challenges that must be considered when operating in them. One major concern in developing countries that pertains to water technologies is affordability (Gebauer & Saul 2014). While philanthropic approaches that rely on donations can be important for development, they lack self-sustainability in economic terms (London 2008). For this reason, Prahalad & Hammond (2002) propose a market-based approach to serve the poor, changing the perception of the poor from receivers of philanthropic aid to customers. This concept creates a value for both the supplying company and the customers served (Prahalad & Hammond 2002) and is known as the BoP approach (Prahalad 2008). Prahalad (2008) describe the BoP as a population with an annual per capita income below US$1,500 (in purchasing power parity (PPP)). While countries such as India, the Philippines and Indonesia have since developed to become lower middle-income countries, the term BoP is still used to describe the poor in these countries (Sánchez *et al.* 2007; Dembek *et al.* 2018).

The SBM concept was chosen for two reasons. First, a sustainability approach is considered to be very important, especially in emerging markets like Viet Nam, where the economic development is often accompanied by detrimental environmental impacts (Niederste-Hollenberg *et al.* 2012). Sustainability here is understood as comprising three dimensions: an economic, a social, and an environmental dimension (White 2013). A sustainable approach is expected to ensure the quality of life of the present generation without impacting that of future ones (World Commission on Environment and Development 1987). Second, as Matos & Silvestre (2013) point out, there is an apparent fit between the SBM concept and the BoP context.

The following section outlines the case and methods applied in this paper. The results describe the special features of the Vietnamese water supply market and potential (BM) responses. The paper closes with the limitations of this study, an outlook and conclusions.

CASE SELECTION AND METHOD

Case description

The next paragraph elaborates some of the country's economic characteristics and the applicability of the emerging market and BoP concepts.

Viet Nam has developed significantly over the last 30 years. It is now regarded as a lower middle-income country (AHK Vietnam 2012; Niederste-Hollenberg *et al.* 2012) and as an emerging market (Le Hau *et al.* 2013). Nevertheless, in 2016, its BoP consisted of 7,754,664 to 27,141,324 persons – according to data of the World Bank Group – revealing that 8.2% of its total population of 94,569,072 were living on US$3.20 and 28.7% on US$5.50 a day (2011 PPP) (The World Bank Group 2019b, 2019c). While this is a very wide range, the income distribution published by the General

Statistics Office of Viet Nam (2018) suggests that the population in the BoP is much closer to the upper bound of this range, as the poorest quintile of the Vietnamese population had an annual income of US$1,223 in 2016 and the annual income of the BoP is defined as less than US$1,500 (Prahalad 2008). The annual income of US$1,233 was calculated based on a monthly income of VND771,000 using a PPP conversion factor of 7,562.939 (The World Bank Group 2019a). This study focuses, in particular, on the less affluent groups of Viet Nam's population, as these are the ones with the poorest access to basic water services (WHO & UNICEF 2017).

Applied methods

This study is based on literature research on SBMs, BM innovation, the BoP concept (Prahalad 2005; Gebauer & Saul 2014; Gebauer *et al.* 2017), and emerging market theory (Sheth 2011; Landau *et al.* 2016; Oyedele 2016). Semi-structured expert interviews are the main method used. We conducted four interviews in 2019 with experts from different fields, i.e. consultants and practitioners. In accordance with Mayer (2013), the sample was purposive and created in advance. Due to the scope of the research project, this study aims at identifying the framework conditions of the Vietnamese water supply market for German (or European) technology providers. For the purpose of our study, we took a European perspective of the problem and only selected European experts for the interviews. In total, we contacted 31 potential interview candidates from 23 different organisations. Fourteen did not respond, six of the originally contacted persons provided contacts to experts they thought to be more suitable, and four declared no interest in or capacity for participation. Three of the contacted experts responded too late to conduct an interview with them. Therefore, four interviews were conducted in total.

The first interview was with the CEO of a foreign NGO operating in the Vietnamese water supply sector, who is referred to as Expert A. The second interview was with Expert B, who has been working in Viet Nam for 4 years as a deputy CEO and a market consultant. The third interview was with the CEO of a German company offering consultancy and planning in the sector, as he has experience of the Vietnamese market (Expert C). Expert D is a German engineer, who has lived and worked in Viet Nam for more than 10 years on water-related issues.

We applied a semi-structured design to allow interviewees the freedom to answer without pre-categorising or limiting their responses. An interview guideline provided some degree of data comparability and structure and served as orientation during the interview (Mayer 2013). The interview questions specifically addressed decentralised water treatment technologies in Viet Nam and the experts were briefed accordingly. The guideline was not used as a static template, but more flexibly as proposed by Meuser & Nagel (2009). The interview guideline was developed based on and in parallel to the literature research, and translates the research question(s) into interview questions. For the interviews with German experts, the interview guideline was translated into German. The interviews were subjected to qualitative content analysis as proposed by Kaiser (2014). The first step was to transcribe the recordings. MAXQDA and Express Scribe (NCH) were used for transcription. Literal transcription was conducted according to the rules of 'simple transcription' developed by Dresing and Pehl and provided by Pädagogische Hochschule Freiburg (n.d.). The introduction, de-briefing, and off-topic passages were not transcribed. Prior to the coding, categories were developed based on the results of the literature research. The categories included all the factors identified during the literature review as specific to or important for emerging markets, the BoP context, SBM design, or Viet Nam. They belonged to the topics 'target group', 'institutional framework', 'stakeholders', 'business models for sustainability', and 'success factors and obstacles'. The transcripts were coded using MAXQDA, and new categories were created when necessary based on the transcripts to ensure the openness of the analysis (Kaiser 2014). As proposed by Kaiser (2014), the transcripts were mainly coded based on paragraphs. In a second step, the statements of each expert were summarised within each category and redundancies eliminated. According to Kaiser (2014), categories that are not used need to be revised. After condensing the material of each interview, the contents of all interviews were merged within the same category and the key messages of the different categories were identified. Finally, the results of the interview analysis were used to verify the results of the literature research with regard to the specific target market – i.e. decentralised water treatment technologies in Viet Nam. The findings are therefore presented in a combined manner.

RESULTS AND DISCUSSION

The utilisation of available water sources in Viet Nam varies across regions (GSO 2008). These include bottled water, tap water, rainwater, water from dug and tube wells, spring water, surface water (without rainwater), and other sources (GSO 2008). While tap water is favoured in urban areas, tube wells, dug wells, and rainwater are more popular in rural areas (GSO 2008). Rainwater is used extensively during the wet season, but in their study of the Tra Vinh province in the Mekong River delta (MRD), Li *et al.* (2016) showed that other water sources experience an upswing during the dry season. The literature shows that most Vietnamese treat their drinking water, for example, by boiling or straining it (Li *et al.* 2016). The differences regarding the utilisation of existing water sources, their treatment, as well as context-specific features need to be kept in mind when designing a BM to introduce innovative water treatment technology in Viet Nam.

Special features and potential (BM) responses

While the attractiveness of the low-income populations in emerging markets is now recognised (Ablaza *et al.* 2007), these markets differ from those of developed countries (Prahalad & Hammond 2002). The literature describes various features specific to emerging markets and the BoP markets, which are shown in Figure 1. We assume that successful operations in these markets must acknowledge and respond to these special features and challenges.

Features specific to emerging markets include a lack of resources including finance, energy, and skilled labour (Sheth 2011) as well as the heterogeneity of the target group regarding economic, geographic, and cultural variables (Rangan *et al.* 2007). The lack of energy and transportation infrastructure might be an obstacle, particularly in rural areas (Prahalad 2005). Landau *et al.* (2016) highlight the role of the institutional environment. Institutions comprise formal and informal laws, constitutions, traditions, cultural patterns (Landau *et al.* 2016), and the organisations themselves (Hodgson 2006). In emerging markets, important institutions tend to be

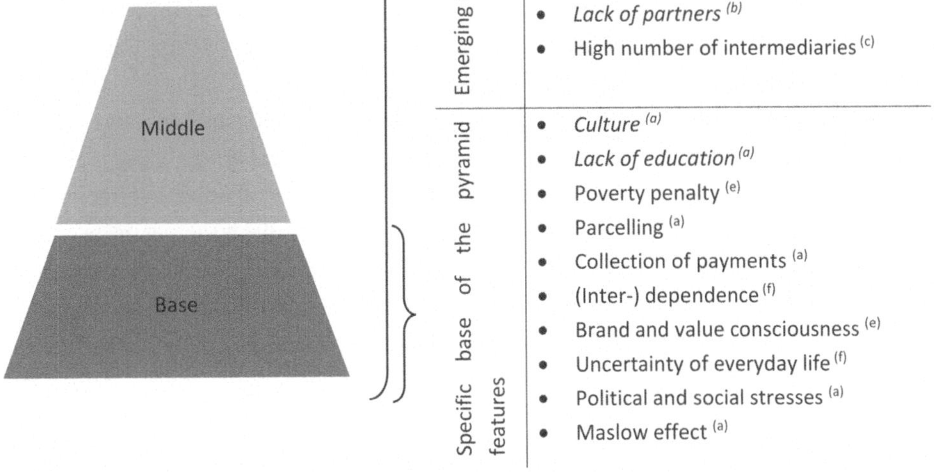

Figure 1 | Features of emerging markets and the BoP. Own illustration based on (a) Ablaza *et al.* (2007), (b) Landau *et al.* (2016), (c) Oyedele (2016), (d) Sheth (2011), (e) Prahalad (2005), (f) Viswanathan *et al.* (2009), and (g) Hoskisson *et al.* (2013).

absent or of poor quality (Khanna & Palepu 1997). Furthermore, suitable partners are often missing (Landau *et al.* 2016), a high number of intermediaries make operations expensive, and brands are rarely used in competition (Oyedele 2016).

In BoP markets, culture, lack of education (Ablaza *et al.* 2007), and uncertainties of daily life (Viswanathan *et al.* 2009) can influence purchase decisions. The poor often pay a poverty penalty, i.e. higher prices due to a lack of accessibility, local monopolies, and intermediaries (Prahalad 2005). Limited financial resources mean that the poor can only afford small quantities, and products need to be parcelled accordingly (Ablaza *et al.* 2007). The collection of payments, especially in post-paid models and in cases of delinquency, can be expensive and a matter of security (Ablaza *et al.* 2007). The relationship of customers and suppliers in BoP markets tend to be characterised by high dependence (Viswanathan *et al.* 2009). Out of necessity and for aspirational reasons, BoP customers tend to pay close attention to values and brands (Prahalad 2005). Political and social stresses may jeopardise the operations of companies in BoP markets (Ablaza *et al.* 2007). Finally, the poor may feel entitled to the satisfaction of basic needs and may engage in theft if these needs are not met (Ablaza *et al.* 2007). This phenomenon is known as the Maslow effect and increases the costs of doing business (Ablaza *et al.* 2007).

These characteristics of BoP and emerging markets were scrutinised during the interviews. The features confirmed in the interviews for decentralised water treatment technologies in Viet Nam appear in *italics* in Figure 1. Features not in italics are general BoP and emerging market characteristics, but were neither confirmed nor excluded during the interviews. Therefore, no statements can be made about their existence in this market. In the following, we address the confirmed features for the Vietnamese water treatment sector in detail and suggest possible (BM) responses in Table 1.

Lack of financial resources

During the interviews, experts pointed out the lack of two resources regarding decentralised water treatment technologies: financial resources, which are addressed in this section; and skilled labour, which is addressed in the following section on *culture and the lack of education*. As heterogeneity was only confirmed as a relevant feature in the field of financial resources (Expert A), it is included in this section.

Constraints on financial resources exist in all markets to some extent, but this problem is much more severe in BoP markets (Ablaza *et al.* 2007). Financial means may be limited and also very volatile (Gebauer *et al.* 2017). A lack of financial resources was confirmed concerning potential customers such as local water producers (Expert A) and communities, districts and provinces (Experts B and C), as well as final consumers. While the interviews showed that shortages of financial resources remain a problem even for decentralised technologies, these technologies still have advantages compared to centralised infrastructure (Sartorius 2007). Financial shortages can be managed by specific BM elements. In urban areas, financial resources are heterogeneous, whereas *heterogeneity* is less common in rural areas (Expert A). More affluent people can afford to buy well-known and trusted brands, while the less affluent population has to trust local retailers from whom they buy most products (Expert A). While price is important for the latter group, the most important selling proposition for the former group is convenience (Expert A). In this *unbranded competition*, existing distribution channels can be utilised to gain the trust of final consumers (Expert A). For example, the organisation of Expert A collaborates with existing local water producers, whereby the distribution channels are maintained and the person(s) from whom the water is bought stay(s) the same.

In response to financial constraints, BM design suggests that customers contribute to value creation (Gebauer *et al.* 2017). In the case of water treatment technologies, the value created for the customer is *access to safe drinking water*. The customer might contribute to installing, operating, and maintaining the technology and to downstream distribution activities, i.e. delivering or collecting the treated water. Other eligible options include the combination of a 'market-based approach' with fund-raising, and local sourcing or assembly, i.e. procuring and assembling components locally (Gebauer & Saul 2014).

Table 1 | Emerging market and BoP features and potential (BM) responses

Features	**Potential (BM) responses**
Lack of financial resources • Customers[b, c, d, e] • Final consumers[b, e]	• Cost reduction: ○ Co-creation[g], e.g. customer contributes to/is responsible for installation, operation, maintenance (self-service) or downstream distribution ○ Local sourcing[f] ○ Life-cycle cost assessment[f] ○ Outsource R&D[f] • Enhancing funding and enlarging the revenue base: ○ Combination of a marked-based approach with fund-raising[f] ○ Customisation[g] ○ Expansion towards middle-income segment[g] ○ Diversification[f]
Unbranded competition • Low-income segment of final customers[b]	• Co-operation with local distributers trusted by the final consumer[b]
Collection of payments	• Deferred payment schemes and micro-credits[f] • Credit and pre-paid cards[f] • Mobile payment[f]
Culture • Relationships[b, c, d] • Word of mouth[c] • Customers' taste[e]	• Collaboration with local partners and intermediaries[c, d] • Utilisation of existing value chains[b] • Gain understanding[b, k]
Lack of education • Basic knowledge[b, d] • Vocational expertise[b, d]	• Education of children[d] • Vocational education and training (of caretakers)[b, d, f] • De-skilling (redesign operational activities to reduce the skills needed to perform them)[e, f] • Maintenance services[b]
(Lack of) infrastructure • Access to (final) customers	• Social networks, television, and direct contact[e] • Collaboration with women's unions and community chiefs[e] • Utilisation of existing distribution channels to the final customers[b]
Market saturation[b]	• Utilisation of existing distribution channels to the final customers[b]
Deficiencies of the institutional environment • Fragmentation of duties[a, e, i] • Flexibility of regulatory framework and importance of local level[b, c, d, e] • Lack of enforcement[b, d, e, i] • Centralised governance[d] • Various illegal practices[b, c, d, e]	• Top-down approaches[d] • Co-operation with local level[d] • (Local) intermediaries[c]
Lack of partners • Lack of high-quality products and services[c]	• Collaboration with local intermediaries[c] • Joint ventures[c] • Import high-quality products[b]

Own table based on (a) Du *et al.* (2016), (b) Expert A, (c) Expert B, (d) Expert C, (e) Expert D, (f) Gebauer & Saul (2014), (g) Gebauer *et al.* (2017), (h) Landau *et al.* (2016), (i) Nguyen (2013), (j) Reichenbach *et al.* (2018), and (k) Viswanathan *et al.* (2009).

Life-cycle cost assessments and outsourcing research and development (R&D) might also be advisable to improve transparency and cost recovery (Gebauer & Saul 2014). More ambitious solutions could include customisation and expansion towards middle-income segments (Gebauer *et al.* 2017). The rationale behind this is the assumption that the low-income segment observes and aspires to the consumption of the middle-income segment (Gebauer *et al.* 2017). To enlarge the revenue basis and reduce dependence on external funding, additional services inside and outside the water sector can be offered in the spirit of diversification (Gebauer & Saul 2014). In the case of community approaches, it is key to reduce the (financial) dependency of the less affluent on the wealthier strata (Funder *et al.* 2012).

Closely related to the lack of financial resources is the *collection of payments*. To secure payment, the BM innovation literature proposes deferred payment schemes, micro-credits, pre-paid and credit cards, or mobile payments (Gebauer & Saul 2014).

Culture and the lack of education

Purchase decisions of the BoP are not only dependent on financial aspects, but may also be guided by culture and the lack of education, which have important implications for doing business with the BoP (Ablaza *et al.* 2007).

In terms of *culture*, the importance of relationships, word of mouth, networks, and trust were pointed out (Experts A, B, and C). Contrary to western culture, contracts are less important and trust is not built by signing a contract, but is based on personal relationships (Expert B). Viet Nam's cultural characteristics make it difficult for foreign entrants to build networks and earn trust (Expert B). However, the importance of trust and relationships can be utilised by collaborating with local partners and intermediaries (Experts B and C), i.e. people or organisations that are able to provide contacts or negotiate on one's behalf. Another option is to utilise the existing value or distribution chains, especially to gain access to the final customer (Expert A). Cultural-based differences to western customers were also described regarding advertisements and product design (Expert D). Due to the lack of relevant intercultural studies, Expert B suggests an orientation towards China, since this country is closely connected to Viet Nam by a long history of colonialism. The cultural and historical differences between Viet Nam's north and south should also be acknowledged (Experts B and C). A deep understanding of the culture, the market, the target group, and their life circumstances is required (Expert A; Viswanathan *et al.* 2009).

Lack of education is problematic at different levels. Knowledge of the importance of safe (drinking) water was described as regionally dependent (Expert A) and rather limited in rural areas (Expert C). Even where awareness does exist, this rarely translates into an adoption of healthy practices. Water treatment is rather haphazard, based on neighbours' practices rather than independent information (Expert D). While the issue of saltwater intrusion is quite obvious, other contaminations are not (Expert B). Due to their long-term effects, it is difficult for the affected population to trace health issues back to contaminated water (Expert A). Thus, people continue to buy cheap bottled water, which might be contaminated by heavy metals or pesticides (Expert A). The Vietnamese population has developed a very critical attitude towards the quality of piped water (Experts B and D). Educating children is a promising avenue to counteract ignorance, as they are able to transfer knowledge to older generations (Expert C). Furthermore, the lack of trust in piped water can be an advantage for decentralised technologies, and the consumption of bottled water has important implications for the distribution of decentralised treated water. Treating the water locally may, in turn, contribute to education about the importance of clean water, as demonstrated by the collaboration of Expert A's organisation with schools.

Regarding the value chain, a lack of skills and expertise was pointed out (Expert A; Expert C), which often causes malfunctions, for example, due to inadequate maintenance (Gebauer & Saul 2014). However, the Vietnamese show a high willingness to learn (Expert B), enabling training (of caretakers) (Expert C). In this context, Expert A points out that compliance with practices is more important than technology itself to ensure high water quality. The interviews showed that vocational training and education are particularly important for decentralised technologies to ensure proper operation and maintenance. De-skilling, i.e. re-designing operational activities to lower the skills needed to perform them, is another approach proposed to eradicate malfunctions (Expert D; Gebauer & Saul 2014). Contrary to cost minimisation approaches, which may result in self-service (Expert D; Gebauer & Saul 2014), the lack of skills and expertise calls for maintenance services, for example, by the technology provider (Expert A).

Lack of infrastructure

Beyond issues of affordability, education, and culture, the relevant infrastructure is necessary to gain access to a potential target group (Sheth 2011). Neither the literature used nor the interviewees addressed a lack of physical infrastructure such as transportation or telecommunications for the Vietnamese market. However, the term infrastructure is understood here as not solely comprising physical constructions, but also the possibility to gather information about and gain access to the target group. A lack of market data was described for the Vietnamese water supply market, and Expert A suggests in-house research as a response. To gain access to the target group, several channels were suggested for decentralised technologies including social networks, television, direct contact, and collaborations with women's unions and community chiefs (Expert D). Expert A described the Vietnamese water supply market as saturated, meaning that many producers are operating on the market resulting in a situation in which new suppliers of drinking water are not needed to satisfy demand. However, these existing suppliers might not be offering what would be considered safe drinking water. Considering the *market saturation*, it might be sufficient to use the existing distribution channels to the final consumers and this approach has the additional advantage of already having the trust of the final customers (Expert A). Utilising existing distribution channels might mean collaborating with local water producers or retailers, and decentralised, modular technologies are particularly suitable for customisation (Sartorius 2007). BMs should be tailored to the diversity of local conditions and the existing structures of the water supply market. Trade fairs were assigned virtually zero importance for gaining market access (Expert B). Gebauer & Saul (2014) suggest a multi-channel approach, optimising the utilised channels, and enhancing the role of distribution centres.

Deficiencies of the institutional environment

Emerging markets tend to lack institutions such as specialised intermediaries, a well-functioning regulatory system and mechanisms to enforce contracts (Khanna & Palepu 1997). These deficiencies can make transactions very expensive or severely impede them (Khanna & Palepu 1997).

Several impediments were described regarding authorities. Firstly, the fragmentation of duties between the Ministry of Natural Resources and Environment (MoNRE), the Ministry of Agriculture and Rural Development (MARD), and the Ministry of Construction results in conflicts (Expert D). A major problem exists concerning MARD and MoNRE (Niederste-Hollenberg *et al.* 2012). When the MoNRE was introduced in 2002 and assigned some of MARD's former functions, ambiguities, gaps (Nguyen 2012), overlaps (Du *et al.* 2016), and conflicts (Molle & Chu 2008) occurred (Niederste-Hollenberg *et al.* 2012), which are still not entirely resolved (Expert D). While various tasks were reassigned from MARD to MoNRE, employees were largely not (Du *et al.* 2016). As a result, insufficient personnel capacity in MoNRE in particular is a major problem, especially on the provincial level (Niederste-Hollenberg *et al.* 2012). In general, the authorities complain about the lack of international experience and knowledge of modern techniques, inadequate training and experience, missing technical equipment and inadequate monitoring networks, inadequate analysis of the monitoring data, and a lack of financial resources (Niederste-Hollenberg *et al.* 2012). Smets *et al.* (2017, p. 170) criticises a 'lack of coordination between government agencies and inadequate information sharing'.

There are four levels of administrative structure in Viet Nam: central, provincial, district, and communal (Tu n.d.). On all of these levels, governmental authorities deal with different topics concerning water (Wegner 2015), and coordination and consultation between them is problematic (Expert D). There is a general lack of financing (Experts B and C). Overall, local levels were described as decisive for the implementation of decentralised water treatment systems (Experts A, B, C, and D). The old saying: 'the power of the emperor stops at the village gate' was mentioned and transferred to modern times: today, the village gate seems to be the gate to the province (Expert B). However, people at the local level might lack the necessary expertise to truly understand the proposed concepts, meaning convincing them will be more difficult (Expert C). Interaction with the national level might thus be more fruitful (Expert C). In addition, Expert C pointed out that Viet Nam is organised as a centralised state, so the national level is very important, and that it has a flat hierarchy, enabling interaction with higher levels. A top-down approach combined with the involvement of local levels was therefore described as promising (Expert C).

Another challenge pertains to the sector's regulatory framework. While the regulatory framework including the rules for drinking water production is described as straightforward (Expert A), the laws are rather flexible (Experts B and C) and their application may be regionally dependent (Expert B). In general, all laws, decrees, and other legal documents are available online and minor obstacles can be overcome via co-operation with local partners (Expert B). A consistently described issue is the lack of enforcement (Experts A, C, and D), including the monitoring of drinking water quality (Expert D) and the prosecution of non-compliant producers of drinking water (Expert A). This deficiency is due to insufficient staffing, knowledge and laboratory equipment, and in some cases, insufficient engagement and corruption (Experts A and D). Nevertheless, Expert A stressed the necessity of strict compliance for foreign actors and predicted stronger enforcement, which might require adapting practices accordingly. Local partners such as intermediaries or customers might possess the necessary practical experience with the regulatory framework (Expert B).

Informal institutions are an additional impediment to operations in emerging markets (Oyedele 2016). The term informal institutions refers, for example, to conventions, norms, and customs, which shape human interactions (North 2009) and may include illegal practices such as corruption (Oyedele 2016). According to Oyedele (2016), informal institutions add to the complexity of emerging markets and need to be taken into account when designing BMs. Some of the practices observed in the Vietnamese water supply market are pointed out in the following.

In the context of socio-cultural practices, traditions and rules, the lack of intercultural studies (Expert B), and data on the Vietnamese water supply market were criticised (Expert A). According to Experts A and D, bottled water is used by the lower income population. The quality of local brands of bottled water was described as very variable, depending on the brand and the season (Expert A). The usage of household water filters by the middle- and high-income urban population was also described (Expert D). Depending on a household's financial resources, they are used to treat all the water consumed by a household or only its drinking and cooking water (Expert D). Furthermore, water boiling was identified as a deeply ingrained habit that is still practised, even for filtered water (Expert D). There is widespread mistrust of piped water (Experts B and D). In line with the literature and the expert interviews, these practices need to be taken into account when designing BMs for decentralised water treatment technologies. The lack of trust in piped water could support decentralised technologies depending on their design. The widespread use of local brands of bottled water suggests collaborations between the suppliers of decentralised water treatment technology and the producers of these local brands.

Several illegal practices were identified, including the following: the distribution of non-functional decentralised water treatment technology (Expert D) and contaminated bottled water (Experts A and D), non-licenced withdrawal of water, illegal discharge of sewage (Expert D), and corruption (Experts A, B, C, and D). Expert B points out the risks of corruption: dependency and its use as leverage. Local intermediaries can function as a firewall (Expert B).

Lack of partners

Organisations operating in emerging markets often face a lack of suitable partners along their value chain (Landau *et al.* 2016). The lack of partners includes suppliers of high-quality inputs such as machinery components and services (Expert B). Intermediaries and joint ventures might be able to counteract this (Expert B); otherwise, the required products need to be imported (Expert A). Collaboration with all kinds of actors is suggested to access valuable resources, competencies, and knowledge (Sánchez *et al.* 2007). The following section addresses the question of key stakeholders. Table 2 shows a summary of the potential partners and their competences. These comprise Civil Society Organisations (CSOs), education and research (E&R)

organisations, international organisations (IOs) and funders, and local intermediaries.

Civil Society Organisations such as water use groups, i.e. groups that coordinate water-related activities, play a dominant role in the country's society (Du *et al.* 2016). CSOs are 'all non-market and non-state organisations outside of the family in which people organise themselves to pursue shared interests in the public domain' (OECD 2009, p. 26). The participation of communities in the water sector has a long tradition and is specified in the Law on Water Resources (Du *et al.* 2016). Recent decisions attempted to clarify the roles of CSOs and increase their incorporation into the political process and society's development (Du *et al.* 2016). They may become active as technical consultants and advocates (Du *et al.* 2016). Nevertheless, implementation lags behind and there is no adequate legal basis for the incorporation of marginalised groups (Du *et al.* 2016).

One of the sector's CSOs is the Vietnam Water Supply and Sewerage Association (VWSA n.d.). The VWSA can assist in establishing contacts to and dealing with local stakeholders (Expert C). CSOs are also active in water resources analysis, technology transfer, educational work (Expert D), and may allow others to share their distribution channels (Gebauer & Saul 2014). Women's unions and community chiefs are able to assist by providing access to potential customers (Expert D).

Table 2 | Potential partners and their competences

Potential partners	Competences
CSOs	• Establish relationships and deal with local stakeholders[d] • Access to customers[e] • Education[a, e] • Sharing distribution channels[f]
E&R organisations	• R&D activities and data[a, b, f] • Training[a] • Local endorsement[b]
IOs and funders	• Technical support[a, c, d, g] • Financial support[a, c, d, g]
Local intermediaries	• Identify potential partners and suppliers[c, d] • Access to customers and potential partners[c, d] • Build relationships and networks[c, d] • Language and cultural expertise[c, d] • Expertise on customer taste[c, d]

Own table based on (a) Du *et al.* (2016), (b) Expert A, (c) Expert B, (d) Expert C, (e) Expert D, (f) Gebauer & Saul (2014), and (g) Reichenbach *et al.* (2018).

Public and private *education and research organisations* play a key role (Du *et al.* 2016). They can provide R&D and data (Du *et al.* 2016; Expert A; Gebauer & Saul 2014) on suitable technologies as well as information on the target market and train staff for the sector (Du *et al.* 2016). The latter is particularly valuable as the lack of skills and expertise has been pointed out as a major impediment for operations and maintenance, and as a risk factor for decentralised technologies (Expert A). Finally, these organisations might engage in local endorsement (Expert A). The organisation of Expert A, for example, works together with schools to provide free drinking water to the students. While this is undoubtedly a philanthropic act, it could also function as an educational activity, creating awareness of the necessity and advantages of safe drinking water and as a marketing activity.

International organisations and funders might be valuable in terms of supporting operations technically and financially (Du *et al.* 2016; Experts B and C; Reichenbach *et al.* 2018). The World Health Organization (WHO), the World Bank Group, the Asian Development Bank (ADB), the Danish International Development Agency, the Kreditanstalt für Wiederaufbau (KfW), and the Japan International Cooperation Agency offer financial and technical support (Du *et al.* 2016; Reichenbach *et al.* 2018). One problem is the lack of overall strategic co-ordination of the projects (Du *et al.* 2016; Reichenbach *et al.* 2018). Furthermore, it is often not possible to upscale pilot projects (Du *et al.* 2016; Reichenbach *et al.* 2018).

According to the OECD (2018c), Viet Nam received net official development aid (ODA) worth US$2,894.8 million in 2016 and US$2,376.4 million in 2017. ODA is '[G]overnment aid designed to promote the economic development and welfare of developing countries. Loans and credits for military purposes are excluded' (OECD 2018b). The largest share of ODA came from Japan, followed by contributions from the International Development Association (IDA), the ADB, and Germany (OECD 2018c). In 2016, US$2,992 million ODA were earmarked for water issues in Asia (OECD 2018a). Viet Nam was the top recipient, receiving US$531 million (OECD 2018a). About 51% of the received

ODA was given by the IDA and 39% came from Japan (OECD 2018a).

Local intermediaries can fulfil several functions. Compared with other members of the Association of Southeast Asian Nations (ASEAN) countries, Viet Nam has a low localisation rate. Most of its machinery is imported from China, Japan, South Korea, and Germany (Expert B). This assessment is supported by Expert A, whose organisation is dissatisfied with the efficiency, level of automation, and remote monitoring possibilities of the machinery sourced in the MRD. Local intermediaries can identify potential partners and suppliers, enable access to customers and potential partners, build relationships and networks, and have good knowledge of the local language, culture, and customer tastes (Experts B and C).

Outlook and limitations

It remains to be determined exactly where and for which water suppliers saltwater intrusion and arsenic contamination are an issue. Against the background of existing solutions and planning, Expert D suggested an assessment of the technically and economically feasible options for the different water suppliers. To design tailored SBMs, we recommend studying the existing BMs in the Vietnamese water supply market, for example, by conducting more interviews with experts from organisations in the sector. We suggest the scope of such a study should go beyond decentralised water treatment technologies, the water supply sector, and the geographic boundaries of Viet Nam, because (the elements of) BMs from different industries might be applicable, too. This is for example proposed in the St. Gallen Business Model Navigator developed by Gassmann *et al.* (2014). A trial-and-error approach should be taken to adjust and customise the SBM elements in practice. In order to ensure the actual state complies with the targeted one, sustainability aspects must be incorporated and assessed when designing and implementing the SBM.

Viet Nam was chosen as the location for the project WaKap, because it is considered representative for many countries in Southeast Asia. Nevertheless, the water markets of Myanmar and Cambodia, for example, may be very different. When trying to replicate a SBM developed for the WaKap technology, factors such as market saturation, different governmental structures, and different cultures need to be taken into account. Studying the framework conditions and stakeholders in these countries could contribute to the design or adoption of SBMs for decentralised water treatment technologies. Another or complementary option is to study the geological conditions in these countries to determine whether the specific treatment system developed in WaKap is suitable for application there.

Finally, we consider the limited selection of experts in this study to be a major limitation, and suggest a larger and more diverse sample to obtain additional insights and different perspectives. The Vietnamese perspective, in particular, would provide essential insights into the framework conditions and stakeholders for the design of (sustainable) BMs of decentralised water technologies. Future research should focus on their perspective. This study also suffers from the typical limitations of qualitative research, such as limitations in the transferability and generalisation of the results.

CONCLUSIONS

In this article, we determined the features that are specific to the BoP markets and emerging markets in the case of decentralised water treatment technologies in Viet Nam from the perspective of European actors. We assume that a successful design of SBMs to ensure access to safe drinking water for the low-income and middle-income population in Viet Nam must consider and respond to these framework conditions. We utilised the existing literature on emerging, BoP, and water markets and conducted interviews with experts to determine these features. Recommendations for designing tailored BM can be derived from them.

The lack of financial resources described in the literature was relevant for the potential customers of decentralised water treatment technologies such as local authorities or water producers. This suggests that low-cost solutions and/or adequate financing schemes should be developed. Unbranded competition, market saturation, and cultural features such as the importance of word of mouth and relationships call for collaboration with local partners and the utilisation of existing value chains. New entrants with decentralised water treatment technologies can exploit what looks like an obstacle at first sight. They can target and collaborate with local partners, especially at the point where customer interaction happens. This can include offering the treated water to local retailers instead of distributing it themselves, or offering their technology or services to existing water producers. This may have implications regarding the technologies themselves, which might need to be adjusted to the scale and conditions of these local water producers.

Apart from and possibly more important than technology, adopting the right practices ensures safe water and a long technology lifetime. Vocational and *on the job* training may therefore yield high returns. De-skilling and providing maintenance services are additional approaches. These points seem particularly important for decentralised technologies as opposed to large-scale centralised operations, as these have different needs regarding on-site operation and maintenance. Furthermore, following the suggestion of targeting local water producers, adequate operation and maintenance skills are key to providing safe water in the long term. Educating children is key to raising awareness, among the adult population as well. Regarding the institutional set-up, Viet Nam's communist, centralised governance structure makes top-down approaches appealing. However, the local levels are also important, especially when implementing decentralised technologies. This is valid not only for authorities, but also for further collaboration with CSOs and local intermediaries. Particularly when targeting local water producers as potential clients of decentralised water treatment technologies, local intermediaries play an important role in finding and interacting with them. The proposed approaches for dealing with illegal practices range from engaging in them to 100% legal compliance. Sage advice here might be to rely on local intermediaries. These know and respect the local rules and can function as a firewall against corruption.

Finally, we want to point out that although the Vietnamese water supply market is complex, access to safe drinking water is vital to human development and worth joining forces to achieve. Governments, CSOs, E&R institutions, and local and multinational enterprises of all sizes should seek to collaborate and support each other to ensure people everywhere have access to the human right of clean water regardless of their country of residence or income level. To make this happen, obsolete ideas about the strict division of tasks between these different actors need to be abandoned. Business needs to be understood as a vehicle of development. Research needs to be fostered that goes beyond technical feasibility towards the implementation and dissemination of sustainable technology. Efforts need to be combined to make SDG 6 and the Agenda 2030 for people, planet, and prosperity a reality.

ACKNOWLEDGEMENTS

The project on which this article is based was funded by the German Federal Ministry of Education and Research (BMBF) under the FONA funding measure WAVE under the grant number 02WAV1413. The successful completion of the work is also due to the close co-operation between the Vietnamese and German project teams and the project management.

DATA AVAILABILITY STATEMENT

Data cannot be made publicly available; readers should contact the corresponding author for details.

REFERENCES

Ablaza Jr., G. C., Aquino, A. T., Beshouri, C., Romano, K. & Zobel del Ayala II, J. 2007 Viable business models to serve low-income consumers: Lessons from the Philippines. In: *Business Solutions for the Global Poor: Creating Social and Economic Value*, 1st edn (V. K. Rangan, J. A. Quelch & B. Barton, eds). Jossey-Bass, San Francisco, CA, pp. 207–222.

Cañas Kurz, E. E., Luong, V. T., Hellriegel, U., Leidinger, F., Luu, T. L., Bundschuh, J. & Hoinkis, J., 2020 Iron-based subsurface arsenic removal (SAR): Results of a long-term pilot-scale test in Vietnam. *Water Research* **181**, S115929.

Delegation der Deutschen Wirtschaft in Vietnam (AHK Vietnam) 2012 *Zielgruppenanalyse Vietnam (Target Group Analysis Vietnam): Städtische Infrastruktur (Urban Infrastructure).*

Dembek, K., York, J. & Singh, P. J. 2018 Creating value for multiple stakeholders: sustainable business models at the base of the pyramid. *Journal of Cleaner Production* **196**, 1600–1612.

Domènech, L. 2011 Rethinking water management: from centralised to decentralised water supply and sanitation models. *Documents d'Anàlisi Geogràfica* **57** (2), 293–310.

Du, T. L. T., Bui, D. D., Quach, X. T. & Robins, L. 2016 Water governance in a changing era: perspectives on Vietnam. In: *Water Governance Dynamics in the Mekong Region* (D. J. H.

Blake & L. Robins, eds). Strategic Information and Research Development Centre, Selangor, pp. 241–276.

Funder, M., Bustamante, R., Cossio, V., Huong, P. T. M., van Koppen, B., Mweemba, C., Nyambe, I., Phuong, L. T. T. & Skielbow, T. 2012 Strategies of the poorest in local water conflict and cooperation: evidence from Vietnam, Bolivia and Zambia. *Water Alternatives* **5** (1), 20–36.

Gassmann, O., Frankenberger, K. & Csik, M. 2014 *The Business Model Navigator: 55 Models That Will Revolutionise Your Business*. Pearson, Harlow, London, New York, Boston, San Francisco.

Gebauer, H. & Saul, C. J. 2014 Business model innovation in the water sector in developing countries. *Science of the Total Environment* **488**, 512–520.

Gebauer, H., Haldimann, M. & Saul, C. J. 2017 Business model innovations for overcoming barriers in the base-of-the-pyramid market. *Industry and Innovation* **24** (5), 543–568.

Geissdoerfer, M., Bocken, N. M. P. & Hultink, E. J. 2016 Design thinking to enhance the sustainable business modelling process – a workshop based on a value mapping process. *Journal of Cleaner Production* **135**, 1218–1232.

General Statistics Office of Viet Nam (GSO). 2008 *Viet Nam: Household Living Standard Survey 2008*.

General Statistics Office of Viet Nam 2018 *Monthly Average Income per Capita at Current Prices by Income Quintile and by Province by Cities, Provinces, Year and Income Quintile*. Available from: http://gso.gov.vn/default_en.aspx?tabid=783 (accessed 15 January 2019).

Hodgson, G. M. 2006 What are institutions? *Journal of Economic Issues* **40** (1), 1–25.

Hoskisson, R. E., Wright, M., Filatotchev, I. & Peng, M. W. 2013 Emerging multinationals from mid-range economies: the influence of institutions and factor markets. *Journal of Management Studies* **50** (7), 1295–1321.

Kaiser, R. 2014 *Qualitative Experteninterviews (Qualitative Expert Interviews): Konzeptionelle Grundlagen und praktische Durchführung (Conceptional Foundations and Practical Execution)*. Springer VS, Wiesbaden.

Khanna, T. & Palepu, K. 1997 Why focused strategies may be wrong for emerging markets. *Harvard Business Review* **75** (4), 41–51.

Landau, C., Karna, A. & Sailer, M. 2016 Business model adaption for emerging markets: a case study of a German automobile manufacturer in India. *R&D Management* **43** (3), 480–503.

Le Hau, N., Evangelista, F. & Thuy, P. N. 2013 Does it pay for firms in Asia's emerging markets to be market oriented? Evidence from Vietnam. *Journal of Business Research* **66** (12), 2412–2417.

Li, L., Li, C. S. & Wichelns, D. 2016 Assessing household willingness to pay for bottled water in rural areas of the Mekong Delta, Vietnam. *Water Resources and Rural Development* **7**, 36–49.

London, T. 2008 The base-of-the-pyramid perspective: a new approach to poverty alleviation. *Academy of Management Proceedings* **2008** (1), 1–6.

Luong, V. T., Cañas Kurz, E. E., Hellriegel, U., Luu, T. L., Hoinkis, J. & Bundschuh, J. 2018 Iron-based subsurface arsenic removal technologies by aeration: A review of the current state and future prospects. *Water Research* **133**, S110–122.

Matos, S. & Silvestre, B. S. 2013 Managing stakeholder relations when developing sustainable business models: the case of the Brazilian energy sector. *Journal of Cleaner Production* **45**, 61–73.

Mayer, H. O. 2013 Interview und schriftliche Befragung (Interview and Written Survey): Grundlagen und Methoden empirischer Sozialforschung (Foundations and Methods of Empirical Social Research). De Gruyter, München.

Meuser, M. & Nagel, U. 2009 Das Experteninterview – konzeptionelle Grundlagen und methodische Anlage (The expert interview – conceptional foundations and methods attachment). In: *Methoden der vergleichenden Politik- und Sozialwissenschaft (Methods of Comparative Politic – and Social Sciences)* (S. Pickel, D. Jahn, H.-J. Lauth & G. Pickel, eds). VS Verlag für Sozialwissenschaften, Wiesbaden, pp. 465–479.

Molle, F. & Chu, T. H. 2008 Implementing Integrated River Basin Management: Lessons from the Red River Basin, Vietnam Working Paper.

Nguyen, T. P. L. 2012 Legal framework of the water sector in Vietnam: achievements and challenges. *Journal of Vietnamese Environment* **2** (1), 27–44.

Nguyen, T. P. 2013 The legal framework of Vietnam's water sector: Update 2013. MPRA. https://mpra.ub.uni-muenchen.de/52996/ (accessed 10 July 2020).

Niederste-Hollenberg, J., Hillenbrand, T., Bark, K., Petry, M., Oldenburg, M., Wrenger, B., Scheifhacken, N., Berendonk, T., Stolpe, H., Brömme, K. & Spoth, K. 2012 *Modernisierungsstrategie für die Deutsche Wasserwirtschaft (Modernisation Strategy for German Water Industry): Maßnahmen zur Stärkung der Präsenz der deutschen Wasserwirtschaft auf internationalen Märkten für Wasserdienstleistungen (Measures to Strengthen the Presence of the German Water Industry on the International Markets for Water Services)*.

North, D. C. 2009 *Institutions, Institutional Change and Economic Performance*. Cambridge University Press, Cambridge.

Organisation for Economic Co-operation and Development (OECD) 2009 Civil Society and Aid Effectiveness: Findings, Recommendations and Good Practice. OECD Publishing. https://doi.org/10.1787/9789264056435-en.

Organisation for Economic Co-operation and Development (OECD) 2018a Development Aid at a Glance Statistics by Region: 4. Asia. http://www.oecd.org/dac/financing-sustainable-development/development-finance-data/Africa-Development-Aid-at-a-Glance-2018.pdf (accessed 10 July 2020).

Organisation for Economic Co-operation and Development (OECD) 2018b Net ODA. Available from: https://data.oecd.org/oda/net-oda.htm (accessed 24 March 2019).

Organisation for Economic Co-operation and Development (OECD) 2018c Recipient Country: Viet Nam. Available from: https://public.tableau.com/views/OECDDACAidataglancebyrecipient_new/Recipients?:embed=y&:display_count=yes&:showTabs=y&:toolbar=no?&:showVizHome=no (accessed 24 March 2019).

Oyedele, A. 2016 Emerging market global business model innovation. *Journal of Research in Marketing and Entrepreneurship* **18** (1), 53–62.

Pädagogische Hochschule Freiburg n.d. Dresing/Pehl: Übersicht über die einfachen und erweiterten Transkriptionsregeln (Overview of Simple and Extended Transcription Rules). Available from: https://www.ph-freiburg.de/fileadmin/dateien/fakultaet1/psychologie/Uploads/wirtz/Henning-Kahmann/Dresing___Pehl_Transkriptionssysteme.pdf (accessed 03 February 2019).

Prahalad, C. K. 2005 *The Fortune at the Bottom of the Pyramid.* Wharton School Publishing, Upper Saddle River, NJ/Harlow.

Prahalad, C. K. 2008 *The Fortune at the Bottom of the Pyramid.* Wharton School Publishers, Upper Saddle River, NJ.

Prahalad, C. K. & Hammond, A. 2002 Serving the world's poor, profitably. *Harvard Business Review* **80** (9), 48–57.

Rangan, V. K., Quelch, J. A. & Barton, B. 2007 *Business Solutions for the Global Poor: Creating Social and Economic Value.* Jossey-Bass, San Francisco, CA.

Reichenbach, J., Böttger, S. & Nöh, C. 2018 *Länderprofil zur Kreislauf- und Wasserwirtschaft in Vietnam (Country Profile of Recycling- and Water Sectors in Vietnam).* https://germanwaterpartnership.de/wp-content/uploads/2019/05/180608_Laenderprofil_Vietnam_Gesamt_Final.compressed.pdf. (accessed 10 July 2020).

Sánchez, P., Rodríguez, M. A. & Ricart, J. E. 2007 Meeting unmet needs at the base of the pyramid: mobile health care for India's poor. In: *Business Solutions for the Global Poor: Creating Social and Economic Value*, 1st edn (V. K. Rangan, J. A. Quelch & B. Barton, eds). Jossey-Bass, San Francisco, CA, pp. 84–91.

Sartorius, C. 2007 *Zukunftsmarkt Dezentrale Wasseraufbereitung und Regenwassermanagement.* Fallstudie im Auftrag des Umweltbundesamtes (Umwelt, Innovation, Beschäftigung Nr. 11/2007). Umweltbundesamt, Berlin.

Sheth, J. N. 2011 Impact of emerging markets on marketing: rethinking existing perspectives and practices. *Journal of Marketing* **75** (4), 166–182.

Silvestre, B. S. & Silva Neto, R. 2014 Are cleaner production innovations the solution for small mining operations in poor regions? The case of Padua in Brazil. *Journal of Cleaner Production* **84**, 809–817.

Smets, S., Lockwood, H., Mansour, G. & Smits, S. 2017 *Sustainability assessment of rural water service delivery models: findings of a multi-country review.* (Water Global Practice water papers). The World Bank, Washington, DC.

Sousa-Zomer, T. T. & Cauchick Miguel, P. A. 2018 Sustainable business models as an innovation strategy in the water sector: an empirical investigation of a sustainable product-service system. *Journal of Cleaner Production* **171**, S119–S129.

The World Bank Group 2019a *GDP, PPP (Current International $).* Available from: https://data.worldbank.org/indicator/NY.GDP.MKTP.PP.CD?locations=VN (accessed 15 January 2019).

The World Bank Group 2019b *Poverty Headcount Ratio at $3.20 A Day (2011 PPP) (% of Population).* Available from: https://data.worldbank.org/indicator/SI.POV.LMIC?locations=VN (accessed 16 January 2019).

The World Bank Group 2019c *Poverty Headcount Ratio at $5.50 A Day (2011 PPP) (% of Population).* Available from: https://data.worldbank.org/indicator/SI.POV.UMIC?locations=VN (accessed 16 January 2019).

Tu, C. T. n.d. *Current Issues in Administrative Boundary Adjustments in Vietnam.* Available from: http://isos.gov.vn/News_Detail/tabid/179/ArticleId/1135/language/en-US/Current-issues-in-administrative-boundary-adjustments-in-Vietnam.aspx (accessed 28 January 2019).

United Nations (UN) n.d. *Sustainable Development Goal 6: Ensure Availability and Sustainable Management of Water and Sanitation for All.* Available from: https://sustainabledevelopment.un.org/sdg6 (accessed 16 May 2020).

Vietnam Water Supply and Sewerage Association (VWSA) n.d. *Vietnam Water Supply and Sewerage Association (VWSA): Guiding Principles and Objectives.* Available from: http://vwsa.org.vn/en/about-us.html (accessed 24 March 2019).

Viswanathan, M., Seth, A., Gau, R. & Chaturvedi, A. 2009 Ingraining product-relevant social good into business processes in subsistence marketplaces: the sustainable market orientation. *Journal of Macromarketing* **29** (4), 406–425.

Wegner, A. 2015 *Domestic Water Supply in Rural Viet Nam: Between Self-supply and Piped Schemes.* doi:10.5445/IR/1000050255.

White, M. A. 2013 Sustainability: I know it when I see it. *Ecological Economics* **86**, 213–217.

Wirtz, B., Pistoia, A., Ullrich, S. & Göttel, V. P. 2016 Business models: origin, development and future research perspectives. *Long Range Planning* **49** (1), 36–54.

World Commission on Environment and Development 1987 *Report of the World Commission on Environment and Development: Our Common Future.* Oxford University Press, Oxford.

World Health Organization/UNICEF 2000 Global Water Supply and Sanitation Assessment Report, Geneva. https://www.who.int/water_sanitation_health/monitoring/jmp2000.pdf. Accessed 10 July 2020.

World Health Organization (WHO) & United Nations Children's Fund (UNICEF) 2017 Progress on drinking water, sanitation and hygiene. 2017 update and SDG baselines. WHO and UNICEF, Geneva, Switzerland.

First received 23 March 2020; accepted in revised form 2 June 2020. Available online 6 July 2020

Assuring water quality along multi-barrier treatment systems for agricultural water reuse

Marius Mohr, Thomas Dockhorn IWA, Jörg E. Drewes IWA, Sybille Karwat, Susanne Lackner IWA, Bryan Lotz, Andreas Nahrstedt, Andreas Nocker, Engelbert Schramm and Martin Zimmermann

ABSTRACT

Based on three pilot- and demonstration-scale projects investigating agricultural irrigation practices with reclaimed water, risks associated with these water reuse practices are highlighted and processes and strategies to minimize associated microbial risks were evaluated. A number of treatment processes and combinations were tested regarding their efficacy for pathogen removal, representing the biggest threat to the quality of products from reuse irrigation practices. In addition, the importance of regrowth potential and different methods for monitoring risks associated with pathogens were discussed. One method for online monitoring is flow cytometry. The results of an exemplary quantitative microbial risk assessment (QMRA) were discussed to determine the significance of microbial risks. Multi-barrier approaches comprised of technical and administrative barriers can reduce the risks of water reuse significantly. Quality management also needs to address all stakeholders involved in a reuse project, starting from source control in the sewershed to marketing of the final products. In addition, environmental risks of water reuse need to be addressed by quality management as well.

Key words | agricultural irrigation, multiple barriers, pathogens, QMRA, water quality management, water reuse

Marius Mohr (corresponding author)
Bryan Lotz
Fraunhofer-Institute for Interfacial Engineering and Biotechnology IGB,
70569 Stuttgart,
Germany
E-mail: *marius.mohr@igb.fraunhofer.de*

Thomas Dockhorn IWA
Sybille Karwat
Technical University of Braunschweig, Institute of Sanitary and Environmental Engineering,
Pockelsstraße 2a, 38106 Braunschweig,
Germany

Jörg E. Drewes IWA
Chair of Urban Water Systems Engineering,
Technical University of Munich,
85748 Garching,
Germany

Susanne Lackner IWA
Department of Civil and Environmental Engineering Sciences, Institute IWAR, Chair of Wastewater Engineering,
Technical University of Darmstadt,
Franziska-Braun-Strasse 7, 64287 Darmstadt,
Germany

Andreas Nahrstedt
Andreas Nocker
IWW Water Centre,
Moritzstraße 26, 45476 Mülheim an der Ruhr,
Germany

Engelbert Schramm
Martin Zimmermann
ISOE – Institute for Social-Ecological Research,
60486 Frankfurt am Main,
Germany

HIGHLIGHTS

- Multi-barrier approaches comprised of technical and administrative barriers can reduce the risks of water reuse significantly.
- A number of treatment processes and combinations were tested regarding their efficacy for pathogen removal (ultrafiltration, ozone, UV, ponds, biological processes, and biofilters).
- Flow cytometry as a method for online monitoring is discussed.
- The importance of regrowth potential is shown.
- An exemplary QMRA is carried out, differentiating between risks from bacteria, viruses, and protozoa.

INTRODUCTION

Global challenges such as climate change, population growth, and an increasing shift toward urbanization are

doi: 10.2166/wrd.2020.039

resulting in an increased demand of water resources for agricultural and industrial purposes. Thus, the gap between the growing need for water supply and the existing limited water resources is widening continuously. With this growing demand, fresh water supplies are becoming an increasingly scarce resource in many regions worldwide including areas with a high population density, e.g. mega cities. Due to perceptible climate change, the lack of available water resources will be of even greater importance worldwide in the near future, as the frequency of extended droughts and heavy rainfalls is becoming a more widespread phenomenon even in moderate climate zones. According to the FAO (2017), agricultural irrigation needs an average of 70% of fresh water supplies worldwide. Thus, water reuse focusing on reducing the amount of conventional fresh water supplies is increasingly considered to avoid conflicts due to local supply shortfalls while implementing a sustainable water resource management with a water quality appropriate to the intended use (Becerra-Castro *et al.* 2015; Becker *et al.* 2017; Maassen 2018; Salgot & Folch 2018; WRI 2019).

MICROBIAL RISK ASSESSMENT AND MANAGEMENT

Water reuse for agricultural irrigation is associated with an elevated risk due to the presence of pathogens in municipal wastewater. Thus, it is paramount that the microbial risk in these applications is properly managed. Microbial risk assessment combines scientific knowledge about water-based pathogens, the effect of natural and engineered barriers, and monitoring strategies. Risk assessment consists of four steps: problem formulation, exposure assessment, dose–response assessment, and risk characterization (WHO 2016; Drewes *et al.* 2019). The combination of scientific findings with mathematical models enables a quantitative microbial risk assessment (QMRA). The risk characterization, therefore, contains information on the probability of an exposed person being endangered by specific pathogenic microorganisms with regard to infection or disease.

The measure of Disability-Adjusted Life Years (DALYs) represents the extent of suffering in the form of the sum of years that a person lives impaired by a disease, Years Living with a Disability, and the years of life lost due to the disease through premature death, Years of Life Lost (Prüss-Üstün 2003; Scheierling *et al.* 2011). DALYs can be calculated for various health hazards and differ depending on the health hazard under consideration. The risk assessment is carried out by comparing the maximum reasonable exposure to infection with the expected risk of infection. With regard to the raw consumption of food irrigated by reclaimed water, the World Health Organization (WHO) determines the maximum acceptable additional burden of an individual from waterborne diseases at 10^{-6} DALY per person per year (WHO 2006, 2016; Drewes *et al.* 2019).

The WHO (2006) recommends that achieving 'health-based targets' for a given reuse application should be determined by means of QMRA. The overall 'health-based target' can be accomplished by determining individual 'health-based goals', which represent the necessary cumulative reductions of pathogen concentrations in raw wastewater by individual barriers, typically expressed in logarithmic reduction steps (Log10) steps.

In order to properly manage the overall risk and to achieve the health-based target, a multiple barrier approach can be adopted which is based on a number of measures that together avoid or reduce the risk of contamination by the use of recycled water (WHO 2016). This is achieved by control measures at various locations along the entire production chain to establish effective barriers against the spread of pathogens (NHMRC 2011; Health Canada 2013).

QUALITY MANAGEMENT IN AGRICULTURE

In the technical debate, it is assumed that the 'fit for purpose' approach entails the production of reclaimed water to a quality that meets the needs of the intended end-use (ISO 20670:2018). In traditional commercial water supply sectors, quality control typically involves monitoring of the quality of the water produced at one or more locations in the process. For this purpose, the supplier either sets quality standards itself or negotiates them with its customers. Another possibility is that the standards are specified by technical norms or legal requirements. In the commercial agricultural business context, quality management is increasingly built up very systematically and comprehensively. However, extending these protocols to water reuse applied in the context of agricultural irrigation has not been developed systematically. A successful water reuse project not only depends on reliable treatment processes employed but also on proper organizational procedures and strict compliance with regulatory requirements while also meeting customer expectations continuously. These different aspects can be addressed under the framework of quality assurance. Quality assurance is comprised of an early identification of hazards or the avoidance of hazards. If this is practised in an organized fashion, it is referred to as quality management. Quality management comprises the initiation and safeguarding of organizational measures that serve to ensure the provision of services and the underlying procedures and processes – and thus also the quality of the final product. Guaranteeing desired quality requirements while also involving all stakeholders in quality management is, therefore, a central prerequisite for the consideration of water reuse in agriculture. The EU Commission has stressed this point as a major obstacle to further growing agricultural water reuse across Europe (Council of the EU 2019).

In this paper, we discuss how effectively water quality can be assured in a multi-barrier approach by analyzing the practices of different agricultural water reuse practices as part of three large-scale research projects ('HypoWave', 'EPoNa', and 'MULTI-ReUse') funded by the German Federal Ministry of Education and Research (BMBF). For these water reuse case studies, an overview of the technical and administrative barriers is given and their effectiveness is assessed. As far as process-related barriers are concerned, we investigated to what extent quality management can improve the process results. In addition, risk management is applied to assess the entire chain of wastewater collection, treatment, and conveyance up to the final application to agricultural products.

MATERIALS AND METHODS

Case studies and analytical methods applied

(A) Treatment technologies tested at pilot-scale in the 'HypoWave' project included an expanded granular sludge bed reactor (EGSB), a sequencing batch reactor (SBR), a biological activated carbon (BAC, empty bed contact time: 3.3 h) filter, and an ozone reactor (OZONE, ozone dose approximately 0.8 mg O_3/mg DOC (dissolved organic carbon)) (Bliedung *et al.* 2020). The feed water to the pilot was primary effluent (influent to the aeration tank, AT) of the wastewater treatment plant (WWTP) Wolfsburg-Hattorf, Germany (a treatment capacity of 10,500 PE). Three parallel treatment trains were employed as illustrated in Figure 1 in order to produce reclaimed water of different qualities.

The analysis of microbial parameters included the determination of the indicator organism *Escherichia coli* (EC) as well as the bacteria groups of total coliforms (C), Enterobacteriaceae (EB), and total viable aerobic count (TC). The aim was to evaluate the microbiological contamination of wastewater before and after using different treatment processes in order to estimate their effectiveness as procedural barriers along the water treatment chain. The methods employed used ready-to-use selective culture media (Petrifilm) of the 3M company followed by incubation and counting of colony-forming units (CFU/ml) according to the manufacturer's specifications: C/EC at $37 \pm 1\,°C$ for

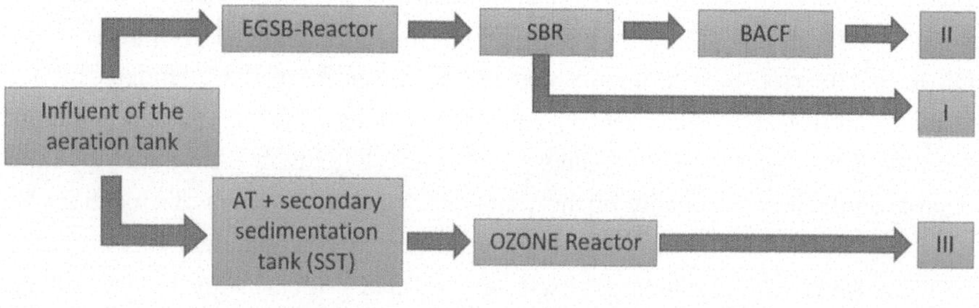

Figure 1 | Flow schematic of the three pilot-scale treatment combinations within the HypoWave project at WWTP Wolfsburg-Hattorf.

24 ± 2 h, respectively, 48 ± 2 h (NMKL method 147.1993); EB at 37 ± 1 °C for 24 ± 2 h and TC at 30 ± 1 °C for 72 h. Dilution was done by deionized water (after 0.2 μm sterile filtration). Three replicates per sample and per bacteria group examined were generated, and corresponding blank samples were collected for each day of analysis. The analysis was based on eight sampling days. Taking the replicates into account, a total of 24 measuring results of each sampling point and each bacteria group was obtained.

How the HypoWave concept can be implemented has been examined in the context of a case study in Germany. The wastewater of the village of Weißenberge (Gifhorn county, approximately 500 inhabitants, 26,400 m^3/a) is currently being discharged into a system of treatment ponds (Figure 2). In case of dry weather conditions, the residence time in the system is more than 90 days. The effluent of the plant still contains relatively high concentrations of nutrients such as nitrogen and phosphorus, but due to the long residence time only low levels of pathogenic microorganisms. In an activated carbon biofilter, the nitrogen, which is largely present as ammonium-nitrogen, will be oxidized to nitrate by aeration and thus be made more readily available to plants. In order to reduce the risk of pathogens, the UV-disinfection process will be used after the biofilter. The treated water will then be used in hydroponic vegetable production, which also will utilize and remove the remaining nutrients. Eggplants, zucchini, tomatoes, peppers, and lettuce are potential crops for this application. For the case study of Weißenberge, a QMRA was carried out.

(B) In the EPoNa project, the town of Outapi in Northern Namibia has been selected. The town has flush toilets and a gravity sewer system in its formal districts. The household wastewater is discharged to waste stabilization ponds (WSP). Population growth and the overflowing of the ponds during the rainy season are challenges for the functionality of the system. Since the mere evaporation of the water is

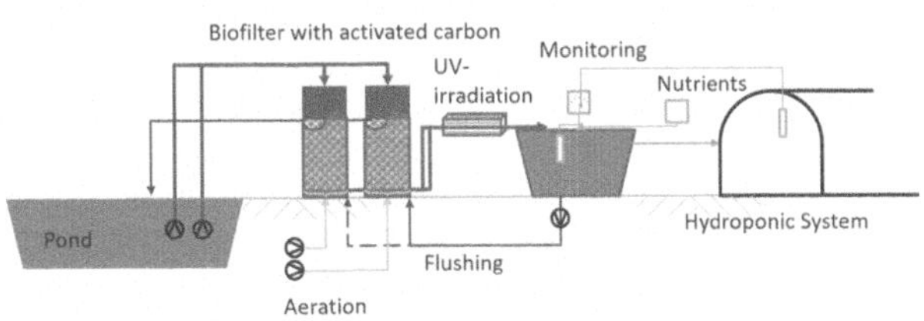

Figure 2 | Flow chart of treatment concept for Weißenberge (HypoWave).

effectively wasting a valuable resource, water reuse for irrigation purposes is considered to be a resource-efficient option in the semi-arid region according to the Namibian Water Resources Management Act (Act No. 24 of 2004).

The WSP system comprises two parallel treatment trains (trains A and B) consisting of four ponds each (one primary facultative pond and three maturation ponds; Figure 3). All eight ponds together have a surface area of approximately 40,000 m^2 and a volume of 55,000 m^3. The treated water reaches an evaporation pond (surface area: 41,000 m^2; volume: 20,000 m^3) as the final stage. The system's average inflow was approximately 753 m^3/d. To enable water reuse, one train (A) of the WSP was equipped with a pretreatment (two options were tested: an up-flow anaerobic sludge blanket reactor (UASB) and a micro sieve) to reduce the organic load (Sinn *et al.* 2019). Additionally, guiding walls helped optimizing the flow conditions in pond A1. A rock filter served as posttreatment in pond A4. These measures aim at reaching an effluent water quality suitable for irrigation purposes to produce animal fodder (Lackner *et al.* 2017).

Microbial samples were analyzed regularly for total coliforms, *E. coli* and enterococci as indicator organisms for pathogens. Total coliforms and *E. coli* were detected by using the IDEXX Colilert-18 test (DIN EN ISO 9308-2:2012), enterococci by the Enterolert 250 and quantified with a Quanti-Tray/2000 system based on the most probable number (MPN) model.

(C) Near the North Sea coast between the Jade bay and the river Weser, drinking water supply is under the regional stress of strong limited water resources with a sufficient low salinity, longer periods with high temperatures (spring, summer, and autumn), and a growing industry. For several decades, the local water supply company Oldenburgisch-Ostfriesischer Wasserverband (OOWV) secures the water supply on the basis of water resources from neighboring administrative districts in competition with the water demand for agricultural irrigation. Long-term forecasts indicate a raise of use conflicts in the near future.

For this reason, the OOWV started working on concepts for a more efficient water supply to deliver their customers with a water quality adapted to their needs. Beside drinking water, this includes different advanced treated water types of a defined quality using tertiary effluent from the municipal WWTP Nordenham (36,000 population

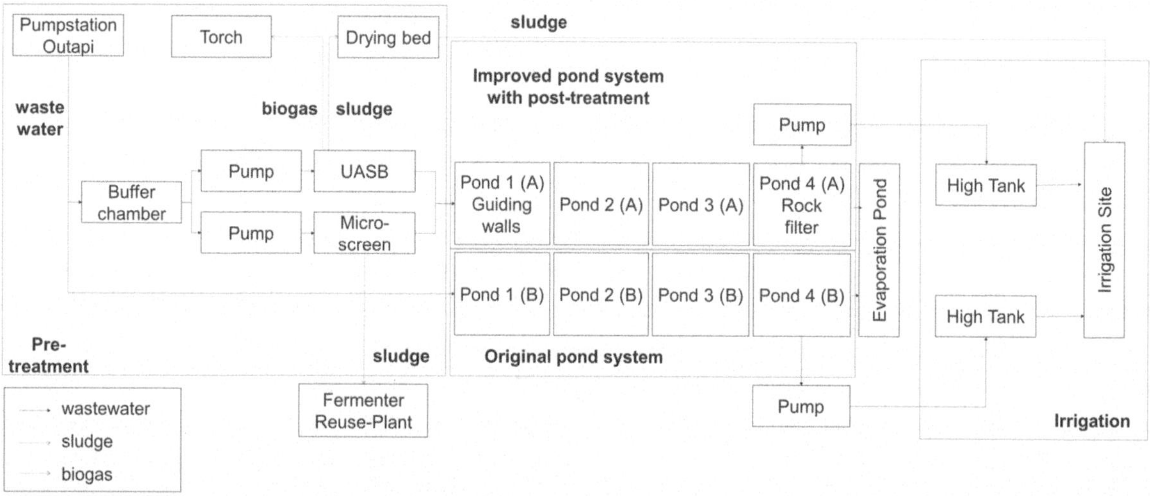

Figure 3 | Flow schematic of the waste stabilization pond system within the EPoNa project (Sinn *et al.* 2019).

equivalents, Q = 3,000–15,000 m^3/d). The different qualities targeted agriculture needs (irrigation for non-food crop) as well as industrial processes for cooling and boiler feed water.

In MULTI-ReUse, a pilot-scale facility with three different trains of modular process units was built to produce three different water qualities (Figure 4): (1) pre-filtration, (optional) powdered activated carbon (PAC) dosage, flocculation, ultrafiltration (UF, each line with a capacity 5.6 m^3/h), and UV disinfection; (2) pre-filtration, flocculation, UF, UV, sand filtration, and granulated activated carbon (GAC) filtration; and (3) pre-filtration (sieve), PAC dosage (optional), flocculation, UF, monochloramine dosage, reverse osmosis (RO, in two trains to compare several low-pressure modules), and UV disinfection.

Log reductions in the treatment processes of the MULTI-ReUse project were monitored by applying traditional culture methods to quantify important microbiological indicator organisms: colony counts at 22 and 36 °C (both DIN EN ISO 6222:1999-07), *E. coli* and coliforms (both DIN EN ISO 9308-2:2014-06), intestinal enterococci (DIN EN ISO 7899-2:2000-11), *Clostridium perfringens* (DIN EN ISO 14189:2016-11), and *Legionella* spp. (DIN EN ISO 11731:2017-05). The culture methods were supplemented

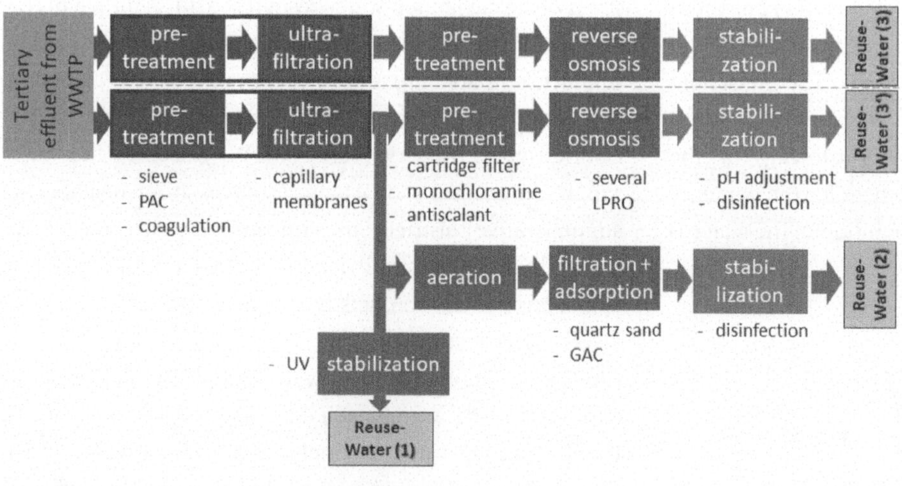

Figure 4 | Treatment train of the pilot plant in the MULTI-ReUse project to produce three different water qualities. The treatment train producing water quality 3 was duplicated for comparison of the effects of process alterations.

by the cultivation-independent assessment of total cell counts (TCC) and intact cell counts (ICC) using flow cytometry in combination with fluorescent staining. Whereas culture values only detect the small proportion of culturable bacteria in the water, flow cytometry allows monitoring of the entire bacterial population independent of the fact whether they form colonies on a given medium under the defined conditions. Samples were transported cooled to the laboratory to ensure that the resulting data represent the microbiological status of the water at the time of sampling (day 0). Obtained values were referred to as $TCC_{day\ 0}$ and $ICC_{day\ 0}$. After analysis of aliquots, water samples were subsequently incubated at 22 °C for 7 days to allow bacterial growth followed by repeated measurement of intact cell count ($ICC_{day\ 7}$). The increase in ICC in these 7 days (i.e. the extent of growth) was determined by the amount of assimilable organic carbon (AOC) present in the sample. $ICC_{day\ 7}$ values represent the maximal concentrations of intact cells supported by the AOC contained in the sample. The offline flow cytometry was supported by online analysis, which allowed high-resolution data. An Online Bacteria Analyzer (OBA Metanor, ONTRONIX AG, Switzerland) was installed after UF which was considered a critical control point.

RESULTS AND DISCUSSION

Multi-barrier systems

Table 1 provides a summary of barriers employed for the projects EPoNa, HypoWave, and MULTI-ReUse. In addition to wastewater treatment, the barriers in the area of wastewater collection and agricultural application are also displayed. In addition to technical barriers, administrative barriers were also used in all projects; for example, in some applications, the risk of hazardous substances in the sewage is reduced by monitoring the connections to the sewer network, and in other applications, the crops that are irrigated are not for human use. In the following section, different technical barriers regarding pathogens are discussed, an exemplary QMRA is executed followed by a brief discussion of quality management and administrative barriers.

Table 1 | Barriers along the water chain in the analyzed projects

Project/case study	Collection	Treatment			Irrigation	Harvest
MULTI-ReUse		Tertiary treatment	UF	Disinfection (UV-radiation)	Energy and industrial crops	
HypoWave Standard case	Controlled sewage collection: No industrial indirect dischargers, no hospitals	Biological treatment (ASP, SBR)	Biofilter/ activated carbon	Disinfection (Ozone, UV-radiation)	Water only in contact with roots, plastic cover between root and shoot of the plants	Specific hygiene measures by the quality assurance system agreed with the trade or purchaser
HypoWave Weißenberge	Controlled sewage collection: No industrial indirect dischargers, no hospitals	WSP (natural UV-radiation, cold temperatures)	Biofilter/ activated carbon	Disinfection (UV-radiation)	Water only in contact with roots, plastic cover between root and shoot of the plants	Specific hygiene measures by the quality assurance system agreed with the trade or purchaser
EPoNa	Controlled sewage collection: No industrial indirect dischargers	Pretreatment line A: Up-flow anaerobic sludge blanket reactor or micro sieve	WSP (natural UV-radiation)	Rock filter in pond A4	Animal fodder	

Performance of treatment processes regarding pathogens

The main goal of these investigations was to highlight how water could be reused in agriculture while properly managing associated health risks. Thus, treatment processes as technical barriers have been tested with a focus on pathogen removal.

The feed water used in the HypoWave project was characterized by the following bacteria concentrations: total aerobic count ranging from 10^6 to 10^7 CFU/ml, EB and total coliforms from 10^5 to 10^6 CFU/ml, and *E. coli* from 10^4 to 10^5 CFU/ml. These concentrations are representative of municipal raw wastewater exhibiting typical *E. coli* concentration of 10^3 to 10^6 CFU/ml and 10^5 to 10^7 CFU/ml for total coliforms, respectively (WHO 2006; Asano *et al.* 2007; Guy *et al.* 2010; DWA 2017; Cornel *et al.* 2018). The efficacy of the different treatment barriers is summarized in Figure 5.

Based on these findings, the removal efficacy of the EGSB appeared to be the lowest in terms of the reduction of all four bacteria groups measured. With a mean *E. coli* removal of 0.09 log units and a maximum of 0.42 log units, its performance was considerably lower than expected. As the EGSB is considered to be a high-efficiency UASB, a mean bacterial removal of at least 1–2 log units was expected (WHO 2006; DWA 2008, 2017). However, during this study, the performance of the EGSB was far from being stable and regrowth in all four bacteria groups has been observed. Contrary to this, the SBR installed after the EGSB removed *E. coli* at least by 1.8 log units and at most by 2.7 log units. This means that even the lower margin of treatment performance of the SBR achieved a log removal efficiency equal to the upper range of the usual bacterial log removal of a conventional activated sludge (CAS) process including secondary clarification, i.e. 1–2 log units, whereas the SBR maximum performance exceeds a CAS process by almost 1 log unit (WHO 2006; DWA 2008, 2017).

With a maximum *E. coli* removal of 2.1 log units and a minimum of 1.1 log units, the purification performance of the BACF is in the range of the usual purification performance of slow sand filtration and to the middle range of fast sand filtration (WHO 2006; DWA 2008, 2017). In the HypoWave project, OZONE performed unusually poor with a mean treatment efficiency far below common values, and even a potential of regrowth regarding EB and total bacteria count was observed. During the ozone process, *E. coli* was removed at least by 0.1 log units and at most by 2.25 log units (Figure 5). This means that the purification performance of the installed OZONE was significantly lower than the usual bacteria removal performance of such a treatment process being 2–6 log units (WHO 2006; DWA 2008, 2017; Seis *et al.* 2016). Moreover, the low purification performance may be due to a deliberately chosen ozone dose which was

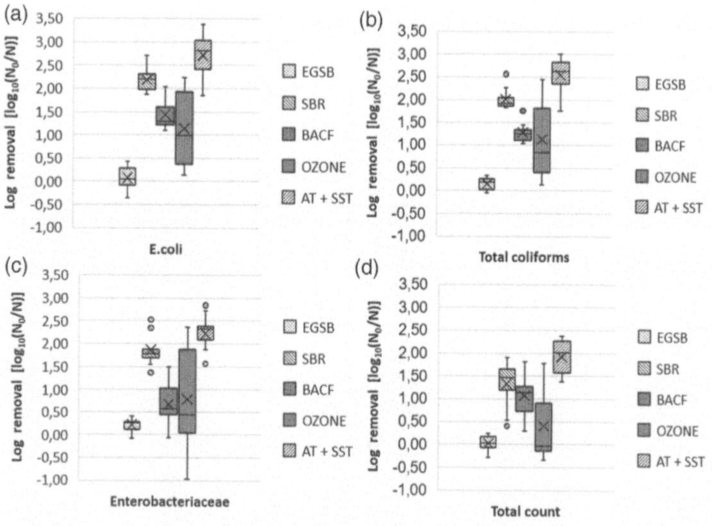

Figure 5 | Performance of each treatment process employed at pilot-scale within the HypoWave project regarding the four measured bacteria groups.

about four times lower than the usual dose for disinfection. A second reason may be the very low concentration of *E. coli* of 10^1 to 10^3 CFU/ml in the influent of ozone. Nevertheless, with a mean *E. coli* removal performance of 2.87 log units, the activated sludge process including secondary sedimentation (AT + SST) shows a performance slightly superior to other data (WHO 2006; DWA 2008, 2017).

In view of fecal coliforms (EC), only irrigation water of treatment II complies with the requirements of the WHO concerning the food production suitable for raw consumption, i.e. $\leq 10^4$ CFU/l. Finally, as *E. coli* is an indicator organism only for bacteria, it has to be stressed that the data cannot be used to make reliable statements on the treatment processes of the pilot reactor regarding protozoa, viruses, and helminths.

The EPoNa project studied WSP for their capabilities to remove pathogens. Properly designed WSP are capable of removing large numbers of fecal bacteria. Aside from sedimentation and decay in the ponds, naturally occurring disinfection by solar UV radiation takes place. However, the effluent quality of WSP often does not comply with common Quality Standards. In the Namibian Code of Practice (DWAF 2012), filtration and disinfection are part of the tertiary treatment steps required for water reuse. However, due to the high amounts of organic solids and ammonia, conventional disinfection (e.g. chlorination) is often not effective and the effluent cannot be reused for unrestricted irrigation. In 2012, the Code of Practice was amended to allow the irrigation of fodder without the requirement of disinfection in exceptional cases. A permit for exemption from the obligation to disinfect wastewater can be granted if it can be proven that the non-disinfected wastewater will not come into contact with any humans or animals.

Data from the sampled WSP indicate that already the overloaded WSP (train B) was able to result in a significant reduction in *E. coli* with 3.90 log units reduction (from the influent to the effluent of B4) and 2.34 log units for enterococci (Figure 6). The upgrade of train A improved the reduction of *E. coli* to 4.29 log units (3.32 log units for enterococci) and concentrations of *E. coli* in the effluent of A4 of $6.9 \times 10^2 \pm 1.0 \times 10^3$ MPN/100 ml. The effect on total coliforms was, however, much less, with a reduction of only 0.94 log units for train A and 0.80 log units for train B. With concentrations of still $2.2 \times 10^4 \pm 5.0 \times 10^4$ MPN/100 ml for *E. coli* in the effluent of B4 (see also Figure 6), the original pond system without any upgrades was not able to meet the WHO guidelines for safe water reuse of 1×10^3 CFU/100 ml for fecal coliforms and application for fodder crop irrigation. Additionally, the reduction of other pathogens might not be as high in the tested technologies, as the effect on enterococci in both lines with and without pre- and post-treatment seemed to be less compared to *E. coli*.

To evaluate the pathogen reduction of the WSP system (with additional pre- and post-treatment) in comparison to only the WSP, the reduction of *E. coli* was analyzed in depth. Pretreatment and upgrading of pond A1 resulted in an *E. coli* reduction of 97% (inflow – A1), compared to 91% reduction without pretreatment. Further reductions by the ponds (from pond 2 to 4) were similar for both, trains A and B, with 99.8%. However, also considering the reuse water storage tanks, concentrations of *E. coli* were still $>1.0 \times 10^3$ MPN/100 ml in line B, and there might even be the danger of regrowth.

Ensure target reduction rates can be achieved by extending the multi-barrier approach to the irrigation system. Besides the WSP and corresponding treatment

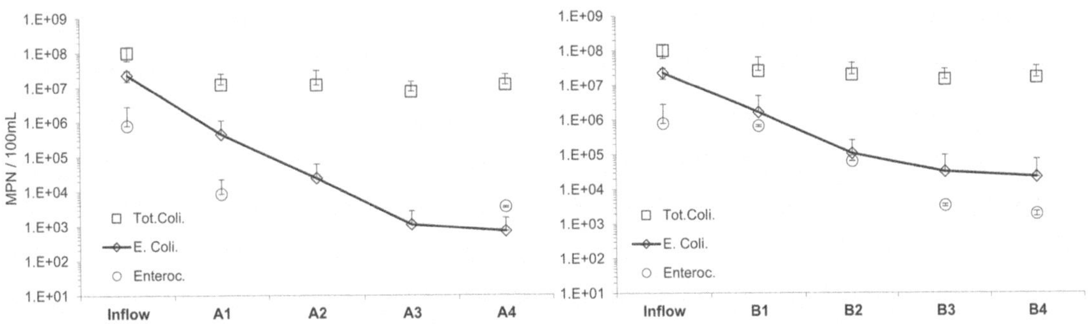

Figure 6 | Reduction of total coliforms, *E. coli* and enterococci through both lines of the WSP (average values include all data with pretreatment (784 days)).

steps, a specific irrigation technology can also act as another barrier. Drip irrigation of low-growing crops would for instance further reduce pathogens by 2 log10 units (WHO 2006). This would bring the values way below the required log removal requirement.

In the MULTI-ReUse project, a membrane integrity concept was applied to assure proper and efficient removal of viruses, bacteria, and protozoa. To validate the initial virus removal efficiency of the UF, modules were tested with MS2 phages yielding a log removal value >4. This indicates a suitable pore diameter distribution and the integrity of the membrane system. During the treatment operation, it was necessary to monitor and ensure this integrity. Single defects are a pathway for feed water bypassing the active membrane (defect flow). But, a critical drop of virus retention will only occur if the ratio of defect flow rate and total flow rate through the active membrane reaches a critical value. For critical defect flow rates in full-scale UF modules, single defects (fiber break, holes in the membrane) must have a size of some micrometer (EPA 2005; DVGW 2007). Defects of this size will also lead to bypassing bacteria. Therefore, the monitoring of a log reduction value (LRV) of 5 or 6 for bacteria retention was used to ensure an LRV >4 for virus retention. With the online measured TCC using flow cytometry, it was possible to monitor an LRV of 3–4 for bacteria nearly in real time. The upper detection limit for the LRV reached is strongly dependent on the filtrate quality, which is affected by the release of bacteria into the filtrate from regrown biofilms in pipes and valves. The culturing method for *E. coli* and coliforms ensures an LRV >6 with a time offset of several hours. To exclude the risk from membrane ageing which can occur in a process under the stress of chemicals and can potentially affect its pore diameter distribution, challenge tests with phages or with (small!) viruses already present in the effluent of a WWTP should be repeated.

To follow the multi-barrier approach, virus and bacteria retention was enhanced by UV inactivation in accordance to the German standards DVGW W 294 Series for drinking water supply and DWA (M)205 for WTP effluent. By meeting these standards, a 4 log inactivation for waterborne pathogens is guaranteed. The flow rate and UV transmission (in a water after UF free of TSS) were monitored online. Furthermore, certified reactors monitor the intensity of UV light inside by an integrated UV sensor.

The hygienic status of treated water was monitored by traditional cultivation of indicator bacteria. UF as the first treatment barrier resulted in complete elimination of *E. coli*, coliforms, intestinal enterococci, *C. perfringens*, and *Legionella* spp. The concentrations of these bacteria in the secondary effluent of a WWTP were subject to strong variations with average values of 79,000 *E. coli*/100 ml, 164,900 enterococci/100 ml, 64,300 *C. perfringens*/100 ml, and >100 *Legionella* spp./100 ml. These indicators were absent also in samples from subsequent treatment steps confirming the hygienic safety of the advanced treated water at the time of sampling. Colony counts in raw water were on average 24,600 CFU/ml at 22 °C and 42,600 CFU/ml at 36 °C and were reduced by UF to 2 and 120 CFU/ml, respectively. UF was therefore seen as a critical control point.

TCC and ICC measurements exhibited strong reductions by UF; however, the UF filtrate side was as expected not sterile. Bacteria colonizing surfaces in the filtrate compartment are the source of freely suspended bacteria in the freshly filtered water. The integrity of the UF membrane was not compromised. Regrowth on the filtrate side could be distinguished from breakthrough events by monitoring relative proportions of low nucleic acid (LNA) bacteria and high nucleic acid (HNA) bacteria using offline flow cytometry. Whereas LNA bacteria tended to be dominant in UF feed water (or at least be present in equal numbers as HNA bacteria), they were absent in the UF filtrate. In case of compromised membrane integrity, it would be primarily LNA bacteria that would be expected in the filtrate as they tend to be smaller in size and have a higher probability of membrane passage. The fact that UF filtrate nearly exclusively contained HNA bacteria indicates clearly that the bacterial population in filtrate originated from regrowth.

Flow cytometry was, therefore, seen as a valuable diagnostic tool to assess whether bacteria on the filtrate side originated from regrowth or from a breach of UF membrane integrity. Online flow cytometry can greatly assist in monitoring the functionality of this important barrier. The use of a fluorescent stain mixture allowing to distinguish between intact and damaged bacteria also provides a continuous monitoring strategy to assess disinfection efficiency in case an oxidative disinfectant (e.g., hypochlorite or monochloramine) is added as an additional barrier (for details see Nocker *et al.* 2020).

Another efficient barrier in MULTI-ReUse was RO as another membrane filtration step to further increase water quality. Both membrane filtration steps led to a profoundly different bacterial community, which was also confirmed by Illumina Sequencing. Sand filtration and granular activated charcoal (GAC) on the other hand greatly increased bacterial diversity and led to biological stabilization of the water. The resulting increase in bacterial numbers greatly diminished the regrowth potential compared to water after UF or RO with filtered, bacteria-stripped water showing high regrowth potentials (Nocker *et al.* 2020). The presence of an indigenous microbial community (as present on sand or GAC filter) has been reported to be antagonistic to the growth of pathogens (Camper *et al.* 1985; Hibbing *et al.* 2010). For drinking water treatment, biological stabilization has been reported to shape the bacterial community composition with lasting effects also during water distribution (Pinto *et al.* 2012; Lin *et al.* 2014; Prest *et al.* 2016). Van Nevel *et al.* (2013) summarized that the bacterial invasion potential in water is determined by nutrient availability and the indigenous bacterial population. The same can also be assumed for reuse water adding another microbiological barrier to the treatment process.

Nutrient depletion as a microbiological barrier is the basis for drinking water treatment when following the philosophy of distributing biologically stable water (instead of employing a disinfectant with residual). Apart from using flow cytometry for process monitoring, this diagnostic method could also provide valuable insight into the change in regrowth potential along treatment. Regrowth was assessed by measuring intact cell concentrations obtained from samples incubated for 7 days at 22 °C. The bacterial concentrations obtained after stagnation represent the maximal numbers supported by the available nutrients contained in the sample. The advantage of this approach lies in the sensitivity in the low nutrient range. After RO, where DOC and traditional AOC (van der Kooij 1992) were near the detection limits of the corresponding methods, the permeate still allowed regrowth to $ICC_{day\ 7}$ concentrations between 10^5 and 10^6 bacteria/ml. The lower end of this range is comparable to bacterial concentrations typically contained in drinking water, the upper range to that of stagnated potable water in household plumbing systems (Hammes *et al.* 2007; Lipphaus *et al.* 2014). As bacterial concentrations in this range are typically not associated with hygienic problems, the quality of the produced recycled water can be considered very good. The critical question is whether the regrowth potential in filtrate or permeate, which has not yet been converted into planktonic suspended biomass (like after GAC), will be deposited in the form of biofilms. Another critical question is whether the regrowth potential can be used by pathogens that accidentally enter the treated water. It is important to realize that the microbial population in water is highly dynamic and changes rapidly over time. The water quality should also be known at the time of water use and not just directly after treatment.

QMRA to assess overall treatment efficacy

The aim of the QMRA undertaken for the HypoWave case study in the county of Gifhorn was to assess whether the suggested concept concerning cultivation and consumption of hydroponically grown food raised by using reclaimed water complies with the maximum additional tolerable health risk specified by the WHO (2006). The QMRA method utilized was based on the approaches of Eregno *et al.* (2017), who followed the provisions of the WHO guideline (2006, 2016) very closely, but included assumptions on hydroponic cultivation such as a retained and ingested exposure volume of 1.0×10^{-7} l of irrigation water per gram of lettuce. In addition, by using lettuce as an example for an agricultural product provided for raw consumption, the focus of the QMRA was on a high exposure to health risks for the consumer. If the residual risk of the Gifhorn concept is below 10^{-6} DALYs, every other hydroponically grown product in the same system could be considered to be hygienically safe as well.

Under the assumption that 100 g of lettuce is consumed in a 2-day rhythm, the risks of infection with Cryptosporidium, Campylobacter, and Norovirus are shown in Figure 7. While the risk of infection with Campylobacter is lowest, the risk of an exposed person becoming infected by a virus seemed to be highest.

In order to verify whether these QMRA results concerning the consumption of hydroponically grown lettuce complied with the maximum additional health risk of 10^{-6} DALY, the DALY value was converted into a maximum tolerable risk of infection per person and year (pppy) for each pathogen according to the WHO Guidelines (2006, 2016). The corresponding values for the reference pathogens Cryptosporidium, Campylobacter and Norovirus – standing for each pathogen group, i.e. protozoa (P), bacteria (B), and viruses (V) – are also shown in Figure 7. Thus, for the three selected reference pathogens, the potential risk of infection for the consumer of lettuce does not cause a higher risk than the WHO (2006) tolerates. For viruses, the calculated risk is very close to the maximum tolerable health risk, though.

Thus, it may be useful to implement further preventive barriers against viruses, i.e. barriers to reduce the risk of virus infection by additional treatment processes, especially when using the water in hydroponics, where plants are in direct contact with irrigation water via their roots (Steele & Odumeru 2004; Uyttendaele *et al.* 2015; Moriarty *et al.* 2019). For example, an UF unit after the BACF could significantly reduce the risk of virus infection of the Weißenberge concept.

Due to a lack of data on the case study Weißenberge, the risk assessment was based mainly on assumptions and thus only shows trends. The assumptions include mean standard pathogen concentrations per liter of raw wastewater for each of the selected reference pathogens (i.e. 5.0×10^3/l regarding Cryptosporidium and Campylobacter and 5.5×10^5/l regarding Noroviruses) and mean standard log reductions to estimate the performance of the treatment methods envisaged in the concept: In the WSP, the LRV is assumed to be 0.5 for protozoa and 1.5 for bacteria and viruses (WHO 2006; Fuhrmann 2013; DWA 2017). Based on the mean *E. coli*-reduction of the BACF in the HypoWave pilot plant, an LRV of 1.44 for all pathogens has been chosen for this treatment process. For the successive UV-disinfection process, an LRV of 3.0 for protozoa and bacteria and of 2.0 for viruses has been assumed (WHO 2006; Seis *et al.* 2016; DWA 2017). Based on the literature on hydroponic plant production based on wastewater (Garland 2000; Ottoson *et al.* 2005) and on results of the HypoWave pilot plant, a mean LRV of 1.0 for all pathogens in the hydroponic system has been used for the calculations.

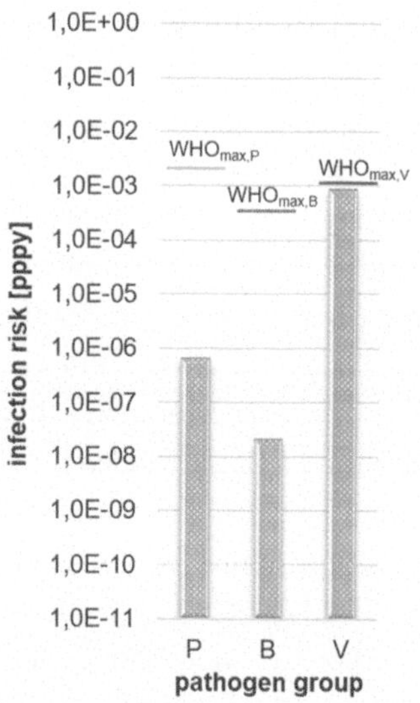

Figure 7 | Mean infection risk per person per year (pppy) for consumption of lettuce for protozoa (P), bacteria (B), and viruses (V) compared with tolerable infection risks of the WHO (2006, 2016).

Nevertheless, the QMRA was applied repeatedly and proved to be a reliable and multifunctional method of risk forecast and risk assessment. Taking into account that the quality of the irrigation water is a decisive factor for the hygienic quality of the agricultural product, it is recommended not to exceed a value of 10^{-5} concerning the infection risk per person per year for all pathogen groups. Risks for the staff working in the greenhouses, e.g. for maintenance and harvesting, have to be taken into account as well.

These results suggest that conventional quality assurance must be supplemented by further measures to ensure that quality targets are reliably achieved in practice. In this context, flow cytometry is seen as a suitable technology for online quality and performance assurance due to its high data resolution and degree of automation. Although microbial signals are not specific and thus do not allow conclusions on the presence of pathogens, critical processes like UF membrane integrity or efficiency of chemical disinfection can be monitored with little time delay. The microbiological consequence of changes in the treatment process can be directly seen in the form of total and intact cell concentrations, the integrity of bacteria, and the ratio between HNA and LNA bacteria. Within the treatment process, flow cytometry further allows the fast assessment of log reduction of total and intact counts independent of their culturability.

Although flow cytometry has proven to be a sensitive tool to detect changes in treatment efficiency and process alterations and can thus be seen as a very useful tool for process monitoring, it does not exclude the necessity for the hygienic assessment of the end product. For the direct assessment of the hygienic status of water, either microbiological hygiene indicators or specific pathogens of interest have to be analyzed. Traditional hygiene indicators (like *E. coli*) have served well for this purpose, especially for fecal borne pathogens.

Quality management as a tool for risk reduction

From the point of view of agricultural production, quality fluctuations in the irrigation water should be avoided as far as possible. For the same reason, downtimes of the water in the supply lines between the producer of the irrigation water and the agricultural producers are to be minimized if possible; it is possible (in the case of longer lines) that there is a monitoring point at the transfer point to the agricultural operator. Experience can lead to a reduction of the actual monitoring. In the same way, attention should also be paid to short retention times in reservoirs. Process stabilization in the field of wastewater treatment facilitates quality assurance of the subsequent water treatment for (agricultural) reuse in a special way. In this respect, it makes the tasks easier if it can be agreed that quality management measures are introduced in the wastewater treatment process, as this will stabilize the process control in the long run, e.g. through higher competence of the operating personnel and inherent process improvements.

The guarantee of desired quality requirements vis-à-vis the cooperation partners is a central prerequisite for the consideration of water reuse in agriculture. In the agricultural business context, quality management is increasingly built up very systematically and at the same time in a far more comprehensive way than is necessary for quality assurance. This approach has not yet been systematically developed for water reuse in agriculture (Schramm *et al.* 2019).

Risk assessments in accordance with QMRA can be incorporated into the quality assurance systems for food production introduced at the European level. This can be used to ensure the quality of the irrigation water. For the European Union, the regulation (in course of legislation) on water reuse will provide a uniform framework for the first time; disinfection barriers will now be mandatory for most applications (Council of EU 2019). The approach of risk-based and process-oriented management of EN ISO 9001 can be used as a basis.

With regard to the scope of quality management, it is necessary to make an explicit distinction between environmental prevention and health care. Environmental policy is primarily interested in environmental precautions. The minimization of environmental risks has so far been given only limited consideration in the existing instruments, as the EU regulation on water reuse makes clear.

CONCLUSIONS

The prime focus of quality management in agricultural water reuse is on the removal of pathogens. A number of technical processes for the elimination of pathogens from water exist and have been applied in the involved research projects. Membrane filtration, e.g. UF, has proved to be a reliable barrier against pathogens. Multi-barrier approaches combine the advantages of single processes and make the reuse concept more resilient against disturbances. Especially aerobic treatment processes and stabilization ponds have a strong additional removal efficiency for pathogens, but also slow filtration processes, UV and ozone help reduce the concentration of pathogens in water. According to the individual reuse case, an appropriate combination of these processes should be chosen. In addition to technical barriers, administrative barriers are important to ensure efficient and safe water reuse. For administering and managing water quality, the question arises as to which institutional arrangements are needed in order to make quality management durable and with which processes it can be successfully coordinated. If municipal wastewater is reused in agriculture, different actors usually carry out quality assurance in parallel. In the long run, it will make sense to coordinate the quality assurance of the various actors along the emerging 'water chain'.

All three projects struggled with the question how to ensure a safe use of the treated wastewater. MULTI-ReUse chose a combination of elaborated treatment steps combined with innovative monitoring to technically minimize the risks. EPoNa reduces the risks from pathogens by upgrading the WSP, thereby adding additional barriers, and restricting the use of the treated water to animal fodder, while HypoWave limits the application of reuse to wastewater collected only in areas without industry and hospitals discharging potentially hazardous substances and stresses the importance of cooperation between the stakeholders involved.

Monitoring can be another barrier, too, especially if the results are available in time to react accordingly. The assessment of the microbiological water quality status directly after treatment should be supplemented by measurement of the regrowth potential. Regrowth potential is currently not assessed on a regular basis, but a standardized procedure for initial testing would prove highly beneficial. Despite standardization, sufficient flexibility should be provided as the relevant conditions (e.g. temperature) of the particular water application have to be considered. Flow cytometry is a valuable diagnostic tool to assess whether bacteria originate from regrowth or from the wastewater. Digital process control systems can contribute to quality management. QMRA is a valuable tool to identify potential threats for a given water reuse case. The barriers can be chosen based on the results of this analysis.

REFERENCES

Asano, T., Burton, F. L., Leverenz Harold, L., Tsuchihashi, R. & Tchobanoglous, G. 2007 *Water Reuse: Issues, Technology, and Applications,* 1st edn. McGraw-Hill, New York, NY, p. 1570. Available from: http://www.loc.gov/catdir/enhancements/fy0668/2006030659-d.html (accessed 05 March 2020).

Becerra-Castro, C., Lopes, A. R., Vaz-Moreira, I., Silva, E. F., Manaia, C. M. & Nunes, O. C. 2015 Waste-water reuse in irrigation: a microbiological perspective on implications in soil fertility and human and environmental health. *Environment International* **75**, 117–135. doi: 10.1016/j.envint.2014.11.001.

Becker, D., Frey, C., Jungfer, C., Krömer, K., Kulse, P., Maaßen, S., Schramm, E., Wencki, K., Zimmermann, B. & Zimmermann, M. 2017 *Marktpotenziale der Wasserwiederverwendung: Anforderungen und Kriterien in Unterschiedlichen Sektoren und Mögliche Zielmerkte für das MULTI-ReUSe-Verfahren (Market Potentials of Water Reuse: Requirements and Criteria in Different Sectors and Possible Target Markets for the MULTI-ReUse Process).* Institute for Social-Ecological Research, ISOE-Materialien Soziale Ökologie 49. Available from: www.isoe-publikationen.de/fileadmin/redaktion/ISOE-Reihen/msoe/msoe-49-isoe-2017.pdf (accessed 27 March 2020).

Bliedung, A., Dockhorn, T., Germer, J., Mayerl, C. & Mohr, M. 2020 Experiences of running a hydroponic system in a pilot scale for resource-efficient water reuse. *Journal of Water Reuse and Desalination* **10** (4), 347–362. doi: https://doi.org/10.2166/wrd.2020.014.

Camper, A. K., LeChevallier, M. W., Broadway, S. C. & McFeters, G. A. 1985 Growth and persistence of pathogens on granular activated carbon. *Applied and Environmental Microbiology* **50**, 1178–1382.

Cornel, P., Mohr, M., Nocker, A., Selinka, H. C., Schramm, E., Stange, C. & Drewes, J. E. 2018 Relevanz mikrobiologischer Parameter für die Wasserwiederverwendung. Factsheet zum WavE-Querschnittsthema 'Risikomanagement in der Wasserwiederverwendung' (*Relevance of microbiological parameters for water reuse. Factsheet on the WavE cross-sectional topic 'Risk management in water reuse'*). Available from: https://www.bmbf-wave.de/_media/WavE_FactSheets_mikrobiol%20Parameter_Final.pdf (accessed 05 March 2020).

Council of the EU 2019 Water Reuse for Agricultural Irrigation: Council Agrees General Approach. Press Release. June 26, 2019.

Department of Water Affairs and Forestry (DWAF) 2012 *Code of Practice: Volume 6 – Wastewater Reuse.* Ministry of Agriculture, Water and Forestry, Windhoek, Namibia.

DIN EN ISO 14189:2016-11 *Water Quality – Enumeration of Clostridium Perfringens – Method Using Membrane Filtration (ISO 14189:2013); German Version EN ISO 14189:2016.* Beuth, Berlin.

DIN EN ISO 6222:1999-07 *Water quality – Enumeration of culturable micro-organisms – Colony count by inoculation in a nutrient agar culture medium (ISO 6222:1999); German version EN ISO 6222:1999.* Beuth, Berlin.

DIN EN ISO 9308-2:2012 *Water Quality – Enumeration of Escherichia Coli and Coliform Bacteria – Part 2: Most Probable Number Method (ISO 9308-2:2012); German Version EN ISO 9308-2:2014.* Beuth, Berlin.

DIN EN ISO 7899-2:2000-11 *Water Quality – Detection and Enumeration of Intestinal Enterococci – Part 2: Membrane Filtration Method (ISO 7899-2:2000); German Version EN ISO 7899-2:2000.* Beuth, Berlin.

DIN EN ISO 9308-2:2014-06 *Water Quality – Enumeration of Escherichia Coli and Coliform Bacteria – Part 2: Most Probable Number Method (ISO 9308-2:2012); German Version EN ISO 9308-2:2014.* Beuth, Berlin.

DIN EN ISO 11731:2017-05 *Water Quality – Enumeration of Legionella (ISO 11731:2017); German Version EN ISO 11731:2017.* Beuth, Berlin.

Drewes, J. E., Zhiteneva, V., Karakurt, S., Schwaller, C. & Hübner, U. 2019 *Risk Management in Water Reuse: International Perspective on Approaches for Germany. Wasserkreisläufe neu Denken*. E. Schweizerbart'sche Verlagsbuchhandlung (Nägele u. Obermiller), Stuttgart, pp. 59–65.

DVGW W 294-Series UV-Geräte zur Desinfektion in der Wasserversorgung (*UV devices for disinfection in the water supply*); Part 1–3. Technische Regeln. June 2006, WVGW, Bonn.

DVGW 2007 *Zur Überwachung der Integrität (Intaktheit) von Membranfiltrationsanlagen (For Monitoring the Integrity (Intactness) of Membrane Filtration Plants)*.WVGW, Bonn, January 2009. Water-Information Nr. 71.

DWA 2008 *Treatment Steps for Water Reuse, German Association for Water, Wastewater and Waste*. DWA, Hennef, p. 31. DWA-Topics

DWA 2017 *Design of Wastewater Treatment Plants in Hot and Cold Climates (EXPOVAL) - DWA-Topics T4/2016 Corrected Version May 2019, German Association for Water, Wastewater and Waste*. DWA, Hennef, p. 75. DWA-Topics

EPA 2005 *Membrane Filtration Guidance Manual*, U.S. EPA, Cincinnati, EPA 815-R-06-009, November 2005.

Eregno, F. E., Moges, M. E. & Heistad, A. 2017 Treated greywater reuse for hydroponic lettuce production in a green wall system: quantitative health risk assessment. *Water* **9** (7), 454. doi: 10.3390/w9070454.

FAO 2017 *Water for Sustainable Food and Agriculture, A Report Produced for the G20 Presidency of Germany*. Food and Agriculture Organization of the United Nations. Available from: http://www.fao.org/3/a-i7959e.pdf (accessed 27 March 2020).

Fuhrmann, T. 2013 *Anwendung und Potenziale von Abwasserteichsystemen im Internationalen Kontext (Application and Potentials of Wastewater Pond Systems in an International Context)*. Dissertation, Technische Universität Darmstadt, p. 236, Available from: http://nbn-resolving.de/urn:nbn:de:tuda-tuprints-38028 (accessed 27 March 2020).

Garland, J. 2000 Graywater processing in recirculating hydroponic systems: phytotoxicity, surfactant degradation, and bacterial dynamics. *Water Research* **34** (12), 3075–3086. doi: 10.1016/S0043-1354(00)00085-3.

Guy, J. L., Pinchas, F. & Asher, B. T. 2010 *Treated Wastewater in Agriculture: Use and Impacts on the Soil Environment and Crops*. Wiley-Blackwell.

Hammes, F., Berney, M., Vital, M. & Egli, T. 2007 *A Protocol for the Determination of Total Cell Concentration of Natural Microbial Communities in Drinking Water with FCM*. TECHNEAU Report Deliverable 3.3.7.

Health Canada 2013 *Guidance on Providing Safe Drinking Water in Areas of Federal Jurisdiction - Version 2*, Water and Air Quality Bureau, Healthy Environments and Consumer Safety Branch, Health Canada, Ottawa, Ontario, Catalogue No. H128-1/05-440-1E-PDF.

Hibbing, M. E., Fuqua, C., Parsek, M. R. & Peterson, S. B. 2010 Bacterial competition: surviving and thriving in the microbial jungle. *Nature Reviews Microbiology* **8** (1), 15–25.

ISO 20670:2018 *Water Reuse - Vocabulary*. International Organization for Standardization, Geneva, Switzerland.

Lackner, S., Sinn, J., Zimmermann, M., Max, J., Rudolph, K. U., Gerlach, M. & Nummer, C. 2017 Upgrading waste water treatment ponds to produce irrigation water in Namibia. *Water Solutions (1), GWF, Wasser - Abwasser* **158**, 82–85.

Lin, W., Yu, Z., Zhang, H. & Thompson, I. P. 2014 Diversity and dynamics of microbial communities at each step of treatment plant for potable water generation. *Water Research* **52**, 218–230.

Lipphaus, P., Hammes, F., Kötzsch, S., Green, J., Gillespie, S. & Nocker, A. 2014 Microbiological tap water profile of a medium-sized building and effect of water stagnation. *Environmental Technology* **35** (5–8), 620–628.

Maassen, S. 2018 *Anforderungen an die Wiederverwendung von Wasser im Sektor Landwirtschaft (Requirements for the Re-use of Water in the Agricultural Sector)*. MULTI-ReUse, Factsheet Agriculture. Available from: https://water-multi-reuse.org/download/1759/ (accessed 27 March 2020).

Moriarty, M. J., Semmens, K., Bissonnette, G. K. & Jaczynski, J. 2019 Internalization assessment of *E. coli* O157:H7 in hydroponically grown lettuce. *LWT - Food Science and Technology* **100**, 183–188. doi: 10.1016/j.lwt.2018.10.060.

NHMRC, NRMMC 2011 *Australian Drinking Water Guidelines Paper 6 National Water Quality Management Strategy*. National Health and Medical Research Council, National Resource Management Ministerial Council, Commonwealth of Australia, Canberra, p. 31.

Nocker, A., Schulte-Illingheim, L., Müller, H., Rohn, R., Zimmermann, B., Gaba, A., Nahrstedt, A., Mohammadi, H., Tiemann, Y. & Krömer, K. 2020 Microbiological changes along a modular wastewater reuse treatment process with a special focus on bacterial regrowth. *Journal of Water Reuse and Desalination* doi: https://doi.org/10.2166/wrd.2020.012.

Ottoson, J., Norström, A. & Dalhammar, G. 2005 Removal of micro-organisms in a small-scale hydroponics wastewater treatment system. *Letters in Applied Microbiology* **40** (6), 443–447. doi: 10.1111/j.1472-765X.2005.01689.x.

Pinto, A. J., Xi, C. & Raskin, I. 2012 Bacterial community structure in the drinking water microbiome is governed by filtration processes. *Environmental Science & Technology* **46**, 8851–8859.

Prest, E. I., Hammes, F., van Loosdrecht, M. C. M. & Vrouwenfelder, J. S. 2016 Biological stability of drinking water: controlling factors, methods, and challenges. *Frontiers in Microbiology* **7**, 45.

Prüss-Üstün, A. 2003 *Introduction and Methods: Assessing the Environmental Burden of Disease at National and Local Levels*, vii, 63. Environmental Burden of Disease Series No. 1, World Health Organization, Protection of the Human Environment, Geneva.

Salgot, M. & Folch, M. 2018 Wastewater treatment and water reuse. *Current Opinion in Environmental Science & Health* **2**, 64–74. doi: 10.1016/j.coesh.2018.03.005.

Scheierling, S. M., Bartone, C. R., Mara, D. D. & Drechsel, P. 2011 Towards an agenda for improving wastewater use in agriculture. *Water International* **36** (4), 420–440. doi: 10.1080/02508060.2011.594527.

Schramm, E., Beythien, U., Dockhorn, T., Ebert, B., Fischer, M., Mohr, M., Wieland, A., Winkler, M. & Zimmermann, M. 2019 Wasserwiederverwendung zur landwirtschaftlichen Nutzung in hydroponischen Systemen: Anforderungen an die Qualitätssicherung (*Water reuse for agricultural use in hydroponic systems: requirements for quality assurance*). *Zentralblatt für Geologie und Paläontologie Teil I, Heft* **1**, 73–82.

Seis, W., Lesjean, B., Maaßen, S., Balla, D., Hochstrat, R. & Düppenbecker, B. 2016 *Rahmenbedingungen für die umweltgerechte Nutzung von behandeltem Abwasser zur landwirtschaftlichen Bewässerung (Framework Conditions for the Environmentally Friendly Use of Treated Wastewater for Agricultural Irrigation).* Report from the German Environment Agency, Dessau-Roßlau, Germany, p. 216

Sinn, J., Cornel, P. & Lackner, S. 2019 Waste stabilization ponds with pre-treatment provide irrigation water – a case study in Namibia. In *Proceedings of the 12th IWA International Conference on Water Reclamation and Reuse*, Berlin, Germany.

Steele, M. & Odumeru, J. 2004 Irrigation water as source of foodborne pathogens on fruit and vegetables. *Journal of Food Protection* **67** (12), 2839–2849. doi: 10.4315/0362-028X-67.12.2839.

Uyttendaele, M., Jaykus, L.-A., Amoah, P., Chiodini, A., Cunliffe, D., Jacxsens, L., Holvoet, K., Korsten, L., Lau, M., McClure, P., Medema, G., Sampers, I. & Rao Jasti, P. 2015 Microbial hazards in irrigation water: standards, norms, and testing to manage use of water in fresh produce primary production. *Comprehensive Reviews in Food Science and Food Safety* **14** (4), 336–356. doi: 10.1111/1541-4337.12133.

Van der Kooij, D. 1992 Assimilable organic carbon as an indicator of bacterial regrowth. *Journal – American Water Works Association* **84** (2), 57–65.

Van Nevel, S., De Roy, K. & Boon, N. 2013 Bacterial invasion potential in water is determined by nutrient availability and the indigenous community. *FEMS Microbiology Ecology* **85**, 593–603. doi: 10.1111/1574-6941.12145.

WHO 2006 *Guidelines for the Safe Use of Wastewater, Excreta and Greywater, Volume 2: Wasterwater Use in Agriculture.* World Health Organization, Geneva, p. 222, Available from: https://www.who.int/water_sanitation_health/publications/gsuweg2/en/ (accessed 05 March 2020).

WHO 2016 *Quantitative Microbial Risk Assessment: Application for Water Safety Management.* World Health Organization, Geneva, p. 204, Available from: https://www.who.int/water_sanitation_health/publications/qmra/en/ (accessed 05 March 2020).

WRI 2019 *17 Countries, Home to One-Quarter of the World's Population, Face Extremely High Water Stress.* World Resources Institute. Available from: https://www.wri.org/blog/2019/08/17-countries-home-one-quarter-world-population-face-extremely-high-water-stress (accessed 27 March 2020).

First received 9 April 2020; accepted in revised form 5 August 2020. Available online 17 September 2020

Experiences of running a hydroponic system in a pilot scale for resource-efficient water reuse

Alexa Bliedung IWA, Thomas Dockhorn IWA, Jörn Germer, Claudia Mayerl IWA and Marius Mohr

ABSTRACT

Within the research project HypoWave, a hydroponic system for plant production was investigated. The hydroponic system was fed with wastewater that had undergone specially adapted treatment. The principal aim was to develop a combined system for water treatment and hydroponic plant production, where water and nutrients were reused efficiently to produce marketable food products. Another goal was to find out whether the reuse of pre-treated wastewater for plant growth in a hydroponic system could also present an additional alternative wastewater treatment step for enhanced nutrient removal. A pilot plant, consisting of various treatment steps such as activated sludge process, ozonation and biological activated carbon filtration, was used to produce lettuce with irrigation water of different qualities. The hydroponic pilot plant was operated in two different modes – flow-through and feed & deplete. This paper focuses on the influence of the various modes of operation and accordingly varying nutrient concentrations (N, P, K) on plant growth. Furthermore, heavy metal content in the various types of treated wastewater and in the produced plants was investigated. In addition, the results of the different modes of operation were verified by mass balances for N, P and K.

Key words | heavy metals, hydroponic system, lettuce, macronutrients and micronutrients, mass balance, water reuse

Alexa Bliedung IWA
Thomas Dockhorn IWA (corresponding author)
Claudia Mayerl IWA
TU Braunschweig,
Institute of Sanitary and Environmental Engineering,
Pockelsstraße 2a, 38106 Braunschweig,
Germany
E-mail: *t.dockhorn@tu-braunschweig.de*

Jörn Germer
Institut für Tropische Agrarwissenschaften (Hans-Ruthenberg-Institut) (490), Fg. Wasserstress-Management bei Kulturpflanzen in den Tropen und Subtropen (490 g),
Universität Hohenheim,
Garbenstr. 13, 70599 Stuttgart,
Germany

Marius Mohr
Fraunhofer-Institut für Grenzflächen- und Bioverfahrenstechnik,
Nobelstr. 12, 70569 Stuttgart,
Germany

HIGHLIGHTS

- Regarding the processes of wastewater treatment as a very specific production of high-quality irrigation water, which is reused in a hydroponic system for growing lettuce.
- Investigation of the effects of various treatment processes (activated sludge process, expanded granular sludge bed reactor, ozonation, biological activated carbon filtration) on the quality and suitability of the produced waters for hydroponic lettuce production.
- Comparison of different modes of operation of the hydroponic system (flow-through and modified circulation system (feed & deplete)).
- Evaluation of nutrient and heavy metal content of the hydroponically cultivated lettuce.
- Verification of the achieved results in regard to the nutrient elimination and nutrient content of the lettuce by mass balances (N, P, K).

doi: 10.2166/wrd.2020.014

GRAPHICAL ABSTRACT

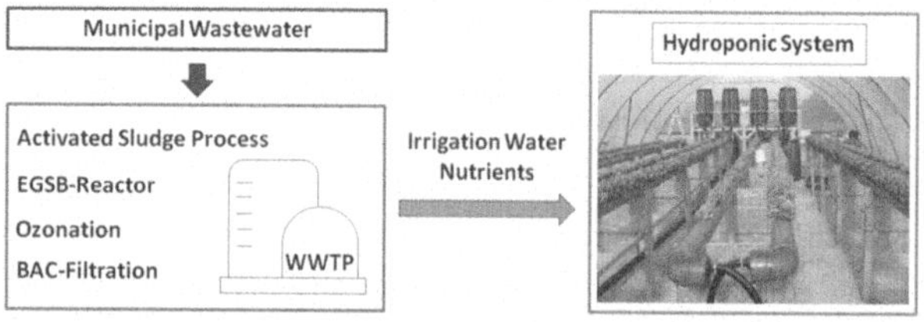

INTRODUCTION

Agriculture, which consumes 70% of the fresh water resources utilized by mankind, is the greatest water user of all (FAO 2017). Climate change, growing population and urbanization are expected to increase the pressure on water resources but also on the availability of arable land and nutrients (WHO 2006a, 2006b; Jiménez & Asano 2008). This makes it increasingly difficult to guarantee food security and to prevent the deterioration of ecosystems. Hence, the investigation of alternative water resources for agricultural irrigation and production is necessary. Countries suffering from significant water stress already use wastewater for irrigation purposes. A great advantage of this option is that not only the water itself is reused but also the nutrients contained in the treated water. However, even treated wastewater contains problematic substances, such as heavy metals, salts, micropollutants or human pathogens. Therefore, treated wastewater could cause salination of the soil and groundwater resulting in losses of plant productivity (WHO 2006b; Marcar *et al.* 2011; Jaramillo & Restrepo 2017). In addition, heavy metals could accumulate in plants if limit values of a certain pollutant are exceeded in the soil and when these plants are consumed by humans, an accumulation of heavy metals in the human body could occur (WHO 2006b). Once a critical concentration is reached in the human body, heavy metals become toxic.

Plant cultivation in hydroponic systems can be an alternative to conventional agriculture. In hydroponic systems, plants are cultivated without soil, where the plant roots either anchor in neutral substrates (e.g. rockwool) or simply extend into nutrient solutions without any substrate at all (Maucieri *et al.* 2019). Such cultivation systems allow very accurate nutrient management and avoid pollution of bodies of water (Jones 2005). The Hoagland solution, for example, is a nutrient solution containing all essential nutrients in balanced concentrations. Depending on the plant to be cultivated or the climatic conditions, the nutrient solution can be optimized. Furthermore, hydroponic systems are more efficient in regard to water and nutrient utilization than conventional agriculture (Sambo *et al.* 2019). In many countries, hydroponic systems already play an important role in agriculture and their contribution to vegetable and fruit production is expected to grow continuously. The most common types of substrate-less hydroponic systems are the Nutrient Film Technique, the Deep Water Culture (DWC) and the Aeroponic System (Maucieri *et al.* 2019).

Considering growing water scarcity and the resulting pressure on finding alternative resources for irrigation water, the combination of hydroponic systems and water reuse constitutes a possible option for the solution of this challenge although this is not a new concept (Tzanakakis *et al.* 2014; Magwaza *et al.* 2020). Previous studies either focused on the decentralized treatment of domestic wastewater (Ayaz & Saygin 1996; Norsträm *et al.* 2003; Vaillant *et al.* 2004; Keller *et al.* 2008; da Silva Carvalho *et al.* 2018) or investigated the treatment of brackish water or industrial wastewater (e.g. brewery wastewater) (Soares *et al.* 2015; Gebeyehu *et al.* 2018; Soares da Silva *et al.* 2018). In addition, all these investigations were carried out in small-scale systems with the focus on treating wastewater with the production of plants as an inevitable byproduct. Therefore, the approach of regarding the processes of wastewater treatment as a very specific production of high-quality irrigation water, which is used in hydroponic systems for the production of products of high quality, is new.

This special focus of the research project, discussed in this paper, is already mirrored in its title 'HypoWave', a combination of the words 'hydroponic' and the official name of the research funding measure 'WavE' (WavE – Zukunftsfähige Technologien und Konzepte zur Erhöhung der Wasserverfügbarkeit durch Wasserwiederverwendung und Entsalzung/Future-oriented Technologies and Concepts to Increase Water Availability by Water Reuse and Desalination) of the Federal Ministry of Education and Research, Germany. The project investigated purpose-oriented treatment and utilization of municipal wastewater in hydroponic systems for efficient water and nutrient reuse. The central element of the project was a pilot plant located on a municipal wastewater treatment plant (WWTP). Technologies such as the activated sludge process (aerobic) or the Expanded Granular Sludge Bed Reactor (EGSB) (anaerobic) were used as basic treatment processes. biological activated carbon filtration (BACF) and ozonation were used as advanced treatment processes. Due to the different combinations of treatment processes, irrigation waters of different qualities for plant production could be produced. Further topics, such as possible operator models in regard to the associated legal framework, feasibility studies and an impact assessment formed a framework for the results of the piloting (Fischer *et al.* 2018; Mohr *et al.* 2018; Ebert *et al.* 2019a, 2019b; Winker *et al.* 2020; Zimmermann & Fischer 2020).

This paper will give an overview from the results of the pilot plant of the research project HypoWave and will focus on the data of the growing periods 2017 and 2018. First, a detailed description of the pilot plant is presented and second, the nutrient ratios of the differently treated wastewaters are described. In addition, the use of wastewater treated especially for the reuse in a hydroponic system for the cultivation of lettuce is discussed, taking into account the impact of heavy metal concentrations. Finally, the used hydroponic system is verified by mass balances together with a recommendation on which type of operation (flow-through, feed & deplete system) is suitable for which type of application, combining adapted wastewater treatment and hydroponic systems. Here, the first mode of operation was a flow-through system. It is a classic open hydroponic system without recycling of the nutrient solution (= treated wastewater) (Maucieri *et al.* 2019). In order to operate the hydroponic system as a recirculation system with treated wastewater (second mode of operation), a circulation system (feed & deplete system) especially optimized for the HypoWave project was used.

In addition to the investigations of the nutrients and heavy metals in the irrigation water and the plants, further investigations regarding pathogens, antibiotic resistances and organic micropollutants were carried out (Blau *et al.* in preparation; Kreuzig *et al.* in preparation; Mohr *et al.* 2020). Knowledge of these parameters is important to evaluate the quality of the products (irrigation water, plants) in regard to guidelines and regulations (DIN 19650:1999-02; WHO 2006b; EC 2018; Schramm *et al.* 2019). The integration of the results into an effective quality management system will be part of a follow-up project.

METHODS

In order to analyse the suitability of various types of treated wastewater for growing lettuce in a hydroponic system installed in a greenhouse, a modularly structured pilot plant was set up at the municipal WWTP Wolfsburg-Hattorf, Germany. The pilot plant consisted of three main modules: basic treatment, quality and hygiene and the hydroponic system (greenhouse). The modules and technologies used have been presented below in more detail as well as the various treatment combinations and modes of operation of the hydroponic system.

Basic treatment

The aim of this module was to reduce the amount of organic compounds of the mechanically pre-treated wastewater by using aerobic and anaerobic processes. Two treatment options were used for the reduction of organic compounds in this research project. Option one being the standard wastewater treatment of the WWTP Hattorf and option two, the combination of an EGSB and a sequencing batch reactor (SBR) within the pilot plant.

The WWTP Hattorf (6,200 PE; 396,000 $m^3\,a^{-1}$ (NMU 2017)) is located in a rural area and the wastewater is collected by a separate sewer system. As there are no industries in the area, only domestic wastewater is collected. Therefore, the incoming wastewater is especially suitable for reuse in hydroponic plant production systems. The treatment chain of the WWTP Hattorf consists of a mechanical treatment step, a compact treatment system for the joint removal of sand and screenings, an aerobic biological stage (activated sludge process) for the reduction of organic compounds and phosphorus and nitrogen removal consisting of biological phosphorus elimination and intermittent denitrification. Furthermore, advanced phosphorus removal is achieved by precipitation in the effluent of the aeration tank. Before the treated wastewater is discharged into the receiving water body, it passes through purification ponds.

For the pilot plant set-up, a pilot-scale EGSB was used. An EGSB is a type of anaerobic treatment, an upgraded upflow anaerobic sludge blanked reactor (UASB) (Meyer 2004). The reactor (ACS-Umwelttechnik GMBH & Co. KG, Rielasingen-Worbingen, Germany) had a volume of 200 L and was fed with the influent of the WWTP Hattorf (after mechanical pre-treatment) at a flow rate of 60 L h^{-1}. The advantage of this treatment combination is the reduction of organic compounds in terms of chemical oxygen demand (COD) without targeting the elimination of phosphorus and nitrogen.

In order to obtain a nutrient-rich effluent with low COD concentrations, an SBR was installed downstream of the EGSB. The exchange volume of the SBR was 240 L per cycle (40–60 L h^{-1}; equal to 1/3 of the volume). By combining the process steps of aeration, settling and decanting, the SBR makes it possible to convert ammonium into nitrate during the nitrification process. The conversion is necessary, since nitrate is more readily available for plants than ammonium (Tochobanoglous *et al.* 2014). At the same time, elimination of phosphorus and denitrification are avoided, thus preserving these nutrients in the irrigation water.

Quality and hygiene

The effluent of the conventional biological treatment process (secondary sedimentation tank (SST)) may still contain organic micropollutants or pathogens. These pollutants in treated wastewater could have a negative effect on the quality of the cultivated product. Therefore, further advanced treatment processes are required. In the HypoWave pilot plant, ozonation and BACF were used as advanced treatment processes. Ozone is one of the principal disinfectants for treating wastewater and a strong oxidant (WHO 2006b; Tochobanoglous *et al.* 2014). Due to the direct oxidation of cell walls, pathogens are killed (Tochobanoglous *et al.* 2014). Additionally, organic micropollutants are eliminated by the oxidative process (Asano *et al.* 2007). BACF is not a common treatment process. One advantage of this technique is the synergetic effect between the adsorption properties of activated carbon, which serves as carrier material and an adapted biofilm (Simpson 2008; Rattier *et al.* 2012). At first, dissolved organic carbon is removed mainly by the physical adsorption process of the activated carbon filtration. Due to the increased retention time, an adapted biofilm grows on the activated carbon material. Formation of a mature biofilm and loading of the activated carbon material takes about three months. This process makes the degradation of hardly biodegradable carbon compounds possible (Simpson 2008; Aktaş & Çeçen 2012; Rattier *et al.* 2012). The operating lifetime of biological activated carbon filters can amount to 10 or even more years (Rattier *et al.* 2012).

Both the ozone reactor and the biological activated carbon filter were used in a pilot scale. The ozone reactor (Xylem Services GmbH, Herford, Germany) was operated at a flow rate of 1.5 m^3 h^{-1} and a dosage of 8 g of ozone m^{-3}. The biological activated carbon filter was filled with 100 L of granulated activated carbon (Epibon A 4×8, 2.36–4.74 mm; Donau Carbon GmbH, Frankfurt, Germany) and was operated at a flow rate of 30 L h^{-1} corresponding to an empty bed contact time of 3.3 h. In order to reduce the maturing time of the biofilm and ensure that biological degradation prevailed during the experiments, the activated carbon was pre-treated with the effluent of the SST at the WWTP Wolfsburg-Hattorf before the experiments were started.

Construction and operation of the wastewater treatment lines

According to the modules 'basic treatment' and 'hygiene and quality', four main treatment lines were developed and used during the experiments. The different treatment processes were combined in the following way:

- Treatment line A: effluent SST WWTP Wolfsburg-Hattorf
- Treatment line B: effluent SST WWTP Wolfsburg-Hattorf – ozone reactor
- Treatment line C: influent aeration tank (I AT) – EGSB – SBR (nitrification)
- Treatment line D: influent aeration tank (I AT) – EGSB – SBR (nitrification) – BACF

The effluent from the hydroponic system was returned to the WWTP (Figure 1).

Hydroponic system

Lettuce (*Lactuca sativa* L. var Hawking RZ; Salanova, Rijk Zwaan Welver GmbH, Welver, Germany) was cultivated hydroponically. The plants thrived in about 7 to 8 cm deep water that was kept in an intermittent circular flow. Due to the high water level maintained in the pipes, the set-up was resilient against water or power cuts. On the other side, the larger volume provided by a barrel in each line increased the necessary exchange intervals, thus facilitating the manual exchange procedure during the feed & deplete mode. The hydroponic set-up consisted of eight treatment lines. Each line was made up of a PVC tube with a diameter of 10 cm and a length of 8 m. The water level in the pipes depended on the length of the roots. Therefore, the air gap was between 20 and 30 mm during the experiments. Pipes were installed without slope and connected to storage tanks. Feeding took place by gravity (year 2017) and pumps (year 2018). Outer lines on each side (two on the south side, one on the north side) served as border lines, which were used to ensure that conditions (e.g. shading, neighbourhood) were identical for all lines/plants. Four lines were operated with the various treated wastewaters and one line was operated as a reference line. In the reference line, a nutrient solution was used with a nutrient concentration corresponding to a 50% Hoagland nutrient solution (Epstein & Bloom 2005). Preliminary tests had shown that 100% Hoagland nutrient solution was too highly concentrated and could damage the plants. Therefore, a diluted 50% solution was used to simulate conventional hydroponic lettuce cultivation as reference.

Modes of operation of the hydroponic system

The experiments were carried out in 2017 and 2018 from April to November. Every growing period lasted about 35 days. Within the project, two modes of operation were investigated. In 2017, the hydroponic system was operated as a flow-through system without any further addition of nutrients. The hydroponic system was operated at an average flow rate of 23.5 L h^{-1} and with 68 lettuce plants per line. During this test period, research investigations covered the analysis of the effluent from the treatment lines A to D.

In a second test period in 2018, a feed & deplete system was used. The treated wastewater ($V_{start} = 185$ L) circulated

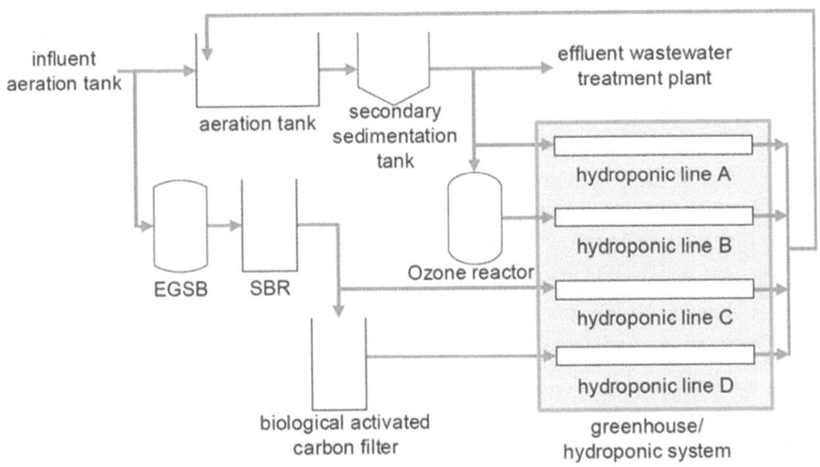

Figure 1 | Scheme of the pilot plant (Bliedung *et al.* 2019b).

in the system until the nutrient concentration of nitrate dropped below a defined threshold. Then the water was exchanged with freshly treated water. This process was repeated until the plants in line B to D reached a target weight of 275 g fresh mass (line A was not in operation in 2018). This target weight was defined, according to standard weight for lettuce sales in supermarkets. In both test periods, 20-day old seedlings were planted in the tubes after nursing and stayed in the system for approximately 35 days depending on the weight and size of the lettuce. In 2018, only 36 plants were inserted in each line, as a result of experiences made in 2017. Based on the nutrient composition of the treated wastewaters, eventually required nutrients were added.

Analyses of treated wastewater types and lettuce samples

The treated wastewater was regularly analysed in regard to its organic content (COD), macronutrients, micronutrients, as well as heavy metals. Forty-two random samples were taken at the effluent of the treatments and the hydroponic system in 2017 and 16 random samples in 2018. A limited number of samples was defined because concentrations are equalized by the different treatment steps of the pilot plant and sampling conditions in the greenhouse were difficult. In addition, in 2018, sampling only took place on those days when the water in the hydroponic lines was exchanged. In the case of the harvested lettuce, fresh and dry matter, the concentration of macronutrients and micronutrients, as well as heavy metals, was examined. Table 1 shows the methods used for analysis.

After the roots were cut off, the fresh weight of the lettuce was measured on the day of the harvest. After weighing, the lettuce was slowly dried at a temperature of 70 °C, to make sure that the material was not destroyed by the heat. The drying process took 3 days. The subsequent analyses were carried out either by the Abwasserverband Braunschweig (Braunschweig, Germany), the Wolfsburger Entwässerungsbetriebe (Wolfsburg, Germany), the TU Braunschweig (Institute of Sanitary and Environmental Engineering TU Braunschweig, Braunschweig, Germany) or the University of Hohenheim (Stuttgart, Germany).

Mass balances

Mass balances were used to check the validity of the system and the plausibility of achieved results. Mass balances were drawn up for the greenhouse covering the load of each parameter in the influent of the hydroponic lines, the effluent of the hydroponic lines and in the harvested lettuce. Loads were calculated on the basis of medians (concentration, mass).

$$F_{\text{Input}}(\text{g}) = F_{\text{Output}}(\text{g})$$

$$F_{\text{IN,water}}(\text{g}) = F_{\text{OUT,water}}(\text{g}) + F_{\sum \text{lettuce}}(\text{g})$$

$$\text{Delta}(\%) = \frac{(F_{\text{Input}}(\text{g}) - F_{\text{Output}}(\text{g}))}{F_{\text{Input}}(\text{g})} \cdot 100$$

Table 1 | Overview of the methods of analysis

Sample	Parameter	Analysis
Water	P, Ca, Mg, K, S, Fe, Mn, Pb, Cd, Cr, Cu, Ni, Zn, Hg	ICP-OES (DIN EN ISO 11885 E22:2009-09)
Water	COD, N_{tot}, NO_3-N, NO_2-N	Cuvette test (Hach-Lange, Düsseldorf, Germany)
Water	NH_4-N	ISE-Sensor NH 500/2; pH/Ion 7310 (Xylem Analytics Germany Sales GmbH & Co. KG., Weilheim, Germany)
Lettuce (shoot)	As, Cd, Co, Cr, Cu, Ni, Pb	ICP-MS (VDLUFA 2011 (vol. VII, cap. 2.2.3.1 (2014)))
Lettuce (shoot)	K, P	ICP-OES (VDLUFA 1995 (vol. II.1, cap. 8.10 (2007)))
Lettuce (shoot)	N	EA (VDLUFA 1995 (vol. II.1, cap. 3.5.2.7 (2000)))
Lettuce (shoot)	Hg	KD-AAS (VDLUFA 2011(vol. VII, cap. 2.2.2.9))

Notation: N_{tot} was analysed in the influent of the aeration tank. For the determination in the irrigation water, the sum of NO_3-N, NO_2-N und NH_4-N was calculated as N_{tot} at the other sampling points.

Nitrogen losses could occur in a hydroponic system on account of microbial nitrification, denitrification or ammonia volatilization (Zerche & Kuchenbuch 1995). During the test periods, it was not possible to measure the loss of nitrogen as gas. Therefore, phosphorus and potassium were balanced as additional control parameters. The growing periods of the lettuce in the hydroponic lines were used as the balancing period and defined system boundaries. The nursing time of 20 days was not considered. Mass balancing in this study was limited to hydroponic treatment line D. Two different runs in two years were chosen for mass balancing, RUN I (14 August to 21 September 2017) and RUN II (19 June to 24 July 2018). For balancing, 24 random water samples of the influent and the effluent of the hydroponic treatment line D were taken during a period of 38 days and 21 lettuce plants per line were considered for RUN I in 2017. For RUN II, five random water samples of the influent and effluent during a period of 35 days were taken and 15 lettuce plants were harvested. Nutrients added before irrigation were also taken into account for mass balancing.

Former research has shown that mass balances of wastewater treatment systems are considered to be closed, if the balance gap is less than or equal to 10% (Mieske 2018). Due to additional sources of error, in this research project, mass balances for hydroponic systems were considered to be closed if the balance gap was less than or equal to 20%. Reasons for this assumption were that the period of nursing (20 days) and the nutrient composition of the roots after harvesting were not considered in the mass balancing. Further, the structure of lettuce is not homogeneous and random samples were taken instead of composite samples on account of the equalization of the treated waters during the various treatment processes.

RESULTS AND DISCUSSION

Treated wastewater

In order to be able to use a larger database, the mean average concentrations of measured values of 2017 and 2018 are presented in the following. To quantify the elimination of organic compounds, the parameter COD was used. In the existing WWTP (processes A and B), COD could be decreased by 96%. This is contrasted by treatment process C, where due to the operation of the SBR, the elimination of the COD only amounted to $\eta = 89\%$. The best performance for COD elimination could be observed for the treatment process D with $\eta = 97\%$. Moreover, approximately 54% of nitrogen could be retained in the treated wastewater by the treatment processes C and D, while the effluent of treatment lines A and B showed a nitrogen content of only 12% compared to the influent of the WWTP. A similar tendency could be observed for phosphorus. Figure 2 shows the concentration of macronutrients in the treated wastewater.

Interestingly, no significant difference in the concentration of potassium, calcium, magnesium and sulphur could be observed in all lines. These findings support the idea that nitrogen and phosphorus are retained in the treated wastewater by treatment processes C and D. This is caused by the higher concentrations of nitrogen and phosphorus in the influent of the aeration tank.

For illustrating data shown in Figure 2 and for emphasizing their significance for the cultivation of lettuce, ratios of nitrogen, phosphorus and potassium were formed (covering both years) and compared to the reference solution. The reference solution had a N:P:K ratio of 10:2.8:12.3. The treated wastewater from line A had a ratio of 10:21:151 and was very similar to the ratio of water from line B (10:20:137). The lack of nitrogen and the considerable surplus of phosphorus and potassium in comparison to the reference solution thus became more obvious and addition of nitrogen and phosphorus is recommended for waters A and B. A close look at the ratios of water from lines C (10:2.3:6) and D (10:2.1:6) shows that there is a much better balance between nitrogen and phosphorus. The concentration of nitrogen in the waters of lines C and D corresponds to 37% of the nitrogen concentration of the reference solution, phosphorus to 29% and potassium to 18%. Potassium, calcium and magnesium were not reduced by any of the wastewater treatment combinations. These findings support the idea that if no specific salt removal process is used, no accidental salt reduction occurs (Marcar *et al.* 2011). However, potassium concentration (24 mg L^{-1}) was low in the treated wastewater, since influent concentrations were already low. In contrast to this, calcium (71 mg L^{-1}) and magnesium (11.5 mg L^{-1}) met the

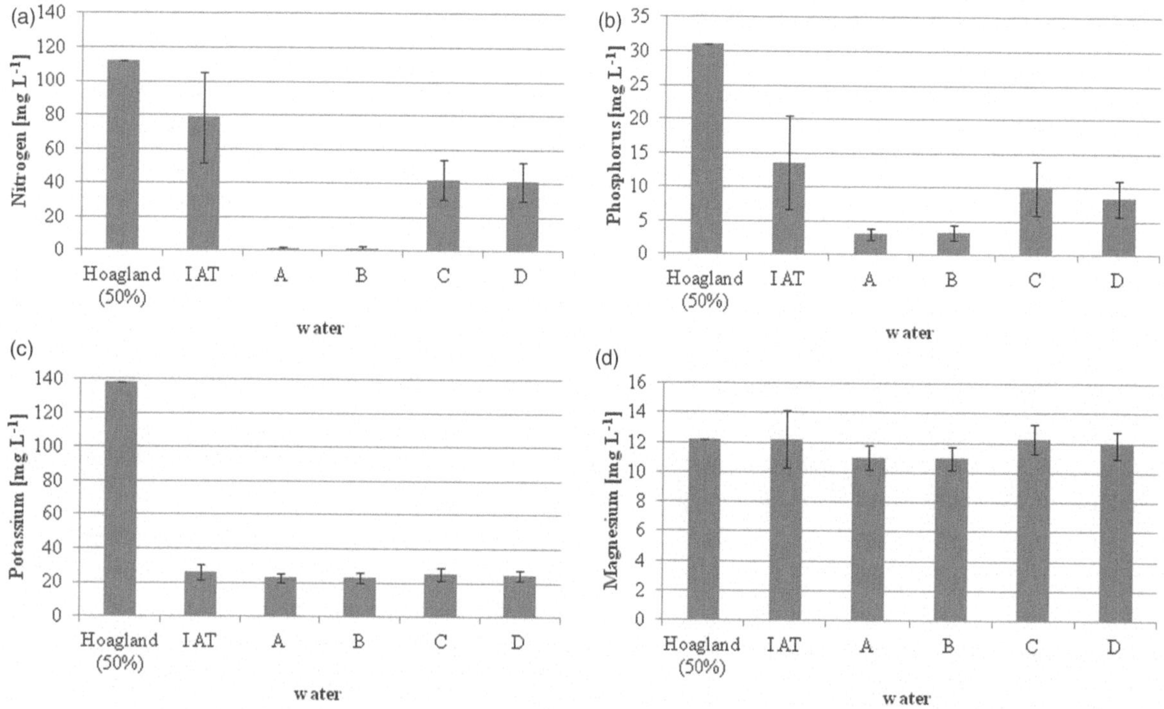

Figure 2 | Average concentrations of N, P, K and Mg in the inflow of the aeration tank (IAT) and the effluent of the differently treated wastewaters (a, b, c and d) in comparison to the reference nutrient solution (2017–2018).

requirements ($C_{Ca} = 80 \text{ mg L}^{-1}$, $C_{Mg} = 12 \text{ mg L}^{-1}$). Like, nitrogen and phosphorus, other nutrients, such as iron ($<0.5 \text{ mg L}^{-1}$) and manganese ($<0.05 \text{ mg L}^{-1}$) were not available in sufficient quantities in the influent of the aeration tank and accordingly in the treated wastewater. Copper, manganese and zinc were reduced by the different wastewater treatment processes. It could be observed that the treatment process D eliminated metals that are important for plant growth. These findings are in agreement with literature where the reduction of heavy metals by adsorption to activated carbon is described (Asano *et al.* 2007).

These results help to understand the composition of nutrients of the treated wastewater and its impact on plant growth. In all treated wastewater, the nutrient composition was not well balanced. A balanced nutrient composition is important, because plant growth is limited by the scarcest resource (Finck 1989). It seems as if nitrogen had the main impact in water from treatment lines A and B, whereas in water from treatment lines C and D, one of the micronutrients seemed to be the critical factor. Hence, to operate the feed & deplete system, nutrients were added to the treated wastewater depending on the results of the water analyses. Nutrient addition included potassium, zinc, sulphur, iron, manganese, copper and molybdenum.

Due to the low concentrations of nitrogen and phosphorus, water from lines A and B was not suitable for a feed & deplete mode of operation. Therefore, water from lines C and D was used for the feed & deplete system. In addition, the effluent of treatment line B was used as well after adding nitrogen and phosphorus to the storage tank in the greenhouse. This altered the average nutrient ratio of water from line B to 10:2.1:37 (N:P:K). Due to the addition of nitrogen in the form of potassium nitrate, the amount of potassium was noticeably higher.

Heavy metals were analysed in the treated wastewater as well. Some of these metals, like iron and zinc, are essential for plant growth. As already mentioned above, these were present in insufficient concentrations. Water from treatment line C contained only 25% of the iron concentration of the reference solution, water from lines A and B 13% and water from treatment line D contained only 3%. A similar tendency could be observed for zinc. The zinc concentration in the water of treatment line C corresponded to

85% of the concentration of the reference solution, for waters from treatment lines A and B to approximately 37% and for water from treatment line D to only 17%. Since iron and zinc concentrations are below values of the reference solution, toxicity is not a problem. Heavy metals, such as mercury, cadmium and chromium, are not needed for plant metabolism. Depending on availability, heavy metals accumulate in plants (FAO 1985; Asano *et al.* 2007). Table 2 presents an overview of the average concentrations of heavy metals in the treated wastewater during the 2 years of investigation. The concentration of mercury in the influent of the aeration tank was below the limit of quantification (0.2 μg L^{-1}), thus mercury could not be detected in any of the treated wastewater samples. The same accounts for cadmium, where the measured values were also below the limit of quantification (0.4 μg L^{-1}).

Since no specific treatment processes for heavy metal reduction were used, water in treatment lines A and B showed nearly the same heavy metal concentrations. Although the treatment combination C eliminates heavy metals, concentrations were still higher than concentrations in lines A and B. It is possible that suspended solids, which could be contained in the effluent of treatment line C, had an impact on this (Hass *et al.* 2011). Due to the activated carbon treatment, water from line D had the lowest heavy metal concentrations. Table 2 shows that the average concentrations of heavy metals varied in 2018. The reason could be that the heavy metal concentrations in the influent of the aeration tank were higher due to less dilution.

Table 2 | Average concentration and deviation of heavy metals given in μg L^{-1}

		Water line A	Water line B	Water line C	Water line D	TrinkwV (2001)
2017	Cd	0.2 ± 0.2	0.2 ± 0.2	0.2 ± 0.2	0.2 ± 0.2	3
2018		0.2 ± 0.2	0.2 ± 0.2	0.2 ± 0.2	0.2 ± 0.2	
2017	Co	1.0 ± 1.0	1.0 ± 1.0	1.0 ± 1.0	1.0 ± 1.0	–
2018		2.0 ± 0.8	1.3 ± 0.9	1.0 ± 1.0	1.0 ± 1.0	
2017	Cr	1.0 ± 1.0	1.0 ± 1.0	1.2 ± 1.5	1.0 ± 1.0	50
2018		1.0 ± 1.0	1.4 ± 0.9	1.9 ± 3.4	1.0 ± 1.0	
2017	Cu	8.1 ± 4.5	8.1 ± 6.0	26 ± 24	3.7 ± 9.3	2000
2018		11 ± 4.2	12 ± 5.0	56 ± 67	9.3 ± 8.8	
2017	Hg	0.1 ± 0.1	0.1 ± 0.1	0.1 ± 0.1	0.1 ± 0.1	1
2018		–	–	–	–	
2017	Ni	1.0 ± 1.0	1.0 ± 1.0	1.2 ± 1.5	1.0 ± 1.0	20
2018		1.0 ± 1.0	1.3 ± 0.9	3.0 ± 3.1	1.0 ± 1.0	
2017	Pb	5.1 ± 2.7	5.1 ± 3.0	5.1 ± 3.6	4.6 ± 2.4	10
2018		8.0 ± 1.6	6.1 ± 2.7	7.2 ± 3.2	6.5 ± 2.8	

Comparing the concentrations of heavy metals in all treated wastewaters to the German regulations of the Drinking Water Ordinance (TrinkwV), it can be seen that overall contaminations are very low and far below all limit values. Nevertheless, it is important to bear in mind that the main influencing factor of the composition of nutrients and heavy metals in the treated wastewater is the composition of the influent of the WWTP. It should be mentioned that results presented above and in the table below support and confirm previous findings of our research work focusing on the results of 2017 (Bliedung *et al.* 2019a, 2019b).

Fresh mass

For both RUN I and RUN II, the fresh mass of the shoots was measured after harvesting along the hydroponic line. Figure 3 shows the fresh mass (shoot) of RUN I (a) and RUN II (b) depending on the position in the hydroponic line. It can be seen in Figure 3 that the lettuce cultivated with the reference solution grew evenly over the entire hydroponic pipe length in both runs. Due to the constantly sufficient nutrient supply, lettuce growth met expectations. The lettuce from the reference line in RUN I had an average weight of 161 g and in RUN II of 553 g, although the duration of the growing period was nearly the same. There are several possible explanations for this effect. Because space per lettuce per line proved to be insufficient in 2017 (68 lettuce plants per line) and limited plant growth, the number of lettuce plants per line was reduced (36 lettuce plants per line) for further investigations. In addition, plant growth is subject to exponential growth; therefore, a small difference in growing period duration could have a large impact on the yield.

The results of the lettuce growth of RUN I are shown in Figure 3(a). The average weight of lettuce grown in line A was 81 g and in line B 103 g. Furthermore, the gradient of the weight along the lines was remarkable. These findings were unexpected, since the plants grew despite the fact that the concentration of nitrogen in wastewaters from line A and B corresponded to only 1.4% of the concentration

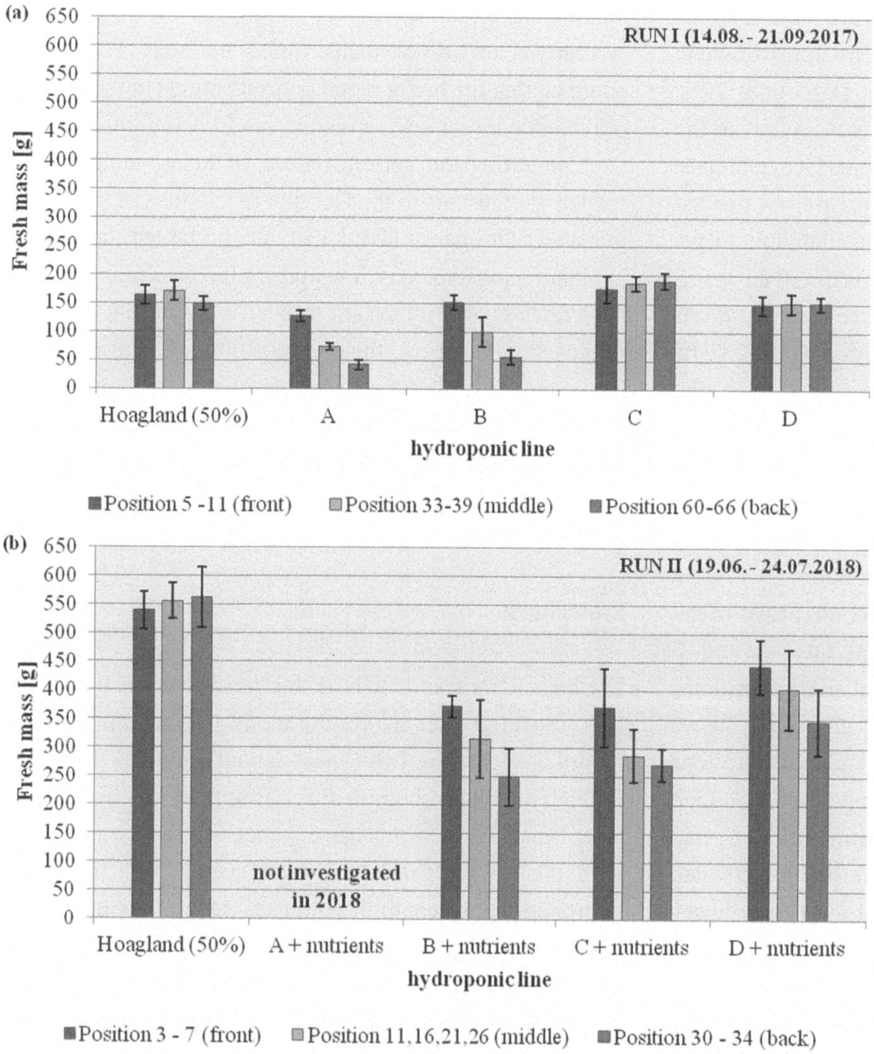

Figure 3 | Fresh mass of lettuce (shoot) for RUN I 2017 (a) and RUN II 2018 (b).

of the reference solution. A possible explanation for this might be the flow-through operation and the related load consistency (45 mg N h^{-1}) during the RUN I. The gradient seems to occur because of an unbalanced nutrient distribution along the lines. The difference in the average weight between A and B could be caused by the additional ozone treatment in line B. The ozone treatment resulted in an increased concentration of dissolved oxygen. Under certain environmental conditions, supersaturated dissolved oxygen has been observed to improve lettuce production (Suyantohadi *et al.* 2010).

In contrast to these results, the weight of the lettuce grown in lines C and D was similar along the entire line. It is noticeable that the lettuce plants in line D (150 g) were about 20% lighter than those in line C (183 g). This difference was attributed to the nutrient composition of the water in lines C and D. Treatment line C and D differ only in the biological activated carbon filter. As already discussed above, the biological activated carbon filter reduced some metals required for plant growth. Lower yield seems to be a consequence of these micronutrient deficiencies. The weight of the lettuce plants from lines D and C corresponded to the lettuce weight in the reference line. Thus, it can be deduced from these investigations that water used in treatment lines D and C is suitable as irrigation water in a hydroponic system operated as a flow-through system.

The results of RUN II are presented in Figure 3(b). The average weight of the lettuce plants grown in line B, C and D was 313, 312 and 398 g. Because of the theoretical calculation of the day of harvest, the lettuce plants were heavier than the defined target weight (275 g). The lettuce plants grown in treated wastewater were 38% lighter on average than those grown in the reference line (553 g). All lettuce plants were harvested at the same time, but the lettuce cultivated with treated wastewater with nutrient addition needed more time to reach the target weight. The most surprising aspect of the data in Figure 3(b) is the gradient of the lettuce weight along the line in all lines. According to the reference line, which was operated as a feed & deplete system, the weight was expected to be similar over the entire length of the lines. A possible explanation might be that the approach flow in the pipes might have varied at the end of the pipe, which could have had an impact on plant growth.

Overall, these results indicate that if the system is optimized (e.g. higher flow rate), it is possible to grow plants with the low concentrations of nutrients of the effluent of a WWTP in a hydroponic system operated as a flow-through system. For optimization, and to run a feed & deplete system, higher nutrient concentrations and a balanced nutrient composition are necessary. Nevertheless, it could be shown that an average concentration of 42 mg nitrogen L^{-1} is sufficient for achieving the required yield (275 g).

Quality of the lettuce

The N content of the shoot biomass corresponded to the biomass production presented above within and between the lines over the two experimental years. In the first year, the shoot N content was between 40 and 50 g kg^{-1} dry matter in the front of all lines and fell to 20 g kg^{-1} towards the end of lines A and B. This suggests that in RUN I, the depletion of N towards the end of these lines A and B limited plant growth. A small, but significant N decrease (less than 5 g kg^{-1}) in the shoot biomass was also observed in the reference line and line D. Accordingly, the shoot P content decreased from 6 to 4 g kg^{-1} along lines A and B. K was higher in the front section of the reference line and lines A and C. However, the K content of all shoots was in the magnitude of 60 g kg^{-1}. In RUN II, the reference shoot biomass contained 65 g N kg^{-1} in the front and 50 g N kg^{-1} in the end of the line, thus distinctly more than in 2017. In all lines fed with treated wastewater, the N content was between 37 and 48 g kg^{-1} without any remarkable difference between the front and end of the lines. The highest P content was measured in the plants from the front of the reference line (10.4 g kg^{-1}) and the lowest in the end of line D (5.4 g kg^{-1}). In all other samples, the P content was between 7.0 and 9.0 g kg^{-1}. Shoot K was 72 g kg^{-1} in the front of the reference line and between 50 and 60 g kg^{-1} in all wastewater lines. The K content fell about 10 to 20 g kg^{-1} in all lines. The shoot biomass nutrient contents are supported by the concentrations in the used water of the hydroponic system (effluent). At the exchange of the water in all lines, the average concentration of nitrogen was <1 mg L^{-1}. In lines C and D, potassium was <1 mg L^{-1} as well. The concentration of phosphorus in the used water of line C and D was between 3 and 2 mg L^{-1}. Since the first change of the irrigation water of RUN II had been scheduled too early, the concentrations (N, P, K) were higher in the effluent (e.g. line C and D 35 mg N L^{-1}).

Mercury was not detected in any of the plant samples of RUN I. As, Cd and Co were occasionally detected with maximum values below 120 µg kg^{-1}, Cr and Pb were detected in all samples with values below 2,100 and 260 µg kg^{-1} respectively. In RUN II, the maximum As value detected was 90 µg kg^{-1}, Co 190 µg kg^{-1} and Cd was below the detection limit in all cases. The highest Cr content was 3,200 and that of Pb 110 µg kg^{-1}.

Mass balances

In order to validate the system, mass balances were carried out exemplarily on the hydroponic line, operated with the effluent of treatment process D. The results of the mass balances are presented in Table 3. It is apparent that the nitrogen, phosphorus and potassium load in the influent, effluent and the lettuce in RUN I are distinctly higher than in RUN II, although the concentrations were similar (see Figure 2).

This difference between the loads is caused by the used amount of treated wastewater. With a flow rate of 23.5 L h^{-1} in total, 21.5 m^3 treated wastewater was used in RUN I. During RUN II, the treated wastewater in the hydroponic line needed to be changed five times. This resulted in an

Table 3 | Mass balances of hydroponic lines operated with treated wastewater from line D

		RUN I (14 August to 21 September 2017)			RUN II (19 June to 24 July 2018)		
		N	P	K	N	P	K
$F_{\mathrm{IN,water}}$	(g)	938	137	498	36	7.8	44
$F_{\mathrm{OUT,water}}$	(g)	946	131	459	5.5	2.8	3.4
F_{lettuce}	(g)	25	3.8	33	33	5.6	34
Delta	(g)	−33	2.2	6	−2.5	−0.6	6.6
Delta	(%)	−3.6	1.6	1.2	−6.9	−7.7	15

applied water volume of 0.93 m^3. Interestingly, the deviation in the balances of RUN I is below ±4%. It is possible therefore that the high flow rate and the comparatively low load in the lettuce plants minimized the error. Compared with this, the deviation of RUN II is higher. However, with a deviation of −6.9, −7.7 and 15%, the balances could still be considered closed. A possible reason for this deviation could be the inhomogeneous structure of the lettuce plants, which makes a homogenous sample difficult. However, altogether the mass balance results validate the research results. The presented data thus can be considered plausible and reliable.

CONCLUSION

This research focused on the question of whether treated wastewater is suitable for growing lettuce in a hydroponic system. In order to determine how plant growth can be optimized, various types of treated wastewaters were used and the system was operated in two different modes: as a flow-through system and as a feed & deplete system. For flow-through operation in 2017, it could be observed that the ratio of nutrients of the various types of treated wastewater differed from the nutrient ratio of the reference solution according to Hoagland. This was especially true for treated wastewater from treatment lines A and B, since a conventional treatment for nitrogen and phosphorus removal was used. Furthermore, these concentrations from line A and B were significantly below the concentration of wastewater used in previous investigations (Ayaz & Saygin 1996; Al-Karaki 2011; Keeratiurai 2013; da Silva Cuba Carvalho *et al.* 2018). Nevertheless, it could be observed that due to the constant flow rate of water in a flow-through system and the subsequently constant fresh supply of nutrients, it was possible to grow lettuce in these lines (A and B) despite the low nutrient concentrations. If the gradient of plant weight in lines A and B is compared to the gradient of plant weight in lines C and D, it can be observed that the system needs to be optimized. Either the number of plants per hydroponic line should be reduced or the flow rate increased to ensure a sufficient nutrient supply when using waters with low nutrient concentration such as waters from treatment lines A and B.

If the hydroponic system is operated as a feed & deplete system, treated wastewater with a nitrogen concentration of >40 mg L^{-1} (e.g. treated wastewater from lines C and D) is suitable. It was shown that this nutrient concentration is sufficient for good plant growth. The target weight (possible sales weight) of 275 g was achieved by all hydroponic lines operated as a feed & deplete system. As da Silva Cuba Carvalho *et al.* (2018) achieved similar weights, the weights of the investigations by Keller *et al.* (2008) were lower. Furthermore, nitrogen, phosphorus and potassium were reduced by the plant uptake (feed & deplete system) by 100, 67–78 and 100%. Therefore, a higher elimination could be achieved than by the studies compiled by Magwaza *et al.* (2020).

Nutrients in all samples, except for N content in the samples from the end of the A and B line in 2017, were within the ranges for soil grown lettuce developed by Hartz *et al.* (2007) according to the Diagnosis and Recommendation Integrated System (Walworth & Sumner 1987). In addition, the nitrogen and phosphorus content of the plants corresponded to the content in lettuce cultivated hydroponically by Bugbee (2004) with a nutrient solution and by da Silva Cuba Carvalho *et al.* (2018) with wastewater.

The detected heavy metal contents do not exceed any current threshold value and are comparable to or lower than reference values in literature. The highest contents of Cd and Pb detected were only a tenth or less of the maximum permissible levels in the EU (Commission Regulation (EC) 1881/2006) and internationally (WHO/FAO 2015). No permissible maximum levels for As, Cr and Co are established, but recommendations on and assessment of daily intake of these elements (EC 2003; Lison 2007; WHO 2010) suggest that these heavy metal contents are of no concern.

Previous studies did not validate the hydroponic system run with wastewater as irrigation water by mass balances. Therefore, to complete the investigations mass balances were performed for two runs (RUN I flow-through and RUN II feed & deplete) of the hydroponic lines operated with treated wastewater from line D in 2017 and 2018. The mass balances were calculated for irrigation water and grown biomass and verified the plausibility of achieved results, thus validating the hydroponic system used.

As a conclusion, it can be stated that it is generally possible to operate hydroponic systems using treated wastewater. Our research showed that wastewaters with high and with low nutrient concentrations are suitable, if the mode of operation is accordingly modified. Both modes of operation – flow-through and feed & deplete – produced satisfying results in regard to plant weight. The advantage of the feed & deplete system lies in the additional water treatment effect. High nutrient concentrations of the influent were reduced due to nutrient consumption by the plants (e.g. <1 mg N L^{-1}). However, the utilization of a flow-through system is possible if a complete WWTP already exists. Then a small amount of the treated wastewater and nutrients could be reused for plant production. Integration of a feed & deplete system is feasible, if the wastewater and the included nutrients are not only utilized for plant production but also used as a treatment step of wastewater. A significant reduction of the nutrients is possible in this process and requirements according to the heavy metals in the irrigation water and the lettuce could be met. A final conclusion in regard to a legal use of the irrigation water could not yet be done. Further parameters such as pathogens, antibiotic resistances, organic micropollutants, boundary conditions, operator models, etc. must be taken into account. Therefore, Winker *et al.* (2020) have investigated this question for Germany and have determined that under certain conditions an implementation of a hydroponic system operated with treated wastewater is possible.

ACKNOWLEDGEMENTS

The authors would like to thank the German Federal Ministry of Education and Research for funding the research project 'HypoWave – Use of hydroponic systems for resource efficient water reuse in agriculture' (grant number: 02WAV1402). The authors gratefully acknowledge the technical support of Wolfsburger Entwässerungsbetriebe AöR, Wolfsburg, Abwasserverband Braunschweig, Braunschweig, aquatune GmbH, Hahnstätten, ACS-Umwelttechnik GMBH & Co. KG, Rielasingen-Worblingen, and Xylem Services GmbH, Herford, all Germany, for operating the WWTP, the pilot system and for carrying out the analyses.

DATA AVAILABILITY STATEMENT

Data cannot be made publicly available; readers should contact the corresponding author for details.

REFERENCES

Aktaş, Ö. & Çeçen, F. 2012 Bioregeneration of activated carbon in biological treatment. In: *Activated Carbon for Water and Wastewater – Treatment Integration of Adsorption and Biological Treatment* (Ö Aktaş & F. Çeçen eds). Wiley-VCH Verlag & Co. KGaA, Weinheim, Germany, pp. 189–235.

Al-Karaki, G. 2011 Utilization of treated sewage wastewater for green forage production in a hydroponic system. *Emirates Journal of Food and Agriculture* **23** (1), 80–94. https://doi.org/10.9755/ejfa.v23i1.5315.

Asano, T., Burton, F. L., Leverenz, H. L., Tsuchihashi, R. & Tochobanoglous, G. 2007 *Water Reuse Issues, Technologies, and Applications*, 1st edn. McGraw-Hill Education, New York, USA.

Ayaz, S. & Saygin, Ç. 1996 Hydroponic tertiary treatment. *Water Research* **30** (5), 1295–1298. https://doi.org/10.1016/0043-1354(95)00284-7.

Blau, K., Jechalke, S. & Smalla, K. in preparation Mikrobiologische parameter (Microbiological parameter). In: *HypoWave – Neue Wege zur Wasserwiederverwendung in der Landwirtschaft (HypoWave – New Ways Towards Water Reuse in Agriculture)* (T. Dockhorn & M. Winker eds). Veröffentlichung des Instituts für Siedlungswasserwirtschaft der Technischen Universität Braunschweig, Braunschweig, Germany.

Bliedung, A., Dockhorn, T., Ebert, B., Germer, J., Mohr, M., Schramm, E. & Winker, M. 2019a *HypoWave-ressourceneffiziente nutzung von gereinigtem abwasser in hydroponischen systemen (HypoWave – resource-efficient use of treated wastewater in hydroponic systems). Zentralblatt für Geologie und Paläontologie* **Teil 1**, 95–104.

Bliedung, A., Dockhorn, T., Fiebig, B., Germer, J., Rossmanith, P. & Wieland, A. 2019b The HypoWave-System – Nutrient and heavy metal flows within an integrated system of adapted wastewater treatment and subsequent water reuse in a hydroponic system. In *Book of Abstracts – IWA Water Reuse 2019*, Berlin, pp. 552–557.

Bugbee, B. 2004 Nutrient management in recirculation hydroponic culture. *Acta Horticulturae* **648**, 99–112. https://doi.org/10.17660/ActaHortic.2004.648.12

COMMISSION REGULATION (EC) No 1881/2006 of 19 December 2006 Setting maximum levels for certain contaminants in foodstuffs. Available from: https://eur-lex.europa.eu/legal-content/EN/TXT/PDF/?uri=CELEX:32006R1881&from=EN (accessed 17 February 2020).

da Silva Cuba Carvalho, R., Bastos, R. G. & Souza, C. F. 2018 Influences of the use of wastewater on nutrient absorption and production of lettuce grown in a hydroponic system. *Agricultural Water Management* **203**, 311–321. https://doi.org/10.1016/j.agwat.2018.03.028.

DIN 19650:1999-02 Irrigation – hygienic concerns of irrigation water; German version.

DIN EN ISO 11885 E22:2009-09 Water quality – determination of selected elements by inductively coupled plasma optical emission spectrometry (ICP-OES) (ISO 11885:2007); German version EN ISO 11885:2009.

Ebert, B., Dockhorn, T., Peters, G., Schramm, E., Teiser, B. & Winker, M. 2019a Operator models for the reuse of municipal wastewater in hydroponic systems: potentials and options for Central and Meditterranean Europe. In *IWA Water Reuse 2019, Book of Abstracts* (International Water Association (IWA), ed.), Dechema, Berlin, pp. 725–730.

Ebert, B., Günkel-Lange, T., Parniske, J. & Schramm, E. 2019b *Weitergehend aufbereitetes kommunales Abwasser als alternative Wasserressource im Hessischen Ried: Eine HypoWave-Fallstudie (Advanced Treatment Municipal Sewage Water as a Alternative Water Resource in the Hessian Ried: A HypoWave-Case Study)*. Fraunhofer IGB, Stuttgart.

Epstein, E. & Bloom, A. J. 2005 *Mineral Nutrition of Plants: Principles and Perspectives*, 2nd edn. Sinauer Associates Inc, Sunderland, Massachusetts, USA.

European Commission (EC) 2003 *Opinion of the Scientific Committee on Food on the Tolerable Upper Intake Level of Trivalent Chromium*. Available from: https://ec.europa.eu/food/sites/food/files/safety/docs/sci-com_scf_out197_en.pdf (accessed 17 February 2020).

European Commission (EC) 2018 *Proposal for A Regulation of the European Parliament and of the Council on Minimum Requirements for Water Reuse*. Available from: https://eur-lex.europa.eu/legal-content/EN/TXT/PDF/?uri=CELEX:52018PC0337&from=DE (accessed 08 May 2020).

Finck, A. 1989 *Dünger und Düngung Grundlagen und Anleitung zur Düngung der Kulturpflanzen (Fertiliser and Fertilisation Basics and Instructions for Fertilising Crops)*. Corrected Reprint of the 1st edn. VCH Verlagsgesellschaft mbH, Weinheim, Germany.

Fischer, M., Beckett, M., Bürgow, G. & Ebert, B. 2018 *Modulares Wasser- und Nährstoffrecycling zur Schnittblumenproduktion in der Gemeinde Raeren (Modular Water and Nutrient Recycling to Produce cut Flowers in the Municipality of Raeren, Belgium)*. Fraunhofer IGB, Stuttgart.

Food and Agriculture Organization of the United Nations (FAO) 1985 *FAO Irrigation and Drainage Paper, 29 Rev. 1*. Food and Agriculture Organization of the United Nations, Rome, Italy.

Food and Agriculture Organization of the United Nations (FAO) 2017 *Water for Sustainable Food and Agriculture – A Report Produced for the G20 Presidency of Germany*. Rome, Italy. Available from: http://www.fao.org/3/a-i7959e.pdf (accessed 13 March 2020).

Gebeyehu, A., Shebeshe, N., Kloss, H. & Belay, S. 2018 Suitability of nutrient removal from brewery wastewater using a hydroponic technology with *Typhy latifolia*. *BMC Biotechnology* **18** (74), 1–13. https://doi.org/10.1186/s12896-018-0484-4.

Hartz, T. K., Johnstone, P. R., Williams, E. & Smith, R. F. 2007 Establishing lettuce leaf nutrient optimum ranges through DRIS analysis. *HortScience* **42** (1), 143–146. https://doi.org/10.21273/HORTSCI.42.1.143.

Hass, A., Mingelgrin, U. & Fine, P. 2011 Chapter 7 Heavy metals in soils irrigated with wastewater. In: *Treated Wastewater in Agriculture Use and Impact on the Soil Environmental Crops* (G. J. Levy, P. Fine & A. Bar-Tal eds). Wiley-Blackwell, Oxford, UK, pp. 247–285.

Jaramillo, M. F. & Restrepo, I. 2017 Wastewater reuse in agriculture: a review about its limitations and benefits. *Sustainability* **9** (10), 1–19. https://doi.org/10.3390/su9101734.

Jiménez, B. & Asano, T. 2008 Water reclamation and reuse around the world. In: *Water Reuse: An International Survey of Current Practice, Issues and Needs* (B. Jiménez & T. Asano, eds). IWA Publishing, London, UK, pp. 3–26.

Jones, J. B. 2005 *Hydroponics A Practical Guide for the Soilless Grower*. CRC PRess, Boca Raton, Florida, USA.

Keeratiurai, P. 2013 Efficiency of wastewater treatment with hydroponics. *ARPN Journal of Agricultural and Biological Science* **8** (12), 800–805.

Keller, R., Perin, K., Souza, G., Cruz, L. S., Zandonade, E., Cassini, S. T. A. & Goncalves, R. F. 2008 Use of polishing pond effluents to cultivate lettuce (*Lactuca sativa*) in a hydroponic system. *Water Science and Technology* **58** (10), 2051–2057. https://doi.org/10.2166/wst.2008.754.

Kreuzig, R., Haller-Jans, J., Bischoff, C., Leppin, J., Germer, J., Mohr, M., Bliedung, A. & Dockhorn, T. in preparation *Removed From Wastewater or Taken up by Lettuce? Micropollutants in Conventional and Advanced Wastewater Treatment and in a Downstream Hydroponic System*.

Lison, D. 2007 Chapter 25 – Cobalt. In: *Handbook on the Toxicology of Metals*, 3rd edn (G. F. Nordberg, B. A. Fowler, M. Nordberg & L. T. Friberg eds). Elsevier, Amsterdam, The Netherlands, pp. 511–528. https://doi.org/10.1016/B978-0-12-369413-3.X5052-6

Magwaza, L. S., Magwaza, S. T., Odindo, A. O. & Miditshwa, A. 2020 Hydroponic technology as a descentralised system for domestic wastewater treatment and vegetable production in urban agriculture: a review. *Science of Total Environment* **698**, 1–13. https://doi.org/10.1016/j.scitotenv.2019.134154.

Marcar, N. E., Theiveyanathan, T. & Stevens, P. D. 2011 Chapter 8 Salinity. In: *Treated Wastewater in Agriculture Use and Impacts on the Soil Environment and Crops* (G. J. Levy, P. Fine & A. Bar-Tal eds). Blackwell Publishing, Oxford, UK, pp. 286–305.

Maucieri, C., Nicoletto, C., van Os, E., Anseeuw, D., Van Havermaet, R. & Junge, R. 2019 Chapter 4 Hydroponic Technologies. In: *Aquaponics Food Production Systems Combined Aquaculture and Hydroponic Production Technologies for the Future* (S. Goddek, A. Joyce, B. Kotzen & G. M. Burnell eds). Springer Open, Cham, Switzerland, pp. 77–110. https://doi.org/10.1007/978-3-030-15943-6_4.

Meyer, H. 2004 *Leistungsfähigkeit Anaerober Reaktoren zur Industrieabwasserreinigung (Performance of Anaerobic Reactors for Industrial Waste Water Treatment)*. PhD Thesis, Institut für Siedlungswasserwirtschaft und Abfalltechnik, Universität Hannover, Hannover, Germany.

Mieske, R. 2018 *Anaerobe Schlammstabilisierung bei Faultemperaturen Unter 35°C – Erweiterung Deutscher Bemessungsregeln (Anaerobic Sludge Stabilisation at Digestion Temperatures Below 35°C – Extension of German Design Rules)*. PhD Thesis, Institut für Siedlungswasserwirtschaft, TU Braunschweig, Braunschweig, Germany.

Mohr, M., Ebert, B., Schramm, E., Germer, J. & Bürgow, G. 2018 *Nutzung des Ablaufs Eines Klärteichs zur Gemüseproduktion im Landkreis Gifhorn (Use of the Effluent of A Treatment Pond for the Production of Vegtables in Gifhorn County)*. Fraunhofer IGB, Stuttgart.

Mohr, M., Dockhorn, T., Drewes, J. E., Karwat, S., Lackner, S., Lotz, B., Nahrstedt, A., Nocker, A., Schramm, E. & Zimmermann, M. 2020 Assuring water quality along multi-barrier treatment systems for agricultural water reuse. *Journal of Water Reuse and Desalination* **10** (4), 332–346.

Niedersächsisches Ministerium für Umwelt, Energie und Klimaschutz (NMU) 2017 *Beseitigung Kommunaler Abwässer in Niedersachsen Lagebericht 2017 (Disposal of Municipal Wastewater in Lower Saxony)*. Available from: https://www.nlwkn.niedersachsen.de/download/121843/Lagebericht_2017.pdf (accessed 21 January 2020).

Norsträm, A., Larsdotter, K., Gumaelius, L., la Cour Janses, J. & Dalhammar, G. 2003 A small scale hydroponics wastewater treatment system under Swedish conditions. *Water Science and Technology* **48** (11–12), 161–167.

Rattier, M., Reungoat, J., Gernjak, W. & Keller, J. 2012 *Organic Micropollutant Removal by Biological Activated Carbon Filtration: A Review*. Urban Water Security Research Alliance Technical Report No.53. Available from: http://www.urbanwateralliance.org.au/publications/uwsra-tr53.pdf (accessed 13 March 2020).

Sambo, P., Nicoletto, C., Giro, A., Pii, Y., Valentinuzzi, F., Mimmo, T., Lugli, P., Orzes, G., Mazzetto, F., Astolfi, S., Terzano, R. & Cesco, S. 2019 Hydroponic solutions for soilless production systems: issues and opportunities in a smart agriculture perspective. *Frontier in Plant Science* **10**, 923. https://doi.org/10.3389/fpls.2019.00923.

Schramm, E., Beythien, U., Dockhorn, T., Ebert, B., Fischer, M., Mohr, M., Wieland, A., Winker, M. & Zimmermann, M. 2019 Wasserwiederverwendung zur landwirtschaftlichen Nutzung in hydroponischen Systemen: Anforderungen an die Qualitätssicherung (Water reuse for agricultural usage in hydroponic systems: requirements to ensure quality). *Zentralblatt für Geologie und Paläontologie* **Teil 1**, 73–82.

Simpson, D. R. 2008 Biofilm processes in biologically active carbon water purification. *Water Research* **42** (12), 2839–2848. https://doi.org/10.1016/j.watres.2008.02.025.

Soares, H. R., de F. e Silva, Ê., da Silva, G. F., Pedrosa, E. M. R., Rolim, M. M. & Santos, A. N. 2015 Lettuce growth and water consumption in NFT hydroponic system using brackish water. *Revista Brasileira de Engenharia Agrícola E Ambiental* **19** (7), 636–642. http://dx.doi.org/10.1590/1807-1929/agriambi.v19n7p636-642.

Soares da Silva, J., da Silva Paz, V. P., Soares, T. M., de Almeida, W. F. & Fernandes, J. P. 2018 Production of lettuce with brackish water in NFT hydroponic system. *Semina: Ciências Agrárias, Londrina* **39** (3), 947–962. http://dx.doi.org/10.5433/1679-0359.2018v39n3p947.

Suyantohadi, A., Kyoren, T., Hariadi, M., Purnomo, M. H. & Morimoto, T. 2010 Effect of high concentrated dissolved oxygen on the plant growth in a deep hydroponic culture under a low temperature. *IFAC Proceedings Volumes* **43** (26), 251–255. https://doi.org/10.3182/20101206-3-JP-3009.00044.

Tochobanoglous, G., Stensle, H. D., Tsuchihashi, R., Burton, F., Abu-Orf, M., Bowden, G. & Pfrang, W. 2014 *Wastewater Engineering Treatment and Resource Recovery*, 5th edn. McGraw-Hill Education, New York, USA.

Trinkwasserverordnung (TrinkwV) 2001 *Verordnung über die Qualität von Wasser für den Menschlichen Gebrauch (German Drinking Water Ordiance – Ordiance on the Quality of Water Intended for Human consumption)*.

Tzanakakis, V. A., Koo-Oshima, S., Haddad, M., Apostolidis, N. & Angelakis, A. N. 2014 Chapter 24 The history of land application and hydroponic systems for wastewater treatment and reuse. In: *Evolution of Sanitation and Wastewater Technologies Through the Centuries* (N. Angelakis & B. Rose eds). IWA Publishing, London, UK, pp. 457–479.

Vaillant, N., Monnet, F., Sallanon, H., Coudret, A. & Hitmi, A. 2004 Use of commercial plant species in a hydroponic system to treat domestic wastewater. *Journal of Environmental Quality* **33**, 695–702. https://doi.org/10.2134/jeq2004.0695.

VDLUFA 1995 *VDLUFA-Methodenbuch Band II.1 Die Untersuchung von Düngemitteln (VDLUFA Methods Book Volume II.1 The Analyses of Fertilisers)*, 4th edn. VDLUFA-Verlag, Darmstadt, Deutschland.

VDLUFA 2011 *VDLUFA-Methodenbuch Band VII Umweltanalytik (VDLUFA Methods Book Volume VII*

Environmental Analysis), 4th edn. VDLUFA-Verlag, Darmstadt, Deutschland.

Walworth, J. L. & Sumner, M. E. 1987 The diagnosis and recommendation integrated system (DRIS). In: *Advances in Soil Science* (B. A. Stewart ed.). Springer, New York, USA, pp. 149–188.

WHO 2006a *WHO Guidelines for the Safe Use of Wastewater, Excreta and Greywater, Volume I Policy and Regulation Aspects*. World Health Organization, Geneva, Switzerland.

WHO 2006b *WHO Guidelines for the Safe Use of Wastewater, Excreta and Greywater, Volume II Wastewater Use in Agriculture*. World Health Organization, Geneva, Switzerland.

WHO 2010 *Exposure to Arsenic: A Major Public Health Concern*. Available from: http://www.who.int/ipcs/features/arsenic.pdf (accessed 11 January 2020).

WHO/FAO 2015 *Codex Alimentarius – International Food Standards*. Food & Agriculture Org. Available from: www.fao.org/input/download/standards/17/CXS_193e_2015.pdf (accessed 20 February 2020)

Winker, M., Fischer, M., Bliedung, A., Bürgow, G., Ebert, B., Germer, J., Mohr, M., Nink, A., Schmitt, B., Wieland, A. & Dockhorn, T. 2020 Water reuse in hydroponic systems – a realistic future scenario for Germany? Facts and evidences gained in a transdisciplinary research project. *Journal of Water Reuse and Desalination* **10** (4), 363–379.

Zerche, S. & Kuchenbuch, R. 1995 Stickstoff- und Kaliumbilanzen bei Anbau von chrysanthemen (*Dendranthema-Grandiflorum*-Hybriden) im plant plane hydroponic verfahren (Nitrogen and potassium balances for the cultivation of chrysanthemums (*Dendranthema-Grandiflorum*-Hybrids) using the plant plane hydroponic method). *Journal of Plant Nutrient and Soil Science* **158** (4), 393–398. https://doi.org/10.1002/jpln.19951580414.

Zimmermann, M. & Fischer, M. 2020 Impact assessment of water and nutrient reuse in hydroponic system using Bayesian networks. *Journal of Water Reuse and Desalination* **10** (4), 431–442.

First received 23 March 2020; accepted in revised form 22 May 2020. Available online 31 August 2020

Water reuse in hydroponic systems: a realistic future scenario for Germany? Facts and evidence gained during a transdisciplinary research project

Martina Winker IWA, Michaela Fischer, Alexa Bliedung, Grit Bürgow, Jörn Germer, Marius Mohr, Andreas Nink, Bea Schmitt, Arne Wieland and Thomas Dockhorn IWA

ABSTRACT

The HypoWave transdisciplinary research project investigated the innovation of water reuse in a hydroponic system, focusing on its applicability in Germany. The methods applied were the operation of a pilot plant for the appropriate treatment and subsequent reuse of water in a hydroponic system, expert interviews, feasibility studies, an impact assessment and a stakeholder dialogue. To identify the concept's advantages and disadvantages, publications specialising in the various disciplines involved and meeting protocols were analysed and the results grouped together in a SWOT (strengths, weaknesses, opportunities and threats) analysis. This revealed that the system has potential for application in Germany subject to certain requirements being met, such as suitable local wastewater treatment conditions and actors looking for new business opportunities within agricultural production. This system is not recommended for the country as a whole, but it does offer an interesting alternative for locations that meet the appropriate conditions. Nevertheless, additional efforts and knowledge are required to promote and operate a new system of this kind. These include the reliable supply of irrigation water with guaranteed plant nutrition, comprehensive quality management to manage potential risks, a good understanding of the cooperation arrangements required and a more detailed examination of energy aspects.

Key words | governance, greenhouse, irrigation water, SWOT analysis, vegetable production, wastewater

HIGHLIGHTS

- Water reuse in hydroponic systems serves two aims: wastewater treatment and regional food production.
- The involvement of different sectors requires new cooperation arrangements between actors; forms of these already exist and can be applied here.
- It can be recommended if certain preconditions are met, e.g. local wastewater treatment facilities, farmers and organisations looking for new forms of agricultural production.

Martina Winker IWA (corresponding author)
Michaela Fischer
ISOE – Institute for Social-Ecological Research, Hamburger Allee 45, 60486 Frankfurt am Main, Germany
E-mail: *winker@isoe.de*

Alexa Bliedung
Thomas Dockhorn IWA
Technical University Braunschweig, Institute of Sanitary and Environmental Engineering, Pockelsstraße 2a, 38106 Braunschweig, Germany

Grit Bürgow
aquatectura – Studios for Regenerative Landscapes, Fritz-Reuter-Straße 33, 13156 Berlin, Germany

Jörn Germer
Institute of Agricultural Sciences in the Tropics (Hans Ruthenberg Institute), University of Hohenheim, Garbenstr. 13, 70599 Stuttgart, Germany

Marius Mohr
Fraunhofer Institute for Interfacial Engineering and Biotechnology, Nobelstr. 12, 70569 Stuttgart, Germany

Andreas Nink
aquatune GmbH, In den Wingerten 13, 65623 Hahnstätten, Germany

Bea Schmitt
team ewen, Hügelstraße 19, 64283 Darmstadt, Germany

Arne Wieland
Xylem Services GmbH, Boschstr. 4-14, 32051 Herford, Germany

doi: 10.2166/wrd.2020.020

GRAPHICAL ABSTRACT

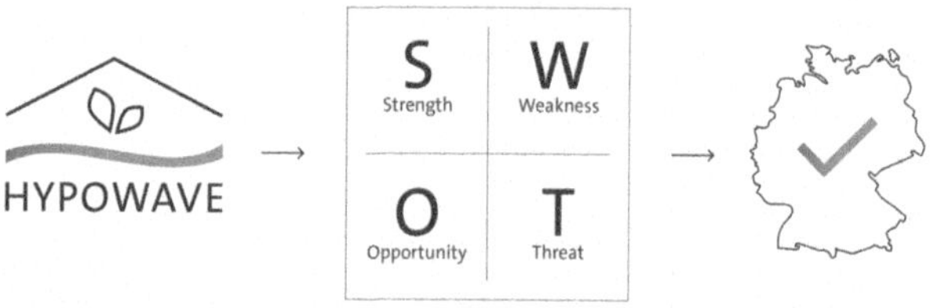

INTRODUCTION

Agriculture is the world's largest consumer of water, and consumption is rising due to increasing production and the effects of climate change (IPCC 2019). Hydroponic systems can contribute to tackling the global problem of water scarcity and an increasing demand for water in food production, because these systems require much less water (Sambo *et al.* 2019). Using treated wastewater for irrigation purposes not only saves drinking water resources, but also allows the nutrients contained in it to be reused for plant fertilisation. This would also help close the water and nutrient cycle. Hydroponic plant cultivation offers two advantages: the removal of nitrogen and phosphorus and an additional water treatment step (Fischer *et al.* 2019). However, the great challenge is to balance these advantages with the effluent quality required. There is also a requirement to meet plant quality standards for heavy metal and organic trace substances, as well as the additional commercial criteria of purchasers such as weight.

Even though Germany has a moderate climate, the hot, dry summers of 2018 and 2019 clearly point to an increasing need for irrigation in agriculture to ensure food security. At the same time, social concerns about globalisation and climate change appear to be stimulating growing interest in regionally produced food. In that regard, hydroponic systems used in conjunction with greenhouses have interesting potential for Central Europe. As well as requiring less water for food production, they also facilitate a higher production yield per area and year-round regional production of Mediterranean vegetables, for example (Hosseinzadeh *et al.* 2017).

Until now, there has not been a thorough investigation of whether the goals of both the wastewater treatment and plant production sectors in Germany can be achieved by this system, particularly the link between centralised wastewater treatment systems and food production. Technical and organisational issues need to be dealt with, as well as questions about product safety and market placement. Furthermore, the required cooperation between the wastewater and agricultural sectors does not yet exist except in the case of the reuse of sewage sludge in agriculture, a practice that is decreasing in Germany. There is a unique cooperation in Braunschweig and Wolfsburg, where treated wastewater is used as irrigation water in agriculture. Overall, this is a fairly typical situation when a socio-technical innovation is being introduced.

These aspects were investigated in the transdisciplinary research project 'HypoWave – Use of hydroponic systems for resource-efficient water reuse in agriculture'. Significant resources were required in the research undertaken to set up and run the largest pilot plant to date for water reuse in hydroponic systems in Central Europe. The plant provided the starting point for joint research. The research results are now available (for details see Bliedung *et al.* 2019, 2020; Ebert *et al.* 2019a; Mohr *et al.* 2019a; Dockhorn *et al.* 2020; Mohr *et al.* 2020; Zimmermann & Fischer 2020). An overall analysis was therefore undertaken that included the various research perspectives and disciplines. This analysis is now presented in this paper. Based on the preconditions outlined above, the research questions were whether it is feasible to apply water reuse in a hydroponic system in Germany and, if so, what key issues need to be addressed to promote a system of this kind.

METHODS

Methods used in the HypoWave research project

The research results analysed in this article were obtained by a team of researchers who have backgrounds in engineering and in natural, agricultural and social sciences. The team also comprised operators of wastewater treatment plants (WWTPs) practising water reuse, an engineering and landscaping office, a software development company, plant engineering companies and a producer of sustainable plastic. This transdisciplinary team was supported by external stakeholders as part of a stakeholder dialogue, as well as by additional external experts.

Within the HypoWave research project, the team undertook different research activities to investigate the concept of water reuse in hydroponic systems in Germany. An overview of the key working steps is provided below. The results and conclusions were sorted in an additional assessment step to provide an overview of the concept's advantages and disadvantages. This methodological step, a SWOT analysis, is presented at the end of this section.

Establishment of a pilot plant to investigate the interaction between wastewater treatment and plant production

For the investigations during the HypoWave project, a pilot plant was implemented and operated over a 3-year period on the site of the Wolfsburg–Hattorf WWTP in Germany (Bliedung *et al.* 2019). The setting covered a total area of 396 m^2 and included a greenhouse measuring 116 m^2. Pre-treated municipal wastewater was continuously processed in an existing activated sludge treatment plant at the Wolfsburg–Hattorf WWTP and also on a technical scale in an anaerobic expanded granular sludge bed (EGSB) reactor and an aerobic sequencing batch reactor (SBR) to eliminate organic compounds and nitrification. Other advanced treatment steps were also undertaken using a biological activated carbon filter or ozonation to reduce organic trace substances and pathogens. Due to the different treatment technologies and steps, various combinations of these were possible to provide treated waters of different qualities, for example with regard to nutrients, trace substances or pathogens. These waters were used for the irrigation of lettuce in eight separate hydroponic lines operated using a slightly modified deep-flow technique. Five lines were used for the main investigations: four hydroponic lines with treated wastewater and one with a reference nutrient solution. The other three lines were used as borderlines and for subordinate investigations. The hydroponic lines consisted of pipes with an inner diameter of 10 cm without any slope. A water level of 7–8 cm was maintained in the pipes.

In this working package, the central scientific and technical task was to establish the steps that are required in wastewater treatment in order to produce high-quality agricultural products within the hydroponic system.

The hydroponic system was operated either as a flow-through system (23.5 L/h) or as an adapted recirculation system (also called 'feed & deplete' with approximately 185 L per exchange; for details, see Bliedung *et al.* 2020c). The differently treated wastewater used as irrigation water had varying compositions. For example, the average concentration of nitrogen ranged from 1.6 to 42 mg/L. Other nutrients were not added to the irrigation water. During a second test period, a recirculation system was used in the hydroponic plant production. Here, the irrigation water had a high initial nitrate concentration (42 mg NO_3-N/L) and was circulated in the system until the nitrate concentration dropped below a defined threshold. The water was then replaced with freshly treated water. This process was repeated until the plants reached a target weight of 275 g FM (fresh matter, FM).

Expert interviews to acquire an understanding of the institutional framework

In the context of this socio-technical innovation, guideline-based expert interviews were used to generate knowledge with the actors involved in this cross-sectoral cooperation. There were three parts to the qualitative expert interviews. The first questions focused on knowledge about the individual and regional challenges faced by the sectors and intersectoral relationships related to the issue of irrigation in agriculture (see also Witzel & Reiter (2012)). Based on this, a second section dealt with perceptions about the technical novelty, enabling the challenges and adaptation requirements to be pointed out, as well as an assessment of whether they matched the problem frames previously articulated. The third and final part highlighted potential cooperation arrangements, responsibilities for certain roles and the challenges of interacting with state agencies and sales structures. Overall, nearly 50 interviews were undertaken.

The experts were identified on the basis of a desktop study. They came from the water and agricultural sectors and their respective consultants and associations, representatives of retailers, NGOs and environmental and consumer protection, as well as governmental institutions and research institutes. The actors' analysis was verified with the project partners, especially practitioners, and at a later point also with the participants of the stakeholder dialogue too. The expert interviews were analysed using a qualitative content analytical method with MaxQDA to identify the key issues.

Feasibility studies to gain knowledge about the concept's adaptability for other locations

Options for implementing the concept were investigated in feasibility studies at four different sites in Germany, Belgium and Portugal (Mohr *et al.* 2019a, 2019b). The aim was to identify beneficial and inhibitory factors in the use of treated wastewater in the hydroponic system. For each feasibility study, a transdisciplinary team was assembled with expertise in wastewater management, plant production, social sciences and landscape design. The team analysed the local framework conditions regarding water supply and wastewater infrastructure, agricultural production and demand for irrigation, marketing of agricultural products and interaction between stakeholders from the water and agricultural sectors through desktop research, literature analysis and on-site visits. Semi-structured interviews with the relevant local experts were conducted to co-create and evaluate the knowledge available at each site. The resulting concept was discussed during a workshop with local stakeholders in order to obtain their evaluation and develop a final local concept for water reuse in hydroponic systems. The concept was then published in a brochure for each study (for details, see the list of references). The transdisciplinary results obtained indicated whether and under which conditions this concept could be implemented locally. During this process, an overall analysis of the feasibility studies was undertaken at a workshop in which all project partners participated. The analysis identified the factors that promoted and impeded the concepts, as well as any similarity in patterns. These overall results can contribute to identifying the concept's advantages and disadvantages, improving them and determining the appropriate framing conditions for the establishment of such systems.

Impact assessment to understand potential niches for the concept

To evaluate the overall sustainability of the concept, a probabilistic graphical model called a Bayesian belief network (BBN) was used to investigate the potential social, ecological and economic implications of the investigated concept and then to identify beneficial system characteristics and supportive measures for the implementation of crop production systems as investigated in the HypoWave project (Zimmermann & Fischer 2020). Qualitative and quantitative data were combined in the BBN (Chen & Pollino 2012), and expert knowledge as well as calculations and project results were included. The various system elements (variables) and their relationships were developed with different members of the HypoWave project team. In a next step, experts evaluated the likelihood of certain social and ecological implications using standardised questionnaires. Economic implications were considered in a model with illustrative calculations for costs and revenues (Wageningen University & Research 2020).

Stakeholder dialogue for additional feedback and guidance

The stakeholder dialogue (Ebert *et al.* 2020) involved practitioners from various state authorities, local NGOs, farmers, other environmental, social and agricultural representatives, wastewater experts and scientists. The dialogue accompanied the project from the outset and was concentrated into three workshops that included a field trip to the pilot plant. A final conference to present the research results was also arranged that was open to the public. The research team presented the HypoWave concept and the intermediate results of the current research process.

During the workshops, stakeholders commented on these inputs but also actively contributed in work phases to identifying relevant actors and framing conditions (first workshop), key topics and issues to be addressed (second workshop), and options for cooperation between the different actors involved (third workshop). The final conference, which offered generous time slots for exchange and discussion, focused on an assessment of the results achieved and general questions to be addressed in future research.

SWOT analysis

An overall assessment of the available research results was undertaken using a SWOT analysis. SWOT stands for strengths, weaknesses, opportunities and threats. 'SWOT analysis is a widely used tool for analysing internal and external environments in order to attain a systematic approach and support for decision situations' (Ghazinoory *et al.* 2011, p. 24). The internal environment, which can be influenced by the system operators themselves (in the analytical framing called 'controllable'), is reflected in the strengths and weaknesses, while the external environment ('incontrollable') is reflected in the opportunities and threats. The methodology was developed in the field of strategic management to provide managers with an analytical tool for developing new business strategies. SWOT analyses have also been applied in research, with agriculture being a prominent field of application (Ghazinoory *et al.* 2011). The SWOT approach selected for the analysis of HypoWave results goes back to Novicevic *et al.* (2004). It merges knowledge (the authors refer to 'intelligence') and planning perspectives and introduces the criteria of controllable and uncontrollable factors, while intelligence is divided into desirable and undesirable factors.

The analysis evaluated the (water) technology, plant production, institutional/organisational settings, and legal, economic and social/societal aspects insofar as these were addressed in the research. The analysis was undertaken using existing project publications and further project outputs, such as the feasibility studies (Fischer *et al.* 2018; Mohr *et al.* 2018; Ebert *et al.* 2019b). It considered the recommendations for practitioners (pre-version) as well as internal documents arising from the project and stakeholder meetings. In addition, there was an overall process of knowledge integration with discussions, (intermediate) results and conclusions regarding the advantages and disadvantages of the HypoWave concept. The process was documented in the protocols of the working groups and project and stakeholder meetings. These documents were also included in the analysis.

RESULTS AND DISCUSSION

Results of the wastewater-driven hydroponic system

To provide a general impression of the developed concept of water reuse in hydroponic systems, a broad overview of the most promising treatment technology, hydroponic management, plant production and governance aspects is given below. The results do not go into great detail because this has been done in different papers published by the researchers and references to these are given in the text. Here, the broad overview is designed to provide guidance on the key issues and clarify the findings of the SWOT analysis. It should also be noted that the text does not discuss the general advantages and disadvantages of wastewater reuse in agriculture. There are many standard textbooks on this subject (e.g. Asano *et al.* 2007; Lazarova *et al.* 2013), and the authors consider the knowledge contained within them as a prerequisite to understanding water reuse in agriculture.

From wastewater to irrigation water

The HypoWave project is the first to investigate the appropriate treatment of municipal wastewater in order to provide irrigation water and its subsequent reuse in a hydroponic system for the production of high-quality products. The most important step in this direction is to identify the suitable combination of treatment steps required.

Based on the results of the pilot tests in HypoWave, a combination of an SBR reactor and a biological activated carbon filter proved to be very efficient in terms of producing high-quality irrigation water. Nitrogen, phosphorus and the required nutrients remained in the treated water, while there was an effective reduction in the chemical oxygen demand (COD), micropollutants and heavy metals

(Bliedung *et al.* 2019, 2020a, 2020b, 2020c; Dockhorn *et al.* 2020; Mohr *et al.* 2020).

Based on the results, the following recommendations can be made. During biological pre-treatment, organic compounds must be significantly reduced in order to avoid biofilm formation or anaerobic conditions in the hydroponic system and not impair plant growth. For this purpose, an anaerobic treatment (e.g. EGSB reactor) can be applied as a first step. However, since temperatures in Central Europe do not allow COD to degrade sufficiently under anaerobic conditions (Urban 2009), a subsequent aerobic treatment for further COD removal is strongly recommended. At the same time, nitrification can take place in this aerobic treatment step since nitrate is the preferred nitrogen source for lettuce and many other plants. To provide the optimal nutrient supply for the plants, the wastewater treatment should not aim to achieve denitrification or eliminate phosphorous in order to maintain the level of water-borne nitrogen and phosphorus nutrients as much as possible.

To reduce pathogens and organic trace substances, ozonation or biologically activated carbon filtration can be used (Bliedung *et al.* 2020b, 2020c; Dockhorn *et al.* 2020; Mohr *et al.* 2020). In the case of ozonation, it is important to adjust the specific ozone doses to minimise the transformation products of organic trace substances and achieve the appropriate elimination of pathogens. However, the combination of both ozonation and biologically activated carbon filtration is also an option. Although this setting involves two treatment steps, the advantage is that transformation products produced during ozonation can be eliminated by subsequent bio-filtration (Knopp *et al.* 2016; DWA-Fachausschuss KA-8.6 2019). However, despite a further reduction of pathogens during water treatment, regrowth can occur in the hydroponic system itself. If necessary, continuous UV disinfection of the (circulated) irrigation water should also be provided.

Plant production

Lettuce plants grown in the hydroponic system and irrigated and fertilised with the irrigation water were produced in comparable quantity and quality to those produced with mineral nutrient solution. Nevertheless, it was evident that nutrient concentrations in the irrigation water were at the lower end of plant requirements, with certain micronutrients in particular being limited (Bliedung *et al.* 2020c). Careful management of the nutrient solution, e.g. the volume and/or addition of the nutrient in deficit, was key for the production of a competitive yield. The nutrient content in the plants was largely comparable with the average found in current data in the literature. Where heavy metals were detected, they were at very low concentrations or below threshold values (Bliedung *et al.* 2020c). The plant quality with regard to organic trace substances and pathogens was also comparable with that of commercially available products (Bliedung *et al.* 2020b; Dockhorn *et al.* 2020).

An investigation of other vegetable varieties and crops in addition to lettuce was undertaken in the feasibility studies, the impact assessment and in additional literature research. Greenhouse crops for the system in question are those suitable for hydroponic systems with a deep-flow technique. Alongside lettuce, tomatoes, aubergines, courgettes and peppers (Mohr *et al.* 2018), as well as cut flowers like chrysanthemums (Fischer *et al.* 2018), have been identified as crops that could potentially be grown in the investigated concept.

There are recent examples in Germany where agricultural cropland has been covered with greenhouses (Kliebisch *et al.* 2009; Moninger *et al.* 2017), and the greenhouse sector in Germany is growing (BMEL 2019). Access to financial resources and limited access to land are currently leading to the establishment of more greenhouses. Another reason is growing customer demand for regional products in supermarkets (Moninger *et al.* 2017). Due to increasing quantities and customer interest, in addition to farmers' markets that were originally supplied, there has been a rise in the distribution of regional products through supermarkets.

An increase in greenhouse crop production supports self-sufficiency, especially since the demand for year-round, regionally produced agricultural products is growing as a result of consumer awareness (BMEL 2017; Meyerding *et al.* 2019). This rise in greenhouse crop production therefore boosts food security and has the potential to also increase farmers' income with the supply of high-value products (Stanghellini 2014). As shown by the results of the impact assessment, compensatory measures should be introduced when greenhouses are constructed on agricultural cropland in order to minimise or mitigate their negative effects on the environment (Zimmermann & Fischer 2020).

Key actors and their cooperation

Agronomic and technical decision-making takes place in a societal context. In social reality, not every apparently rational choice is valued as an appropriate solution (for differentiation, see Ebert *et al.* 2019a). Therefore, it is important to have knowledge of the norms and values as well as the logics and desires relevant to decision-making in both the wastewater and farming sectors. The newly established production system involves heterogeneous actors who were either socialised in the service-orientated public sector or operate within the logic of competition prevailing in small- and medium-sized private entities. Whereas the former, here operators of WWTPs, understands and frames problems by taking collective matters into account, such as preserving water bodies in this case, the expert interviews with farmers showed that their farm's continued existence and their identification with their products and production practices (e.g. plant production in soil) are very salient issues for them. In this context, nature plays a role when it comes to long-term perspectives for securing the existence of farms as private entities.

With this as the starting point, both sectors need to adapt the system to suit their own needs. The key actors involved at an early stage were the wastewater associations, their supervisory authorities and (organised) farmers. Horticultural producers and farmers had also expressed an interest in the system as an additional production branch. They should now search for advisors and potential purchasers. The system can be operated on the basis of individual private-law contracts. With regard to the above-mentioned rural applications of water reuse in hydroponic systems, the project showed that established enterprises are willing to establish complementary branches within their companies and become water service providers. Furthermore, regional organisations, such as irrigation associations and cooperatives, are a major group for creating opportunities and establishing such a concept by integrating it into their institutions or establishing new ones (Mohr *et al.* 2018; Ebert *et al.* 2019b). Cooperatives, in particular, contributed to sectorial and cross-sectorial understanding. The cooperation of actors in such heterogeneous sectors as wastewater treatment and plant production depends on mutual recognition and a basic understanding of belief systems behind the apparently rational choices of actors (e.g. pluralistic core beliefs on precaution, and distributional or intergenerational justice) (Jenkins-Smith & Sabatier 1994; Shinn 2004). Awareness of this was particularly relevant in order to secure joint quality management (Mohr *et al.* 2020; Schramm *et al.* 2019). For the actors, this might even justify the establishment of specialist water and nutrient management services (e.g. new independent service providers for irrigation water) (Ebert *et al.* 2019a).

Despite exceptional cases in decentralised continental European applications, the operation of water reuse in hydroponic systems requires collective action. Maintaining autonomy is an important rationale articulated in the expert interviews with farmers. Nevertheless, farmers are confronted with new dependencies involving other actors once more collective belief systems enter the picture. This process of establishing new modes of cooperation is accompanied by the ambiguous positions of other stakeholders such as NGOs, the large supermarket chains, government agencies, the scientific community and financial providers (Ebert *et al.* 2017). Hence, water reuse in hydroponic greenhouses takes place in a societal context in which collective actors, as well as individuals, make decisions in a situation of uncertainty. Contracts, organisational arrangements and (quality) management plans agreed between organisations can contribute to reducing this uncertainty and create signalling effects for decision-makers, e.g. with respect to responsibility and trustworthiness.

Potential locations for implementation

The feasibility studies, stakeholder discussions, expert interviews and impact assessment showed that it is possible to implement HypoWave, especially in rural areas where the wastewater infrastructure needs to be adapted and land for setting up greenhouses is more readily available (Fischer *et al.* 2018; Mohr *et al.* 2018). At the same time, settlements are usually smaller and produce less wastewater. This means that the amount of irrigation water and available nutrients is lower, and thus the space requirements for greenhouses are smaller. Moreover, industrial wastewater is less likely to be found and easier to identify in the wastewater of small settlements. As nutrients are not very expensive (<0.5% of the operational costs of the greenhouse production system) and water for agriculture is still cheap in many parts of Germany (e.g. Lower Saxony: € 0.007/m^3 for groundwater used in irrigation; Niedersächsisches Ministerium für Umwelt Energie Bauen und Klimaschutz 2020), the economic feasibility greatly depends on the synergies between advanced wastewater treatment (reducing nutrient concentrations) and the production of marketable products. Thus, the concept is interesting in areas that have a need to limit further the discharge of nutrients (nitrogen and phosphorous) into the environment to prevent eutrophication in receiving water bodies.

Furthermore, in peri-urban contexts, water reuse in hydroponic systems can be established for WWTPs that plan to add a fourth treatment stage due to administrative requirements, e.g. the removal of pharmaceutical residues. If this kind of additional treatment is implemented, the water quality of the effluent can become suitable for irrigation purposes. Being located in areas with a growing agricultural irrigation demand as well as an increasing drinking water demand (for example, expanding metropolitan regions such as Rhine Main), this concept offers opportunities to substitute the groundwater currently being used for irrigation purposes (Ebert *et al.* 2019b). Established local structures in horticulture as well as the demand for regional food strongly support this approach on a regional level. However, the need to integrate the demands of several farmers as well as requirements regarding irrigation water quality can lead to the decoupling of nutrient reuse (Ebert *et al.* 2019b). In this regional approach, reused water of high quality can contribute to structural policy, for example by keeping agricultural production attractive, to services of general interest, such as regional food supply, and to precautionary measures, for example preventing trace substances from entering the water cycle. The amount of water that can be used for irrigation greatly depends on the minimum ecological flow required by the receiving waters.

Results of the SWOT analysis

The results of the disciplinary research and the transdisciplinary integration process are presented here using the logic of a SWOT analysis. A summary of the results is provided in Table 1, which shows the topics identified during the SWOT analysis. The table also locates the topics within the analysis. As an additional measure, additional guidance is provided by the categories 'material' and 'immaterial'. Both categories provide thematic subcategories: 'material' includes the subcategories 'water treatment', 'plant production' and 'environment', while 'immaterial' comprises 'operation and management', 'economy', 'actors and institutions (law)' and 'values and standards (sustainability)'.

The analysis of the HypoWave material also highlighted additional topics and issues that could not be grouped into one of the four SWOT fields. These aspects are presented and discussed in the text below the respective tables.

The concept focuses on two aims at the same time: the treatment of wastewater and the service provision of irrigation water. The system provides a dual service and addresses the need for resource-efficient handling of water and nutrients, as well as a secure water source for agriculture, especially in situations with water shortages. As the results show (see also Table 2), both objectives can be achieved. However, a profound knowledge of wastewater treatment and agricultural production and understanding among the actors involved is crucial to its success. Required forms of cooperation exist, of which the actors are aware, and can be implemented. This is a huge asset because it implies security within the establishment of such a sociotechnical innovation, which by its nature involves numerous insecurities because there is no experience of it to date. The environmental benefits of the system are also due to its resource efficiency, the reuse of already extracted water, and being sealed off from the environment, thus minimising risks such as groundwater pollution.

A sensitive point in water reuse is always the handling of the risks originating from raw wastewater containing heavy metals, pathogens and pharmaceutical residues. As the results showed, all three of them were successfully eliminated by the chosen wastewater treatment and agricultural production process (Table 2). Nevertheless, the safe

Table 1 | Summary of the location of the identified topics within the SWOT analysis

Analytical framing		Desirable		Undesirable	
		Strength		**Weaknesses**	
		Material	**Immaterial**	**Material**	**Immaterial**
Internal aspects	Controllable	Water treatment – Water quality – Phosphorus recovery – Pharmaceutical residues, heavy metals, pathogens – Measurement and control technology	Economy – Profitability – Circular economy	Water treatment – Contamination – Salinity – Pharmaceutical residues, heavy metals, pathogens – Wastewater quality	Operation and management – System approach (2) – Staff – Existing experience (2)
		Plant production – System configuration – System approach – Weather-independent production	Actors and institutions – Forms of cooperation – Secure water supply	Plant production – Control systems – Nutrient supply	Economy – Economic incentives – Barriers to entry
		Environment: – Closed production system – Discharge water quality		Environment – Conversion of land	Values and standards (sustainability) – Change of perspective
		Opportunities		**Threats**	
		Material	**Immaterial**	**Material**	**Immaterial**
External aspects	Uncontrollable	Water treatment – Protection of water resources – Wastewater quality	Operation and management – Knowledge	Environment – Discharge water quality	Operation and management – Existing horticulture sector – Marketing experience
		Plant production – Regional food production (2) – Water scarcity – Use of artificial intelligence	Economy – Income and employment – Competition in resource efficiency		Economy – Market competition – Entry point to market
		Environment – Usable land	Actors and institutions – Experience (2) – Regional food production – Political will (2)		Actors and institutions – Reaction of media – Permits – Certifications
					Values and standards (sustainability) – Public acceptance

Only subcategories that provide information are listed. Where topics are addressed more than once, this is indicated in brackets.

Table 2 | SWOT analysis of water reuse in hydroponic systems for German conditions: Strengths

Strengths	
Material	**Immaterial**
Water treatment • Treatment of wastewater to gain irrigation water is possible in qualities up to DIN 19650, even for the irrigation of vegetables eaten raw. • The concept can also address upcoming requirements of phosphorus recovery. • Pharmaceutical residues, heavy metals and pathogens can be reduced sufficiently within wastewater treatment or in the case of pathogens by implementing additional hygienic barriers. • Parts of the necessary instrumentation and control infrastructure already exist. For instance, most WWTPs are equipped with online measuring devices to detect ammonium, nitrate and phosphorus. ***Plant production*** • Plant growth functions in hydroponic systems fed with this irrigation water. Moreover, the supply of irrigation water via a tube includes an additional barrier regarding microbial risks as the water is only touching plant roots. • Depending on the design of the irrigation system, two goals can be achieved: nutrient efficiency (flow-through system) and water efficiency (recirculating nutrient solution). • Different greenhouse crops like fruits/vegetables and ornamental flowers can be produced, while the high-yield production is not exposed to seasonal weather patterns like droughts in summer or frost in winter as long as illumination and heating is provided. ***Environment*** • The system reduces environmental risks, such as groundwater pollution, as the hydroponic system itself is closed to the environment. • The hydroponic system can be balanced in a way that its effluent flow contains very low concentrations of nutrients. Thus, it can be discharged into sensitive water bodies without further treatment.	***Operation and management*** • First economic studies indicate that hydroponic greenhouses can be operated economically (Fischer *et al.* 2018; Mohr *et al.* 2018; Zimmermann & Fischer 2020). • As nutrients recycled from wastewater are used, this concept is an example of a circular economy. It helps to reduce the dependency on artificial fertilisers and reduces the need of nutrient elimination and therefore the cost of sewage treatment (Fischer *et al.* 2019) ***Actors and institutions (law)*** • There are forms of cooperation that can be used to organise and manage such a system among the involved actors: (a) private contracts and (b) regional organisations. The categories such as liability, obligations and quality assurance, requiring special care in the respective agreements, are known. • Quantitatively, the system contributes to secure supply streams that mitigate groundwater competition and reduce the risk of illegal overextraction, which is not currently monitored or measured. ***Values and standards (sustainability)*** –

production and high quality of the products (water and plants) have to be guaranteed at all times. As the system is complex, many aspects have to be considered (for details, see Table 3) and different actors are involved, a risk management concept is highly recommended for the whole wastewater processing and plant production process that addresses these weaknesses and how they are dealt with.

A minimisation of the risks mentioned, such as pathogenic contamination of plants or pharmaceutical residues contained in the products (for details, see Table 3), can also be achieved by further processing of the vegetables produced (e.g. cooking), when this is economically justifiable and the expense and workload involved in processing remains manageable. Where this is possible, the farmer gains greater independence from fresh food contractors. This provides leeway with regard to other market segments and contracts, as well as marketing channels.

Additionally, there needs to be clarification that the absolute dependency of the two coupled subsystems, wastewater treatment and agricultural production, only exists in cases where an optimum in water and/or nutrient efficiency is to be achieved. Otherwise, water reuse in hydroponic systems can be designed in such a way

Table 3 | SWOT analysis of water reuse in hydroponic systems for German conditions: Weaknesses

Weaknesses	
Material	**Immaterial**
Water treatment • Proximity of plant production to the area of wastewater treatment/irrigation water preparation can cause an additional airborne loading of plants with pathogens via aerosols. • Salinity of treated wastewater should not exceed the plant requirements. Removal of excess salinity, for example by reverse osmosis, could be advisable. In such cases, the management of residues might also be necessary. • The additional treatment of wastewater also cannot guarantee that there are no pharmaceutical residues in the products because there are too many of them and their range is continually changing, making holistic analyses impossible. • Technical challenges presented by a wide range of influent water situations (dry weather and heavy rainfall). ***Plant production*** • Adequate plant nutrition on a minimum feeding level needs close attention and control systems. • Certain nutrients need to be added artificially to secure adequate plant nutrition. ***Environment*** • If a conversion of agricultural cropland to a greenhouse area takes place in an uncontrolled way, there will be negative ecological impacts such as the reduction of biodiversity and groundwater recharge, if no countermeasures are undertaken. Moreover, the landscape quality might decrease (Bürgow *et al.* 2019).	***Operation and management:*** • The system is highly complex as two goals have to be achieved: wastewater treatment and plant production. This requires a high level of competence and monitoring. • Due to this complexity, quality assurance plays a major role and requires competent and dedicated staff in a field where untrained staff are often employed. • So far, there is no comprehensive reference and therefore no experience of a combined system like this. • The conversion of traditional farming to greenhouse production requires new knowledge and experience, and also needs the establishment of new professional networks, including sales structures. • There is a dependency between wastewater treatment/irrigation water production and plant production. At the moment, the two systems are combined; individual decisions in one of the two subsystems always have an influence on the other and need to be coordinated. ***Economy*** • Synthetic fertilisers are cheap and, in areas with sufficient freshwater resources, there is usually little economic incentive for water reuse. • Barriers to entry in this new business field are relatively high initial investment costs for the greenhouses and the dependency on partnerships with WWTPs. So far, there is no experience of such partnerships in Germany. ***Actors and institutions (law)*** – ***Values and standards (sustainability)*** • The operator of a WWTP has to change perspective from disposal to provision and deliver a product that meets qualitative standards and quantitative requirements.

that only parts of the treated water are used as irrigation water in hydroponic production and the rest is discharged, for example. By designing the wastewater treatment in this way, a more adaptive and robust management of the whole system can be achieved. This might also be important in periods when hydroponic production is halted for cleaning or other maintenance activities.

Developments in the field of digitalisation and artificial intelligence are an additional driver, as listed in Table 4. The newly evolving field between water provision and agricultural production might be an interesting business area for companies in this sector. Closer cooperation between the (waste)water and agricultural sector is just a matter of time due to the already ongoing changes in water availability and requirements. At the same time, new tools might greatly alleviate the coordination and interaction between the two sectors, especially when it is developed as a learning system that helps to steer the overall processes with real-time interpretation of the recorded data.

New systems like this will not only advance sustainable development in agriculture and increase competition for more resource-efficient production (Table 4), but could be implemented as a wastewater-driven hydroponic system for urban areas too. Developing areas like this are often located on the border between urban and rural areas where agricultural land becomes part of an urban region and farmers have to leave. Water reuse in a hydroponic system could be

Table 4 | SWOT analysis of water reuse in hydroponic systems for German conditions: Opportunities

Opportunities	
Material	**Immaterial**
Water treatment • The (re)cycling of water in smaller loops lowers the impact on nature as water withdrawal is minimised, while increasing resource efficiency. Natural water resources remain unspoiled or are available for uses other than the ones for which treated wastewater is used. • Wastewater that does not contain industrial influences and water collection in a separated sewer system are good starting points for setting up such a system. ***Plant production*** • Demand for regionally produced vegetables/fruits, also outside their seasons, is often higher than the production. • Water scarcity will become more severe due to climate change; thus, the demand for water reuse and water-efficient agriculture will increase in the coming years. This is especially true as the limits of current agricultural irrigation using mostly groundwater and surface water sources will be reached. • Rapid developments in artificial intelligence help to regulate and coordinate the two aims of plant production and nutrient reduction in the water. ***Environment*** • Due to the system being closed, any area can be used to set up the production system, including unpopular areas such as industrial wastelands. In such cases, the system can upgrade the landscape and additional efforts be made to improve its ecological characteristics.	***Operation and management*** • There is already knowledge around managing hydroponic systems in Germany, and this can be obtained in the form of consultancy by interested operators. ***Economy*** • Farmers' income increases compared with that from traditional agriculture. Jobs are also created since horticulture production is more labour-intensive. The HypoWave concept can offer new business options. • New systems like this can advance sustainable development in agriculture and increase competition for more resource-efficient production. ***Actors and institutions (law)*** • The experience and knowledge of water reuse in agriculture among actors and institutions might also spill over to new fields of water reuse. • The existence of regional organisations such as irrigation associations and cooperatives are an asset as they can be a starting point for setting up the cooperation required for such a system. A major advantage is that they know how to mediate between collective and individual interests. • Food retailers and supermarket chains might be interested in supporting new regional production systems due to existing demand for regionally produced food. • Preservation of nature and precaution can be positively influenced through political will and an executive habitus. • Political will can support a sustainable transformation of plant production by taking responsibility for structural planning and policy. ***Values and standards (sustainability)*** –

a way of farming continuing in areas like this and as a historical reference to the area's origins. It could, for example, also serve to integrate inhabitants through environmental education and joint social activities within farming.

One threat identified in the SWOT analysis (for details, see Table 5) is that consumers might be worried about the product safety of vegetables and fruits. Here, an additional potential option investigated in HypoWave is the production of ornamental flowers. A product like this could offer an alternative by implementing the system in the non-food sector. Nevertheless, it should be clear that by heading in the direction of cut flowers, market opportunities and competition with potentially international operators also have to be considered and parts of the SWOT analysis will have to be redone.

Key topics for further development of the innovation

It is striking how few *threats* there were in the 'material' category (Table 5). Nevertheless, this corresponds to the research questions targeted in the HypoWave project. Alongside the analysis of institutional aspects, the research addressed the key tasks of wastewater treatment and plant production. System configurations and processes were optimised during the 3-year project.

The HypoWave pilot testing proved that the combination of a nitrifying SBR reactor and a biologically activated carbon filter was very efficient in terms of producing high-quality irrigation water. Nevertheless, this combination was only tested on lettuce plants. Hence,

Table 5 | SWOT analysis of water reuse in hydroponic systems for German conditions: Threats

Threats	
Material	**Immaterial**
Water treatment – ***Plant production*** – ***Environment*** • Water reuse can be critical for water bodies with low water levels. The reduction of the discharge can affect river ecosystems, especially in summer.	***Operation and management*** • The attitude of the horticultural sector towards the new production system is unclear regarding concerns or competition. • There is no experience of marketing production systems like this. ***Economy*** • The food sector is very sensitive about product quality and at the same time highly competitive. It might be difficult for single producers to receive binding and stable contracts for purchase. • The operator needs to identify a good and reliable entry point into marketing his products and relies on a respective partner in food trading if there are no plans for self-sale, e.g. farmers' markets. ***Actors and institutions (law)*** • Uncertainty and unpredictability about the media's views on the topic. There are different options for how the media will deal with it. This can range from promoting the positive image of regional food production and closed nutrient cycles to telling horror stories about product quality due to potential risks from the origin of the irrigation water. • The system is not fully subject to self-governance of public and private entities. Permits from local and regional state authorities are a necessity and might be difficult to get due to lack of experience and knowledge. This can present a challenge for the planning and implementation process. • It is not currently possible to receive organic certification for such crops, as organic products have to be produced in soil. Labelling as a regional product is possible, but dependent on the strength of the region's identity and integration. ***Values and standards (sustainability)*** • Consumers might be scared of the fact that the products are grown with irrigation water from wastewater and they might not accept the system or its products.

further development is required regarding the scaling-up of the technology, including treatment capacity, determining ideal operating parameters for various crops and therefore developing a robust technology for continuous irrigation water supply. Moreover, the integrated management of water treatment and plant production, including specific plant requirements, needs to be developed further.

The results also revealed a minimisation of risks resulting from heavy metals, trace substances and pathogens (Table 2). This is key to the successful promotion of the system (see Table 3, for example). The HypoWave results also indicated that this aim can be achieved and initial measures were identified (see Table 2, for example). At the same time, the analysis clearly showed the demand for high quality regarding irrigation water, discharge and the food produced. To address all these requirements, alongside the need for risk management and quality assurance, joint quality management is also required along the entire production chain to secure adequate quality of the water, discharge and food products. This will also help it to gain acceptance among consumers and address any criticism in the media (Table 5).

A clear asset of the system is existing forms of cooperation, such as irrigation associations and cooperatives, which can be used to establish water reuse in hydroponic systems (Table 2). However, cooperation experience regarding water reuse in hydroponic systems does not yet exist in Germany, and actors can only rely on experience transferred from adjacent subjects (Table 4). Therefore, the close support of actors entering this field is important to identify open questions and pitfalls, as well as develop ideas about how to design such cooperation. Municipal actors and civil servants cannot provide support with their existing knowledge and experience, but instead have to explore the innovation themselves to reach appropriate decisions (Table 5). An important measure in this regard can be the early involvement of the necessary stakeholders from the outset to identify any issues and concerns as early as possible and work together on solutions.

The energy consumption of water reuse in hydroponic systems was not a focus of this research project and therefore was only examined incidentally in two of the feasibility studies (Fischer *et al.* 2018; Mohr *et al.* 2018). However, energy aspects have an important impact, especially when the hydroponic system is combined with greenhouses. Fischer *et al.* (2018), for example, show that effluents in the hydroponic system could be treated in a short-rotation plantation, using fast-growing trees as one source for the required heating system.

Moreover, the production system will also be of great interest to many other countries where water scarcity is already more severe and the need for food production and economic development is high on the agenda. This was also shown in a feasibility study undertaken in Portugal (Germer *et al.* 2020). Hence, the concept also offers opportunities for the system, its technical components and the knowledge acquired to be exported.

Limitations of the analysis

A SWOT analysis offers overall guidance and supports decision-making. However, the aspects formulated in this paper might look different under specific local conditions, and the varying influence of local drivers, such as water shortage, demands on irrigation water or the role and influence of local institutions and actors. Due to the system's novelty, the decision about implementing water reuse in hydroponic systems currently involves uncertainty, especially when legislation is changing (e.g. EU legislation on water reuse) or certification processes are uncertain, such as organic certification of soil-less production systems.

Moreover, it is clear that the presented results depend a great deal on the specific knowledge and information made available by the project, contributed by the project partners, interview partners and stakeholders involved. A different selection of knowledge carriers might have had an effect on the results, since they contribute and act according to their own values, norms, motivations and reasoning in decision-making.

Last but not least, as mentioned above, a SWOT analysis provides overall orientation, which requires complex systems and their interactions to be presented in a less complex way. Nevertheless, this analysis provides guidance and can support decision-making. Especially when it comes to cases of (inter-)organisational cooperation, including institutional design, a SWOT analysis is not adequate for all the details to be shown. That is why the concept only has a limited explanatory force when explaining intersectoral cooperation based on collective action among heterogeneous actors.

CONCLUSIONS

This study revealed that water reuse in hydroponic systems is a realistic scenario for Germany in the future. The analysis showed that, on the basis of changing climate conditions and upcoming water shortages, Germany also needs to prepare and search for additional and new forms of food production – a position clearly shared by the stakeholders involved. Early promising results indicate the conditions and circumstances under which water reuse in hydroponic systems might be attractive. They also emphasise that this is no overall solution for everybody everywhere, but is an interesting innovation when considering certain preconditions such as the local wastewater treatment situation and interested farmers or organisations such as associations looking for new forms of agricultural production. Nevertheless, additional efforts have to be made to advance the system with regard to the robust preparation of irrigation water in combination with reliable plant nutrition, joint quality management to handle potential risks, good understanding of the cooperation and configurations required and a more detailed investigation of the energy aspect – all of which were at the heart of the overall analysis conducted by the HypoWave project.

ACKNOWLEDGEMENTS

The authors would like to thank the German Federal Ministry of Education and Research for funding the research project 'HypoWave – Use of hydroponic systems for resource-efficient water reuse in agriculture' (grant no. 02WAV1402). The authors also wish to thank all their interview partners, partners in the feasibility studies and stakeholders for their time, their interest and their participation in the HypoWave research project. These results would not have been achieved without them. Last but not least, the authors of this article would like to thank the whole HypoWave consortium for the joint research and for the constructive and fruitful discussions.

DATA AVAILABILITY STATEMENT

All relevant data are included in the paper or its Supplementary Information.

REFERENCES

Asano, T., Burton, F., Leverenz, H., Tsuchihachi, R. & Tchobanoglus, G. 2007 *Water Reuse: Issues, Technologies, and Applications*. Metcalf & Eddy Inc., McGraw Hill, New York.

Bliedung, A., Dockhorn, T., Fiebig, B., Germer, J., Rossmanith, P. & Wieland, A. 2019 The HypoWave-System: nutrient and heavy metal flows within an integrated system of adapted wastewater treatment and subsequent water reuse in a hydroponic system. In: *IWA Water Reuse 2019, Book of Abstracts*. Internationl Water Association (IWA), Dechema, Berlin, pp. 552–557.

Bliedung, A., Dockhorn, T., Fiebig, B., Mohr, M., Peters, G., Rossmanith, P., Teiser, B. & Wieland, A. 2020a 4. Abwassertechnische Untersuchungen (4. Investigations of the wastewater). In *HypoWave – Einsatz hydroponischer Systeme zur ressourceneffizienten landwirtschaftlichen Wasserwiederverwendung – Gemeinsamer Schlussbericht des Verbundvorhabens (HypoWave – Use of Hydroponic Systems for Resource Efficient Water Reuse in Agriculture – Joint Final Report of the Joint Project)*. pp. 51–63

Bliedung, A., Dockhorn, T., Germer, J., Kreuzig, R., Haller-Jans, J., Bischoff, C., Leppin, B. K., Jechalke, S. & Smalla, K. 2020b 6. Produktqualität (6. Quality of the product). In *HypoWave – Einsatz hydroponischer Systeme zur ressourceneffizienten landwirtschaftlichen Wasserwiederverwendung - Gemeinsamer Schlussbericht des Verbundvorhabens (HypoWave – Use of Hydroponic Systems for Resource Efficient Water Reuse in Agriculture – Joint Final Report of the Joint Project)*. pp. 88–121.

Bliedung, A., Dockhorn, T., Germer, J., Mayerl, C. & Mohr, M. 2020c Experiences of running a hydroponic system in a pilot scale for resource efficient water reuse. *Journal of Water Reuse and Desalination* **10** (4), 347–362. https://iwaponline.com/jwrd/article/doi/10.2166/wrd.2020.014/76358.

BMEL 2017 *Deutschland, wie es isst: Der BMEL-Ernährungsreport 2018 (Germany, How it is Eating: The Nutrition Report 2018 of the BMEL)*. Berlin.

BMEL 2019 *Ertragslage Garten- und Weinbau 2019: Daten-Analysen (Earning Position Horticulture and Viniculture: Data Analysis)*. Berlin.

Bürgow, G., Brüll, A., Fischer, M. & Winker, M. 2019 Der Landschaftsbaukasten: Flächensensibles Wasserrecycling durch hydroponische Gewächshäuser (Landscape model kit: water recycling in hydroponic greenhouses). *Stadt & Grün* **12**, 28–34.

Chen, S. H. & Pollino, C. A. 2012 Good practice in Bayesian network modelling. *Environmental Modelling & Software* **37**, 134–145. http://dx.doi.org/10.1016/j.envsoft.2012.03.012.

DIN 19650:1999-02 *Bewässerung: Hygienische Belange von Bewässerungswasser (Irrigation: Hygiene Parameters for Irrigation Water)*. Beuth Verlag, Berlin.

Dockhorn, T., Bliedung, A., Blau, K., Jechalke, S., Kreuzig, R., Smalla, K., Fischer, M., Winker, M., Gebhardt, J., Nink, A., Wieland, A., Rossmanith, P., Germer, J., Mohr, M. & Peters, G. 2020 Hydroponische Systeme als ressourceneffizienter Beitrag zur Wasserwiederverwendung – Erfahrungen aus dem Verbundprojekt HypoWave (Hydroponic systems as resource efficient input for water reuse – Experiences from the research project HypoWave). In *Conference Proceedings, 53. Essener Tagung 2020*, Aachen. GWA Gewässerschutz-Wasser-Abwasser Band 252, S. 21/1-21/15.

DWA-Fachausschuss KA-8.6 2019 DWA-Themen Aktivkohleeinsatz auf kommunalen Kläranlagen zur Spurenstoffentfernung – Verfahrensvarianten, Reinigungsleistung und betriebliche Aspekte - (T1/2019) (DWA-Topics use of activated carbon at municipal wastewater treatment plants for the removal of trace substances – process variants, purification efficiency and operational aspects), 1st edn., DWA, Hennef.

Ebert, B., Schramm, E. & Winker, M. 2017 Building new nexus: wastewater reuse in agriculture from a multilevel network perspective. In: *Book of Abstracts: Third European Conference on Social Networks (EUSN2017) Main, Germany, 26–29 September 2017*. Johannes Gutenberg Universität, Mainz, pp. 295–296.

Ebert, B., Dockhorn, T., Peters, G., Schramm, E., Teiser, B. & Winker, M. 2019a Operator models for the reuse of municipal wastewater in hydroponic systems: Potentials and options for Central and Mediterranean Europe. In: *IWA Water Reuse*

2019, Book of Abstracts. Internationl Water Association (IWA), Dechema, Berlin, pp. 725–730.

Ebert, B., Günkel-Lange, T., Parniske, J. & Schramm, E. 2019b *Weitergehend aufbereitetes kommunales Abwasser als alternative Wasserressource im Hessischen Ried: Eine HypoWave-Fallstudie (Advanced Treated Municipal Sewage Water as Alternative Water Resource in the Hessian Ried: A HypoWave-Case Study)*. Fraunhofer IGB, Stuttgart.

Ebert, B., Schramm, E. & Winker, M. 2020 *Gemeinsam Innovationsentscheidungen Identifizieren: Methoden und Konzepte für Stakeholderdialoge und die Entwicklung sektorübergreifender Kooperationsmodelle am Beispiel landwirtschaftlicher Wasserwiederverwendung (Identifying Joint Decisions of Innovation: Methods and Concepts for Stakeholder Dialogues and the Development of A Sector Overarching Kooperation Models Along the Example of Agricultural Water Reuse) ISOE-Studientext 25*. ISOE – Institute for Social-Ecological Research, Frankfurt am Main.

Fischer, M., Beckett, M., Bürgow, G. & Ebert, B. 2018 *Modulares Wasser- und Nährstoffrecycling zur Schnittblumenproduktion in der Gemeinde Raeren, Belgien (Modular Water and Nutrient Recycling to Produce cut Flowers in the Municipality of Raeren, Belgium)*. Fraunhofer IGB, Stuttgart.

Fischer, M., Schramm, E. & Winker, M. 2019 How business models can help to introduce nutrient recycling in hydroponic greenhouses. *Watersolutions* (2), 58–65.

Germer, J., Ebert, B. & Mohr, M. 2020 *Concept for Sustainable Wastewater Treatment and Water Reuse in the Alentejo, Portugal: A HypoWave Case Study*. Fraunhofer IGB, Stuttgart.

Ghazinoory, S., Abdi, M. & Azadegan-Mehr, M. 2011 Swot methodology: a state-of-the-art review for the past, a framework for the future. *Journal of Business Economics and Management* **12** (1), 24–48. http://dx.doi.org/10.3846/16111699.2011.555358.

Hosseinzadeh, S., Verheust, Y., Bonarrigo, G. & Van Hulle, S. 2017 Closed hydroponic systems: operational parameters, root exudates occurrence and related water treatment. *Reviews in Environmental Science and Bio/Technology* **16** (1), 59–79. http://dx.doi.org/10.1007/s11157-016-9418-6.

IPCC 2019 *Climate Change and Land: An IPCC Special Report on Climate Change, Desertification, Land Degradation, Sustainable Land Management, Food Security, and Greenhouse Gas Fluxes in Terrestrial Ecosystems*. Available from: https://www.ipcc.ch/site/assets/uploads/sites/4/2020/08/200730-IPCCJ7230-SRCCL-Complete-BOOK-HRES.pdf

Jenkins-Smith, H. C. & Sabatier, P. A. 1994 Evaluating the advocacy coalition framework. *Journal of Public Policy* **14** (2), 175–203. http://dx.doi.org/10.1017/S0143814X00007431.

Kliebisch, C., Altmann, M., Landsberg, A. & Looije, M. 2009 *Masterplan Agro-Park/Gartenbaugebiet (Pilotprojektion für den Kreis Kleve): Gutachten (Master Plan Agro-Park/Horticulture Area (Pilot Projection for the Municipality of Kleve): Expertise*. Bonn.

Knopp, G., Prasse, C., Ternes, T. A. & Cornel, P. 2016 Elimination of micropollutants and transformation products from a wastewater treatment plant effluent through pilot scale ozonation followed by various activated carbon and biological filters. *Water Research* **100**, 580–592. https://doi.org/10.1016/j.watres.2016.04.069.

Lazarova, V., Asano, T., Bahri, A. & Anderson, J. 2013 *Milestones in Water Reuse. The Best Success Stories*. IWA Publishing, London, New York.

Meyerding, S. G. H., Trajer, N. & Lehberger, M. 2019 What is local food? The case of consumer preferences for local food labeling of tomatoes in Germany. *Journal of Cleaner Production* **207**, 30–43. http://dx.doi.org/10.1016/j.jclepro.2018.09.224.

Mohr, M., Dockhorn, T., Drewes, J., Karwat, S., Lotz, B., Nahrstedt, A., Nocker, A. & Schramm, E. 2020 Quality management along the water chain: completing multi barrier systems in agricultural water reuse. *Journal of Water Reuse and Desalination* **10** (4), 332–346. https://iwaponline.com/jwrd/article/doi/10.2166/wrd.2020.039/76804.

Mohr, M., Ebert, B., Schramm, E., Germer, J. & Bürgow, G. 2018 *Nutzung des Ablaufs eines Klärteichs zur Gemüseproduktion im Landkreis Gifhorn (Use of the Effluent of A Treatment Pond for the Production of Vegtables in Gifhorn County)*. Fraunhofer IGB, Stuttgart.

Mohr, M., Günkel-Lange, T., Fischer, M. & Bürgow, G. 2019a Water reuse in hydroponic systems: results from four European feasibility studies. In: *IWA Water Reuse 2019, Book of Abstracts*. Internationl Water Association (IWA), Dechema, Berlin, pp. 96–102.

Mohr, M., Schramm, E., Ebert, B., Germer, J. & Bürgow, G. 2019b Nutzung des Ablaufs einer Teichkläranlage zum Anbau von Gemüse im hydroponischen System im Landkreis Gifhorn – Ergebnisse einer Fallstudie (Usage of a pond waste water sewage plant's flow to grow vegetables in a hydroponic system in the municipality of Gifhorn – Results of a case study). *Zentralblatt für Geologie und Paläontologie* **Teil 1** (1), 65–72.

Moninger, S., Schmidtell, G., Kraft, E. & Reß, S. 2017 *Knoblauchsland: Agrarstrukturelles Gutachten (Knoblauchsland: Agristructural Survey)*. Würzburg.

Niedersächsisches Ministerium für Umwelt, Energie, Bauen und Klimaschutz 2020 *Wasserentnahmegebühr (Fee for Waterwithdrawl)*. Available from: https://www.umwelt.niedersachsen.de/startseite/themen/wasser/grundwasser/wasserentnahmegebuhr/wasserentnahmegebuehr-weg-8621.html (accessed 12 July 2020).

Novicevic, M. M., Harvey, M., Autry, C. W. & Bond, E. U. 2004 Dual-perspective SWOT: a synthesis of marketing intelligence and planning. *Marketing Intelligence & Planning* **22** (1), 84–94. http://dx.doi.org/10.1108/02634500410516931.

Sambo, P., Nicoletto, C., Giro, A., Pii, Y., Valentinuzzi, F., Mimmo, T., Lugli, P., Orzes, G., Mazzetto, F., Astolfi, S.,

Terzano, R. & Cesco, S. 2019 Hydroponic solutions for soilless production systems: issues and opportunities in a smart agriculture perspective. *Frontiers in Plant Science* **10**, 923–923. http://dx.doi.org/10.3389/fpls.2019.00923.

Schramm, E., Beythien, U., Dockhorn, T., Ebert, B., Fischer, M., Mohr, M., Wieland, A., Winker, M. & Zimmermann, M. 2019 Wasserwiederverwendung zur landwirtschaftlichen Nutzung in hydroponischen Systemen: Anforderungen an die Qualitätssicherung (Water reuse for an agricultural usage in hydroponic systems: requirements to ensure quality). *Zentralblatt für Geologie und Paläontologie* **Teil 1**, 73–82.

Shinn, T. 2004 Paradox oder Potenzial: Zur Dynamik heterogener Kooperation (Paradox or potential: About the dynamic of heterogene cooperations). In: *Kooperation im Niemandsland (Cooperation in no Man's Land)* (J. Strübing, I. Schulz-Schaeffer, M. Meister & J. Gläser, eds). VS Verlag für Sozialwissenschaften, Wiesbaden, pp. 77–101.

Stanghellini, C. 2014 Horticultural production in greenhouses: efficient use of water. *Acta Horticulturae* **1034**, 25–32. http://dx.doi.org/10.17660/ActaHortic.2014.1034.1.

Urban, I. 2009 *Anaerobe Kommunalabwasserbehandlung – Einsatz und Bemessung von UASB-Reaktoren (Aneorbic Treatment of Municipal Wastewater – Implementation and Dimensioning of UASB Reactors)*. Dissertation (PhD Thesis), Institut für Siedlungswasserwirtschaft und Abfalltechnik. Leibniz Universität Hannover, Hannover, Germany.

Wageningen University & Research 2020 *Nieuwe update van Kwantitatieve Informatie (KWIN) voor de Glastuinbouw uitgebracht (New Updates of Quantitative Information (KWIN) for Greenhouse Horticulture), Version 2013–2014*. Available from: www.wur.nl/nl/show/Kwantitatieve-Informatie-Glastuinbouw-2016-2017-KWIN-Glastuinbouw.htm (accessed 17 March 2020).

Witzel, A. & Reiter, H. 2012 *The Problem-Centred Interview: Principles and Practices*. Sage, Los Angeles, Ann Arbor, MI.

Zimmermann, M. & Fischer, M. 2020 Impact assessment of water and nutrient reuse in hydroponic systems using Bayesian Networks. *Journal of Water Reuse and Desalination* **10** (4), 431–442.

First received 27 March 2020; accepted in revised form 8 September 2020. Available online 15 October 2020

Microbiological changes along a modular wastewater reuse treatment process with a special focus on bacterial regrowth

Andreas Nocker, Lorenz Schulte-Illingheim, Hubert Müller, Anja Rohn, Barbara Zimmermann, Anil Gaba, Andreas Nahrstedt, Hooman Mohammadi, Yannick Tiemann and Kerstin Krömer

ABSTRACT

Water reuse is becoming an increasing necessity due to depleted water resources or increased water demand. A treatment process on a pilot scale was designed to produce different water qualities for different applications in industry or agriculture. We report here microbiological changes along the modular process using treated municipal wastewater effluent as raw water. Treatment technologies included coagulation, ultrafiltration (UF), reverse osmosis (RO), quartz sand, activated granular activated carbon (GAC) filtration and disinfection. Elimination of traditional hygiene indicator bacteria was already achieved by ultrafiltration as the first barrier. Profound changes by each treatment step also applied to the microbiome. Total and intact cell concentrations as quantified by flow cytometry underwent a strong decline after UF and RO, whereas biological stabilization was achieved through quartz sand filtration and GAC passage. Interestingly assimilable organic carbon (AOC) was still present even after RO at levels that allowed substantial regrowth of bacteria. Overall, UF and RO led only to a 0.43 and 0.78 log decrease in intact cells concentrations in stagnated water after regrowth compared with 6.5 log intact cells/ml in the stagnated raw water. Temperature was shown to be an important parameter determining the microbiome of the regrown population. Regrowth could, however, be efficiently suppressed by monochloramine.

Key words | bacterial regrowth, flow cytometry, process monitoring, water recycling, water reuse

Andreas Nocker (corresponding author)
Lorenz Schulte-Illingheim
Anja Rohn
Barbara Zimmermann
Anil Gaba
Andreas Nahrstedt
IWW Water Center,
Muelheim an der Ruhr,
Germany
E-mail: *a.nocker@iww-online.de*

Hubert Müller
University of Duisburg-Essen,
Campus Essen, Universitätsstraße 5,
45141 Essen,
Germany

Hooman Mohammadi
Yannick Tiemann
Kerstin Krömer
Oldenburgisch-Ostfriesischer Wasserverband (OOWV),
Georgstraße 4, 26919 Brake,
Germany

HIGHLIGHTS

- A versatile water reuse process was established producing different water qualities (MULTI-ReUse).
- Elimination of traditional hygiene indicator bacteria was already achieved by ultrafiltration as the first barrier. This predestined ultrafiltration as a critical control point for high-resolution process monitoring by online flow cytometry that produced data on total and intact cells in high temporal resolution.
- Offline cytometry allowed the assessment of the change in regrowth potential along the treatment process.
- Despite strong reductions in bacterial concentrations by UF and RO, the suppression of bacterial numbers was not sustainable as the regrowth potential (as determined by offline flow

doi: 10.2166/wrd.2020.012

cytometry) was only reduced moderately (on a log scale) despite a strong reduction of DOC (on a linear scale).

- Flow cytometry proved a very sensitive diagnostic tool both in an offline and online format.

GRAPHICAL ABSTRACT

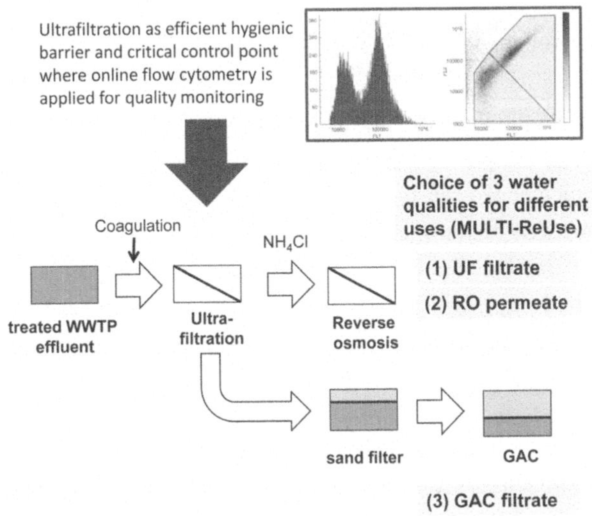

INTRODUCTION

Climate change leads to prolonged dry weather periods and extending semi-arid regions and poses an increasing stress for cities, industries and agriculture in many regions worldwide (Ormerod 2016; Faour-Klingbeil & Todd 2018). Also, in Germany that has previously not been associated with water shortage, local water limitations have become more obvious in the recent dry summers. To avoid excessive stress for groundwater sources, water reuse is increasingly debated especially for industrial uses and irrigation (Drewes & Horstmeyer 2015; Schmid & Bogner 2018; Drewes *et al.* 2019). Given the different types of water uses and the different corresponding water quality requirements, water treatment concepts should ideally be flexible and adjustable to needs (Becker *et al.* 2019). Further challenges consist in the allocation of variable demand-driven water volumes at competitive prices.

To gain experience with how to accomplish different quality requirements in such a scenario, a project named 'MULTI-ReUse' was launched. A coastal region with competing water uses (demand from water-intensive industries versus demand for drinking water) was chosen for a demonstration. As groundwater use would come with the risk of saltwater intrusion due to its direct proximity to the North Sea coast, this region in the northwest of Germany does not have its own drinking water supply. Instead, drinking water is transported over long distances. Whereas relatively low water qualities would be sufficient for most industrial applications and agriculture, high-quality drinking water is used in many cases. Water reuse would, thus, be an immediate alleviation on drinking water demand, groundwater resources elsewhere and energy transport costs. Aims of the project consisted of the (1) optimization and validation of modular treatment trains to achieve different water qualities, (2) improvement of process monitoring strategies, (3) assessment of economic and ecological aspects and socio-cultural acceptance and (4) development of marketable treatment solutions for globally relevant and typical reuse applications.

For this purpose, a German multicenter research consortium developed, demonstrated and evaluated a combination of different water treatment technologies to polish treated municipal wastewater treatment plant (WWTP) effluent to achieve different water qualities. The project was performed in collaboration with the local water supplier Oldenburgisch-Ostfriesischer Wasserverband (OOWV) in Lower Saxony. The treatment train of a pilot plant included, among other technologies, coagulation, ultrafiltration (UF), reverse osmosis (RO), quartz sand filtration, granular activated carbon (GAC) filtration and disinfection. Using a modular approach, wastewater was treated to different quality standards for different applications: (1) particle-free water for washing and cooling processes, (2) RO treated water free of ions and micropollutants for industrial processes and special uses and (3) particle-free, biologically stable water with low content of organic micropollutants for agricultural irrigation and groundwater infiltration.

While other members of the consortium worked on other aspects, this study aimed at mapping microbiological changes along the treatment train. Cultivation approaches were supported by cultivation-independent techniques to assess microbiological changes along the treatment process and the effect of different treatment steps. Applied monitoring technologies included flow cytometry (offline and online) to detect changes in total and intact cell concentrations (TCC and ICC) and high-throughput sequencing to capture changes in the microbiome along the treatment process. Both diagnostic methods have been shown to be useful tools to capture microbiological changes along treatment processes (Prest *et al.* 2014; Van Nevel *et al.* 2017; Hull *et al.* 2019). For flow cytometric analysis, the analysis of freshly taken water samples that reflect the microbiological status of the water at the time point of sampling (day 0) was supplemented with the analysis of the same samples after an incubation for 7 days (day 7) at defined temperatures. The day 7 values capture the regrowth potential of the bacteria contained in the corresponding water sample and reflect the planktonic bacterial concentrations that are obtained within the time stagnation time (Gillespie *et al.* 2014). The bacterial levels obtained after the 7 days are supposed to correlate with the assimilable organic carbon (AOC) and nutrient levels contained in the sample (Farhat *et al.* 2018). Indirectly the approach, therefore, also allows conclusions on the efficiency of nutrient removal.

In summary, this study aimed at monitoring changes in the numbers of bacteria along a modular water reuse treatment process and the composition of the bacterial population. The analysis was not limited to the microbiological status of the water directly after sampling, but also tried to cover the changes in that water after stagnation.

METHODS

Water recycling pilot plant

The pilot plant assembled by De.EnCon GmbH and IWW was accommodated in two containers located on the site of a municipal WWTP in the village of Nordenham in Lower Saxony. The region was chosen as it is characterized by competing water demands between drinking water supply and industrial uses. The treated wastewater effluent served as raw water for the pilot plant. The optimization phase over approximately 1 year was followed by a 6 months demonstration phase. Only data from the demonstration phase are shown. The treatment technologies included, among others, coagulation ($FeCl_3$), ultrafiltration (UF, flux of 60 L/(m^2 h) with 4.8 m^3/h per UF module, 90.9% recovery), low-pressure reverse osmosis (RO, flux of 16 L/(m^2 h) with 0.5 m^3/h per RO line, each with $3 \times 4''$ modules, 75% recovery), quartz sand filtration and GAC filtration, UV disinfection and optional disinfection with monochloramine to reduce RO membrane biofouling (Figure 1).

Two UF modules and two RO lines, each with three modules and recycling of concentrate, were operated in parallel to test different process conditions during an initial test

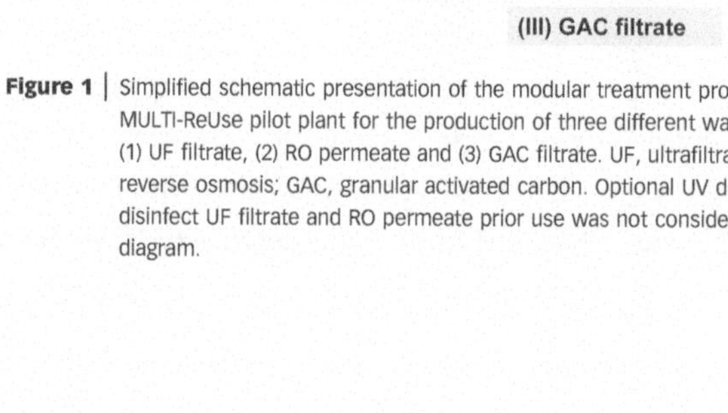

Figure 1 | Simplified schematic presentation of the modular treatment process of the MULTI-ReUse pilot plant for the production of three different water qualities: (1) UF filtrate, (2) RO permeate and (3) GAC filtrate. UF, ultrafiltration; RO, reverse osmosis; GAC, granular activated carbon. Optional UV disinfection to disinfect UF filtrate and RO permeate prior use was not considered in this diagram.

phase. Operating conditions were more constant during the demonstration phase that forms the basis for the data shown here. Monochloramine was generated to give a final concentration of 1 mg/L and added to the filtrate storage tank of UF 1, whereas UF 2 filtrate was not chemically disinfected. A detailed process and operation description is given elsewhere (Nahrstedt *et al.* 2020).

Preparation of AOC-free glassware

An alkaline solution of potassium permanganate was made by dissolving (A) 30 g of permanganate (cat. nr. 105082, Merck KGaA, Germany) and (B) 100 g of sodium hydroxide pellets (cat. nr. 106498, Merck KGaA, Germany) separately in each 500 ml deionized water. The latter is strongly exergonic and was allowed to cool down. The two solutions were subsequently combined in a 1 L bottle and stored in the dark prior to use. 50 ml borosilicate glass vials (cat. nr. 7612150, Th. Geyer GmbH & Co.KG) used for flow cytometry samples and other borosilicate glassware that should be AOC free were filled approximately one-third of the total volume with this alkaline solution and sealed with appropriate caps. Glassware was subsequently shaken in regular intervals for at least 15 min to ensure contact of the entire surface area with the solution. The permanganate solution was decanted (for recycling purposes) and glassware and caps were rinsed three times with tap water and twice with deionized water. Dried glassware and caps were wrapped in aluminum foil. Glassware was muffled at 280 °C for at least 8 h or overnight, caps were heated at 180 °C during that time.

Sampling procedure

Water samples were taken from designated sampling valves after disinfection with isopropanol (70%) followed by desiccation for at least 2 min (following guidelines by DIN EN ISO 19458) and flushing with 2 L of water. For cultivation, water was sampled using 500 ml polypropylene sampling bottles (cat. nr. 295348, Aqua Laborservice, Wertheim, Germany) containing thiosulfate. For flow cytometry, water was sampled in 50 ml AOC-free borosilicate glass vials containing thiosulfate and sealed with a cap with Teflon sealant. Vials were filled up to a 20 ml mark. The final thiosulfate concentration in the sample was 18 mg/L, an effect of thiosulfate on the regrowth potential was experimentally excluded. For sequencing, water was sampled in 1 L glass bottles (SCHOTT, Jena, Germany). All sampling bottles were sterile. Samples were transported cold (6 ± 2 °C) in styropor containers containing cooling elements and processed within 24 h.

Cultivation

Culture analysis was performed by the laboratory of the Public Health Department of Aurich. The following organisms were quantified: colony counts 22 and 36 °C (both DIN EN ISO 6222:1999-07), *Escherichia coli* and coliforms (both DIN EN ISO 9308-2:2014-06), intestinal enterococci (DIN EN ISO 7899-2:2000-11), *Clostridium perfringens* (DIN EN ISO 14189:2016-11) and *Legionella* spp. (DIN EN ISO 11731:2017-05).

Offline flow cytometry

Fluorescent dyes used in this study were SYBR Green I (10,000× stock, Invitrogen™) and propidium iodide (PI, 1 mg/ml, Invitrogen, Thermo Fischer). Water samples were processed undiluted or 10× diluted with 0.1 μm filtered mineral water (Evian, Evian-les-Bains, France) in case the total signal exceeded approximately 5.000 signals/s. SYBR Green I was diluted to a working stock concentration of 100× using dimethylsulfoxide (DMSO, Sigma-Aldrich) and stored at −20 °C until use. Aliquots (250 μl) of water samples were transferred into 96-well plates (cat. nr. 601808, HJ-Bioanalytik GmbH). To determine total cell concentrations (TCC), 200 μl sample aliquots from this plate were transferred into the wells of a second 96-well plate with pre-aliquoted 2 μl of the 100× SG working stock solution followed by thorough mixing by pipetting up and down several times using a multichannel pipette. To determine ICC, 200 μl sample aliquots from the first plate were transferred into the wells of a second 96-well plate with pre-aliquoted 2.4 μl mixture of a 100× SG I and PI in the ratio of 5:1 again followed by thorough mixing. Final concentrations of fluorescent dyes were 1× SG and 3 μM PI. Staining was performed at 37 °C for 13 min in an incubator. Data were collected using an ACEA NovoCyte® benchtop instrument equipped with a 488 nm laser (OLS OMNI Life Science, Bremen, Germany). A NovoSampler® Pro enabled analysis on a 96-well plate basis. Data were analyzed using the instrument-specific NovoExpress software and a gating procedure similar to the one described by Gatza *et al.* (2013).

Online flow cytometry

Online flow cytometry was performed using an Online Bacteria Analyzer (OBA, ONTRONIX AG, Switzerland). A stain solution with a total volume of 100 ml was prepared by mixing 2 ml PI (1 mg/ml), 0.1 ml SYBR Green I (10,000×), 10 ml DMSO and 87.9 ml ultrapure water to give concentrations of 10× SYBR Green I, 30 µM PI and 10% DMSO. This solution was kept in a designated storage compartment in the instrument at 10 °C for a maximum of 4 months. This stain solution was mixed with the water samples (volume of 90 µl) at a ratio of 1:9 to obtain final concentrations of SYBR Green I of 1× and of PI of 3 µM. Incubation of the dye-sample mixture was performed at 37 °C for 13 min. Measurements were typically taken every 2 h. When analyzing samples containing monochloramine, the stain solution was dosed with thiosulfate to a concentration of 180 mg/L to ensure a final thiosulfate concentration in the water sample of 18 mg/L for disinfectant neutralization.

Illumina sequencing

Genomic DNA was extracted from filtered water samples using the Aquadien™ DNA Extraction and Purification Kit (cat. nr. 3578121, Bio-Rad Laboratories GmbH, Feldkirchen, Germany). Illumina sequencing (MiSeq run nano V2, 250 bp paired-end) was performed by Microsynth AG (Balgach, Switzerland) after amplification of 16S rRNA genes using standard universal primers 341F_ill (5′-CCTACGGNGGCWGCAG-3′) and 802R_ill (5′-GACTACHVGGGTATCTAATCC-3′; Klindworth *et al.* 2013) amplifying the V3/4 region of the 16S rRNA gene. We received de-multiplexed raw reads from the sequencing company. After removing indices with bbduk (http://jgi.doe.gov/data-and-tools/bbtools/), reads were processed in mothur (version 1.43.0) following the MySeq SOP (Schloss *et al.* 2009; Kozich *et al.* 2013). After trimming, de-noising and chimera removal, sequences were clustered into operational taxonomic units (OTUs) using a 97% similarity cutoff. To compare the microbial diversity across samples, we sub-sampled and rarefied the dataset to the lowest number of sequences per sample, which was 4,072 sequences. This led to a total amount of 2,919 individual OTUs in all samples. Species richness, Simpson and Shannon indices as a measure for microbial Alpha diversity and Bray Curtis as a measure for Beta diversity were calculated in RStudio (version 1.1.463 and R version 3.5.2) using the package Phyloseq (version 1.24.2) (McMurdie & Holmes 2013). Phyloseq in combination with ggplot (version 3.2.1) and Inkscape (version 0.92) (http://www.inkscape.org/) were used for plotting and illustration.

RESULTS AND DISCUSSION

Assessment of the hygienic status

The most important criterion of water is its hygienic safety. Traditional bacterial hygiene indicators and bacterial colony counts as quantified by culture methods were abundant in treated wastewater effluent serving as raw water for the pilot plant (Table 1). Microbiological raw water quality underwent strong variations as indicated by minimum and maximum values. Independent of these variations, UF was shown to be highly efficient at removing all studied hygiene indicators with only one sporadic finding of one intestinal Enterococcus in a single sample out of $n = 17$. This finding suggests that a very good hygienic status was already accomplished after the first barrier. Also, total colony counts dropped substantially although the filtrate compartment was as expected not sterile.

Hygiene indicators were also found absent in the subsequent treatment process. Nevertheless, the water contained an autochthonous bacterial population that underwent profound changes along the treatment train in terms of numbers and composition. Assessment of colony counts was supported by flow cytometry to quantify the entire bacterial waterborne population independent of culturability.

Microbiological changes monitored by offline flow cytometry

Analysis of water (transported cold to the laboratory and analyzed within 24 h after sampling) provided the bacterial concentration reflecting the time point of sampling (day 0).

Table 1 | Traditional bacterial hygiene indicators and total colony counts in treated WWTP effluent (=raw water) and in UF filtrate

		Raw water		UF filtrate	
		Average	Maximum/Minimum	Average	Maximum/Minimum
E. coli	MPN/100 ml	79,100	198,600/20,600	0	0/0
Enterococci	MPN/100 ml	164,900	2,200,000/600	0	1/0
Clostridium perfringens	MPN/100 ml	64,300	810,000/4.200	0	0/0
Legionella spp.	MPN/100 ml	>100	>100/0	0	0/0
Colony counts 22 °C	CFU/ml	24,600	96,000/5,800	2	10/0
Colony counts 36 °C	CFU/ml	42,600	290,000/6,900	120	700/0

Average numbers are shown for the pilot plant demonstration phase between June 2018 and June 2019 ($n = 17$) with corresponding maximum and minimum values owing to changes in raw water quality.

Whereas bacterial concentrations in the UF feed water were comparable with the raw water (approximately 10^7 suspended cells/ml), UF resulted in a substantial decrease in TCC and ICC by an average of approximately 3.5 log units or 99.97% (Figure 2(a)). Bacterial concentrations tended to increase again within a UF filtrate storage tank and during passage to the RO module. As expected, bacterial concentrations in RO permeate were very low and near or under the detection limit of the method (approximately 100–1,000 cells/ml depending on the cleanliness of the flow cytometer at the time point of sample analysis).

Biological stabilization of UF filtrate via quartz sand filter and GAC passage led to a recovery of bacterial concentrations to approximately 10^6 cells/ml by enrichment of flow-through water with bacteria sessile on these filters. Lowest bacterial concentrations were, thus, contained in water qualities II and I, followed by water quality III. This was in agreement with colony counts that showed the same tendency although on a lower part of the log scale (Figure 2(b)). Comparing bacterial numbers obtained with flow cytometry and culture, between 0.001% (22 °C, quartz sand filter runoff) and 5.9% (36 °C, UF filtrate after storage tank) of intact bacteria were culturable (Figure 2(c)). Overall the percentages of culturable bacteria were thus subject to substantial variations over the treatment train. With the exception of the raw water, more bacteria were culturable at 36 °C than at 22 °C (Figure 2(b) and 2(c)). In the raw water, the proportion of culturable bacteria at the two incubation temperatures was comparable with 0.361% (at 22 °C) and 0.351% (at 36 °C). Based on flow cytometry and culture data, biologically stabilized water after GAC (=water quality III) contained 0.8 log units less bacteria than the raw water based on ICC data, but 3 log units less bacteria based on colony counts. Judgment of treatment efficiency, therefore, greatly depended on the chosen diagnostic method. Given the strong variation of the proportion of culturable bacteria over the treatment process, cultivation-independent quantification by flow cytometry gave a better overview of overall changes in bacterial numbers.

Distinction between membrane integrity breach and regrowth

Especially in the first weeks of the operation of the pilot plant, treatment suffered from technical process interruptions. During that time, bacteria were found in significant numbers in the UF filtrate leading to the question about their origin. Although a retention of *Pseudomonas diminuta* of 9.7 log units was measured for the type of UF membrane used (according to the ASTM standard F838-05, data not shown), a breach could not be excluded. Apart from mere quantification, offline flow cytometry proved useful in this respect. Bacteria in UF filtrate predominantly belonged to the high nucleic acid (HNA) population (data not shown). Low nucleic acid (LNA) bacteria as abundant in UF feed water, on the other hand, were absent in the filtrate. UF filtrates from modules with impaired integrity, on the other hand, had been reported to contain primarily LNA bacteria that have a higher probability to pass through due to their smaller size (Buysschaert *et al.* 2018). In fact, HNA bacteria can already be separated from LNA bacteria by means of a 0.4 μm filter (Proctor *et al.* 2018). The pore size of the UF

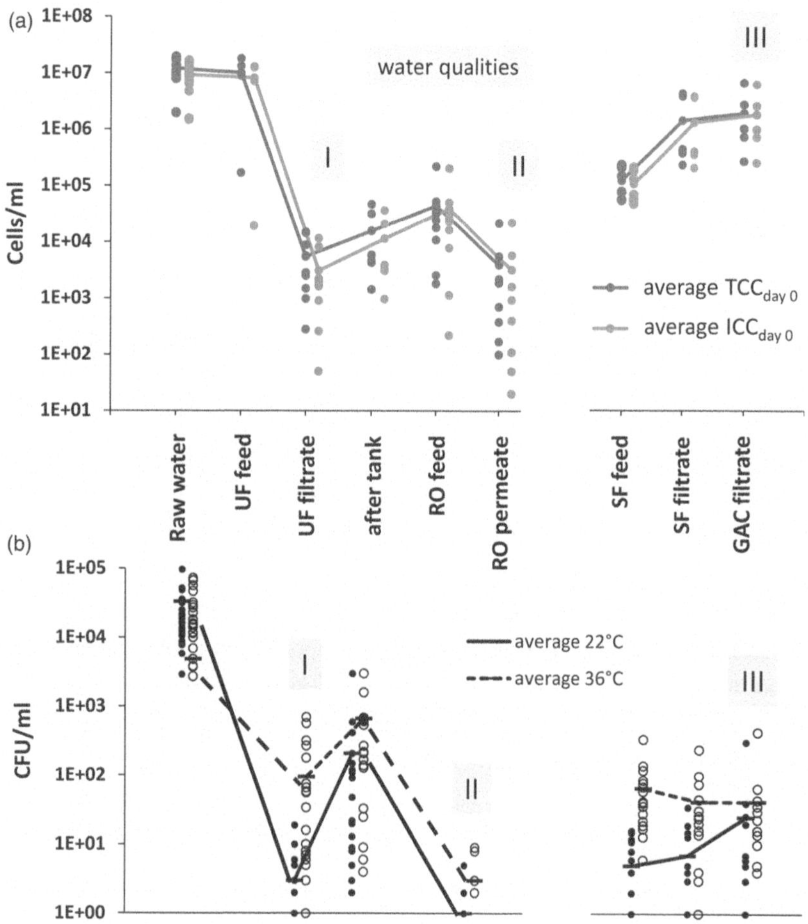

(c)

% culturable	Raw water	UF filtrate	after tank	RO permeate	SF feed	SF filtrate	GAC filtrate
22°C	0.361	0.098	1.853	0.031	0.005	0.001	0.002
36°C	0.351	3.103	5.920	0.094	0.069	0.004	0.003

Figure 2 | Changes of (a) total cells, (b) culturable bacteria and (c) proportion of culturable bacteria at different steps of water treatment. (a) Total and intact cell concentrations at the time point of sampling (day 0) as determined by flow cytometry. (b) Colony counts at 22 and 36 °C as determined by cultivation. Samples were transported cold to the laboratory and analyzed within 24 h so that the concentrations reflect the microbiological *status quo* of day 0. Individual dots show individual sampling events, lines combine arithmetic mean values. The line is interrupted as the sand filter is fed with UF filtrate. (c) Average relative proportions of culturable bacteria based on comparison of A and B. TCC, total cell concentrations; ICC, intact cell concentrations; CFU, colony-forming units; UF, ultrafiltration; RO, reverse osmosis; SF, quartz sand filter; GAC, granular activated carbon.

membrane used here was 0.02 μm. The observation that UF filtrates in our study were dominated by HNA bacteria made us exclude the possibility of a breach in membrane integrity and assign their presence to regrowth during water stagnation. This view was supported by the finding that the composition of the bacterial community in the filtrate was profoundly different from the one in the feed water (see section Changes of the microbiome along the treatment process). Bacterial concentrations in our study tended to be elevated when residence time in filtrate compartments was high or when the filtration process was intermittently interrupted leaving time for regrowth. This also applied to the intermittent presence of small numbers of bacteria in the RO permeate, which is in line with the growth potential on the permeate side reported in earlier studies (Park & Hu 2010). As for UF, we hypothesize the bacteria that colonize the surfaces of the filtrate or permeate compartment to be the source of bacteria in the bulk water.

Changes of the microbiome along the treatment process

Changes in microbial community composition were monitored using Illumina sequencing of 16S rRNA gene amplicons. Great microbial diversity (Shannon index: $H' = 5.96$; Simpson index: $D = 0.99$) was observed for the raw water with an observed richness of 1,068 unique OTUs belonging to 34 different detected bacterial phyla (Figure 3(a)). Diversity ($H' = 2.23$; $D = 0.77$) and observed richness (186 OTUs) were strongly reduced by UF. The resulting bacterial community was very distinct from the raw water and suggested *Pseudomonas* to be among the dominant genera (Supplementary Figure S1). As members of *Pseudomonas* spp. are well known to be sessile and to form biofilms (Tolker-Nielsen *et al.* 2000), the result might reflect enrichment of the filtered and thus bacteria-stripped water by new bacteria attached to the surfaces on

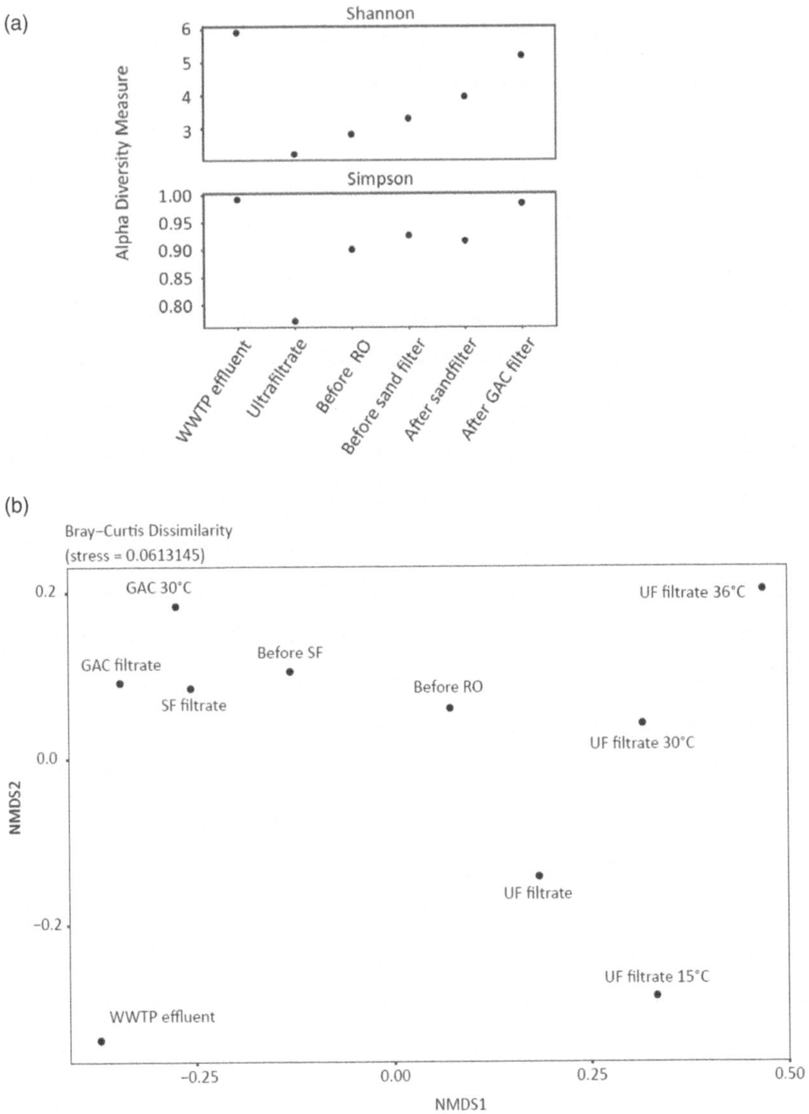

Figure 3 | Relative changes of the bacterial microbiome along the treatment process based on 16S rRNA gene sequencing by Illumina. (a) Change of Shannon and Simpson alpha diversity indices at different points of the treatment train. (b) Bray-Curtis dissimilarity index for water directly after different treatment steps and in part after regrowth. UF filtrate was incubated at 15, 30 or 36 °C for 7 days, GAC filtrate was incubated at 30 °C. No results were obtained for the RO permeate due to the very low numbers of bacteria and insufficient DNA concentrations obtained from this sample. GAC, granular activated carbon; RO, reverse osmosis; SF, quartz sand filtration; UF, ultrafiltration.

the filtrate side. The bacterial community underwent further change within the UF filtrate tank and on its way to RO, consistent with a moderate increase in cell numbers (Figure 2). In case of the sand filter and GAC passage of UF filtrate, diversity and species richness (407 and 600 OTUs after sand filter and GAC, respectively) increased significantly in agreement with biological stabilization. Overall, the bacterial microbiome was shown to be highly dynamic and to undergo various substantial changes along the treatment train reflected in a Bray-Curtis dissimilarity plot (Figure 3(b); heat map shown in Supplementary Figure S1). The bacterial communities in water qualities I, II and III appeared highly distinct from the one of the raw water.

The bacterial community can undergo further change during stagnation as shown for UF filtrate incubated over 7 days at either 15, 30 or 37 °C. The temperature of incubation strongly influenced the community composition of stagnated UF filtrate (heat map shown in Supplementary Figure S2). Stagnation of GAC filtrate led to less dissimilarity probably because the original sample was closer to biological stability.

It has to be pointed out that sequence analysis was only performed on a single set of samples, so variations over the duration of the demonstration phase cannot be excluded.

Process monitoring by online flow cytometry

Online flow cytometry was used to monitor ICC in high resolution with data points generated every 2 h. ICC in August and September 2019 are shown as an example. In this time period, the effect of monochloramine was tested to suppress biofouling on the RO membrane. For this purpose, monochloramine was added to UF-1 filtrate (prior to the filtrate storage tank) to obtain a final concentration near 1 mg/L, whereas UF-2 filtrate did not receive disinfectant (Figure 4).

Monochloramine was initially dosed nearly continuously over 22 h/day. Dosing time was subsequently gradually reduced with the intention to define the necessary dosing time interval that still achieves a microbiological effect. ICC values were around 2 log units lower for UF-1 filtrate compared to UF-2 demonstrating the effect of monochloramine on the integrity of bacteria contained in the water. Reduction of dosing times to 19, 16 and 13 h/day still achieved a comparable disinfection effect whereas with 10 h/day a risk for rising bacterial concentrations was given. Further reduction of the dosing time to 7 h/day led to an increase in ICC values. As a moderate ICC increase was also seen for UF-2, part of the increase in ICC could also have been caused by changes in the raw water quality with the nutrient content of treated wastewater effluent typically being unknown. Independent of the precise cause, a monochloramine dosing time below 13 h/day was seen unable to suppress such scenarios. Overall, the data resolution of online flow cytometry and the ability to compare two parallel processes proved very useful for process control as it enabled to directly measure the biological consequences of changing process parameters or natural water quality alterations. Like for every flow cytometer, regular

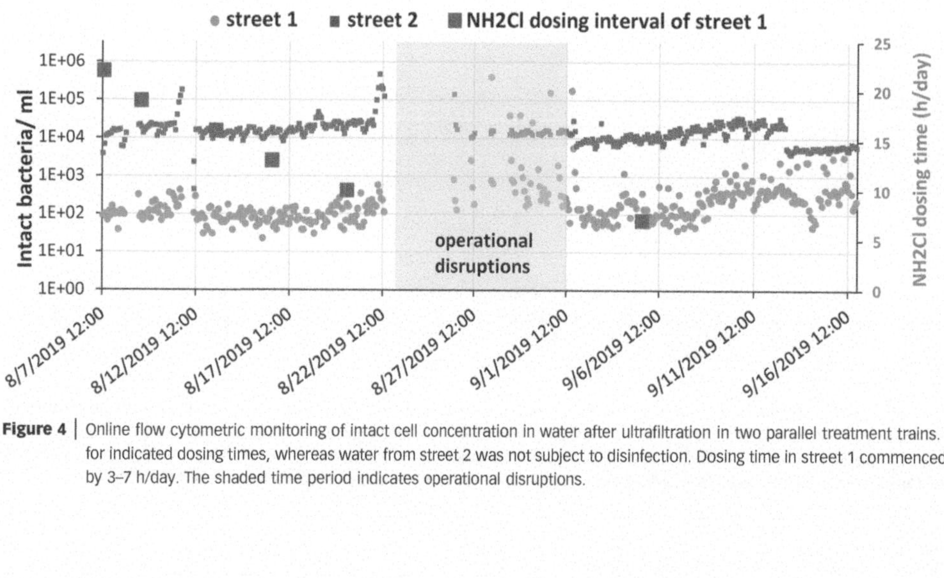

Figure 4 | Online flow cytometric monitoring of intact cell concentration in water after ultrafiltration in two parallel treatment trains. Water from street 1 received monochloramine for indicated dosing times, whereas water from street 2 was not subject to disinfection. Dosing time in street 1 commenced with 22 h/day and was gradually reduced in steps by 3–7 h/day. The shaded time period indicates operational disruptions.

optimized clean intervals were necessary to prevent overestimation of bacteria due to internal fouling events and to ensure correctness of data. The greatly improved process control does, of course, not eliminate the need for traditional hygienic testing as flow cytometry results as applied here do not give information about the species composition of the water.

Regrowth potential after different treatment steps

Flow cytometric day 0 values of water samples reflecting the microbiological status at the time point of sampling were supplemented with so-called day 7 values. The latter was generated by re-measurement of samples that were stored for 7 days at 22 °C. The underlying idea was that in the case that AOC and other nutrients are present in the water, the autochthonous bacteria can propagate and recover to a cell concentration that is supported by the given AOC/nutrient level. Day 7 values should, thus, represent the full regrowth potential of the water (Figure 5) and are shown together with average TCC and ICC values measured at day 0 (reproduced from identical average values from Figure 2).

As often observed for raw waters, bacterial $TCC_{day\ 7}$ and $ICC_{day\ 7}$ concentrations in treated WWTP effluent moderately declined over the incubation time of 7 days. This decline owes itself to the nearly complete conversion of LNA into HNA bacteria. As the total carbon pool remains identical, it supports less HNA bacteria (which tend to be bigger in size) than the total number of LNA and HNA bacteria that were present in the original sample. In other words, the pool of small LNA bacteria converts into a smaller number of bigger HNA bacteria. The same effect could be seen in UF feed samples that were nearly identical with raw water.

Strong deviations between day 0 and day 7 cell numbers were observed in the subsequent treatment train. Whereas flow cytometry of day 0 samples showed a strong reduction of bacterial concentrations following UF and RO filtration (Figure 2(a) and reproduced in Figure 5), the decline was much less pronounced in stagnated day 7 samples (Figure 5). Regrowth was not surprising as the downstream filter compartments were not sterile. Compared to raw water

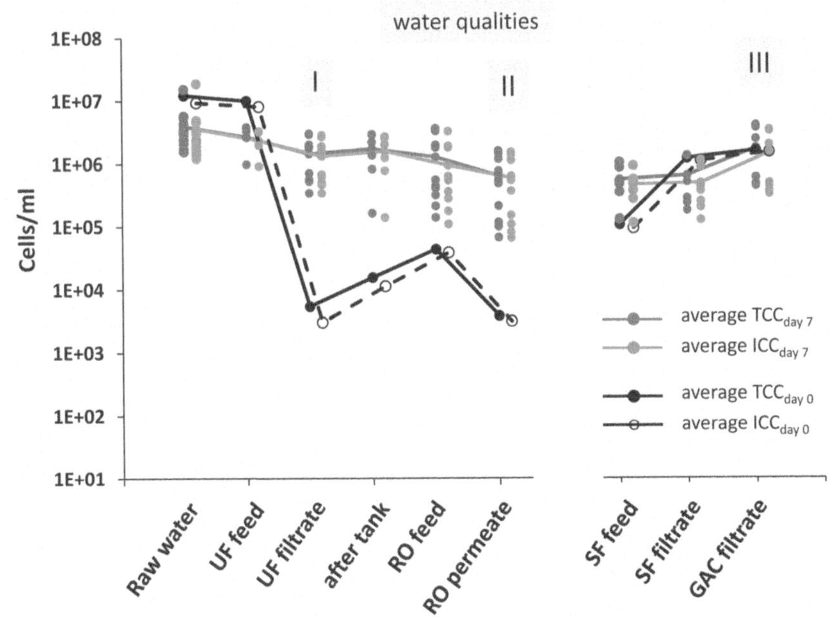

Figure 5 | Change of (a) total and intact cell concentrations after 7 days of incubation at 22 °C over the treatment train. $TCC_{day\ 7}$ and $ICC_{day\ 7}$ values were determined by flow cytometry and represent the maximal total and intact cell concentrations that are supported by the nutrients contained in the corresponding water. Individual dots show individual sampling events, lines combine arithmetic mean values. The line is interrupted as the sand filter is fed with UF filtrate. Average $TCC_{day\ 0}$ and $ICC_{day\ 0}$ values (identical with Figure 2) are shown for easy comparison of day 7 and day 0 values. TCC, total cell concentrations; ICC, intact cell concentrations; UF, ultrafiltration; RO, reverse osmosis; SF, quartz sand filter; GAC, granular activated carbon.

(stagnated under the same conditions), intact cells ($ICC_{day\ 7}$) only declined by 0.43 and 0.63 log units in regrown UF filtrate and regrown RO permeate, respectively.

Strong regrowth as seen for UF filtrate and RO permeate can imply the risk of excessive biofilm formation, biofouling, clogging of tubes and nozzles (Prest *et al.* 2016). An in-line candle cartridge pre-filter installed to reduce fouling potential of the subsequent RO membrane needed to be exchanged regularly due to visible biofilm formation and increased filter resistance (data not shown). During the time when monochloramine was dosed, the filter installed in street 2 (that did not receive disinfectant) was much more prone to biofouling and needed to be exchanged more often than the filter in street 1 that was dosed with monochloramine. Interestingly, reducing disinfectant dosing time to 10 h and especially to 7 h shortened the lifetime of the cartridge filter, which was in line with online flow cytometry results.

Comparison of $ICC_{day\ 7}$ values with traditional AOC

As bacterial concentrations are limited by AOC, the data suggest AOC removal or regrowth potential removal in the given ranges. DOC in comparison was on average reduced by approximately 17% (10.7 mg/L) and 96% (<0.5 mg/L) after UF and RO, respectively, compared to the average DOC concentration in raw water (12.8 mg/L). To address the comparability of the $ICC_{day\ 7}$ values with traditional AOC assessment (van der Kooij 1992), both parameters were compared for a selected set of identical samples on one particular day. As the two parameters have different scales, their changes relative to the values from raw water (100%) are shown as percentages together with the absolute values (Table 2). Whereas UF led to a reduction of both parameters in samples taken prior to a storage tank, values increase in samples after that tank. As the water after UF was supplemented with chloramine at the time point when samples were taken and as tank surfaces with biofilm can enrich the water with nutrients, an increase in regrowth potential (after neutralization of monochloramine) along tank passage was conceivable. In RO permeate, traditional AOC was negligible with 0.08 µg ac-C eq/L, whereas $ICC_{day\ 7}$ values were still 6% of those in stagnated raw water. Both parameters increased after GAC passage.

Despite the very low AOC value in RO permeate, the sample contained sufficient nutrients to support regrowth to a concentration of 1.1×10^5 intact bacteria/ml corresponding to approximately 6% relative to the $ICC_{day\ 7}$ value of raw water on that day. This example illustrates that traditional AOC assessment has a detection limit. Whereas the parameter is useful at measurable concentrations, AOC values near zero on a linear scale can still translate to a substantial number of bacteria on a log scale. Even under extremely oligotrophic conditions bacteria can reach concentrations in the range of 10^5–10^6 cells/ml (Egli 2010). It has to be acknowledged in this context that the traditional AOC assay is optimized for potable water. To improve the assay's suitability for reuse applications, the two bacteria *Pseudomonas fluorescens* P17 and *Spirillum* sp. NOX forming the basis of the drinking

Table 2 | Comparison of ICC values obtained after 7 days of incubation at 22 °C ($ICC_{day\ 7}$) and AOC values obtained with the traditional method according to van der Kooij (1992)

	$ICC_{day\ 7}$		AOC (van der Kooij)		
	Intact bacteria/ml	Value relative to raw water (%)	µg ac-C eq/L	Value relative to raw water (%)	DOC* mg/L
Raw water	1.9×10^6		358		12.8
UF filtrate prior tank	1.5×10^6	79	174	49	10.7
UF filtrate after tank	2.7×10^6	142	213	60	n.d.
RO permeate	1.1×10^5	6	0	0	<0.5
GAC filtrate	5.0×10^5	26	180	50	8.7

*The given DOC values represent average values over the entire demonstration phase and give an estimate over the change along the treatment process. They cannot be directly compared to the indicated $ICC_{day\ 7}$ values or AOC that were determined on a specific day for a direct comparison of those two parameters.

Changes of absolute values over the treatment process are shown for one specific sampling day together with relative changes in comparison with the raw water on that day. Average DOC values during the entire demonstration phase are shown as a reference.

water assay have been supplemented by Zhao *et al.* (2013) with three additional strains together with other modifications such as increased inoculums densities and higher incubation temperatures (25 °C). Although this modified AOC assay has not been applied in our study, the required long starvation times and the fact that even strains with the ability to biodegrade a relatively wide range of organic compounds still have a limited metabolic spectrum (that might not be universally adequate for every reclaimed water at every treatment step) make us consider the presented ICC_{day7} approach as a good alternative. Ideally, the size of the bacteria might be considered in the future as the carbon footprint depends on the size. The $ICC_{day\ 7}$ values represent the maximal bacterial concentrations supported by the given AOC. The metabolic spectrum of the complex autochthonous bacterial population can be assumed to be well adapted to the specific water type. Representation of the regrowth potential as an $ICC_{day\ 7}$ value on a log scale ensures high sensitivity in a range that is not covered by the traditional AOC assay due to its relatively high detection limit. The example shows that a traditional AOC value near 0 μg ac-C eq/L could theoretically translate into any ICC over 5 log units. In other words, a log scale showing actual bacterial concentrations provides more sensitivity in the low nutrient range than a linear scale showing traditional AOC.

Nutrient sources and regrowth control

Whereas the presence of AOC required to support regrowth was not surprising for UF filtrate as nutrients are not retained by UF, the finding of regrowth potential in RO permeate with the $ICC_{day\ 7}$ value in the range between 10^5 and 10^6 cells/ml raised the question about the origin of the nutrients. Average DOC levels <0.5 mg/L in RO permeate were comparable with RO-treated drinking water (Buysschaert *et al.* 2018). Passage of cationic and neutral hydrophilic organic molecules of small size had been reported especially under low flux conditions (Albergamo *et al.* 2019). Another possible source of nutrients might be migration or leaching of assimilable nutrients from polymeric pipe materials into permeate water (Wen *et al.* 2015). To control possible regrowth, the addition of a disinfectant and maintaining an appropriate residual concentration has been proposed (Farhat *et al.* 2018). In our study, regrowth of the autochthonous bacterial flora in UF filtrate could be suppressed for at least 5 days and for ≥14 days by addition of 1 or 2 mg/L free chlorine, respectively (data not shown). If UF filtrate was further processed by RO, monochloramine was preferred due to its lesser reactivity. When dosed at approximately 1 mg/L into UF filtrate, monochloramine in RO feed was around 0.1 mg/L which did not incur measurable damage to the RO membrane. The membrane was found to be partially permeable for monochloramine with remaining residual being able to suppress regrowth for 7 days.

Overall, the necessity for regrowth control with a residual has to be decided on a case-to-case level as bacterial concentrations around 10^6 cells/ml are not uncommon also in stagnated potable water in house plumbing systems (data not shown). Typically, no problems are associated with such systems. Where regrowth control does, however, become relevant is if hygienically relevant bacteria can multiply.

Hygienic challenges

Concerns related to pathogen growth in reclaimed water have been expressed in earlier studies (Jjemba *et al.* 2010; Weinrich *et al.* 2010). Reclaimed water from treated effluent was reported to be susceptible to regrowth of opportunistic pathogens like *Aeromonas*, *Legionella*, *Mycobacterium* and *Pseudomonas* (Jjemba *et al.* 2010) adding to a relatively high biofilm formation potential and problems from smell and color (Oesterholt *et al.* 2007). The major culprit for these problems is seen in high concentrations of biodegradable organic carbon as the source of regrowth (Weinrich *et al.* 2010), but also water after extensive treatment was susceptible. Learbuch *et al.* (2019) demonstrated growth potential of *Legionella pneumophila* (concomitant with biofilm growth) in remineralized RO permeate as a result of PE or PVC-P materials used in the permeate compartment. Also in our study, growth of *L. pneumophila* artificially introduced in low numbers into aliquoted UF filtrate and RO permeate was obtained, although not on a reproducible basis. Uncontrolled day-to-day variations in the nutrient content and composition released from the WWTP, presence and absence of biofilm particles or amoeba and other factors might have played a role. The possible regrowth potential of pathogens in different water qualities will need further investigation but warrants the addition of disinfectant with a residual for safety reasons.

Another way to minimize the risk for pathogen regrowth might be obtained by biological stabilization. Least regrowth was obtained after biological stabilization of water (Figure 5) as water passage through the sand and GAC filter resulted in bacterial enrichment with a highly diverse population. It remains to be addressed whether the small regrowth potential applies also to pathogens. GAC filters with an established microflora were shown to have a 'protective' effect in regard to filter colonization by bacterial pathogens such as *Salmonella typhimurium*, *Yersinia enterocolitica* and enterotoxigenic *E. coli*, whereas sterile GAC filters became rapidly colonized by these pathogens (Camper *et al.* 1985). In other words, competition with the indigenous microbial community can be antagonistic to the growth of pathogens in the water. For drinking water treatment, biological stabilization has been proposed to shape the bacterial community composition, which is subsequently only moderately modified during water distribution (Pinto *et al.* 2012; Prest *et al.* 2016). This is in line with our finding that stagnation of GAC filtrate resulted in less dissimilarity than stagnation of UF filtrate (Figure 3(b)).

CONCLUSIONS

Microbiological monitoring of the given treatment process demonstrated an efficient removal of hygienically relevant bacteria with ultrafiltration being an efficient microbiological barrier. The latter predestined UF as a critical control point for high-resolution process monitoring by online flow cytometry. The bacterial microbiome was exchanged multiple times with the most profound changes induced by membrane filtration. Despite profound changes in bacterial concentrations along with treatment, the reduction was not sustainable as the regrowth potential (as determined by offline flow cytometry) was only reduced moderately despite a strong reduction of DOC. Potential growth of pathogens is seen as the greatest hygienic risk and might warrant a disinfectant residual if water is at risk to stagnate for extended times.

ACKNOWLEDGEMENTS

Kathrin Wiede and Dietmar Pütz from IWW are gratefully acknowledged for sample preparation and technical analysis as well as for help in data analysis. We extend our thanks to Mrs Katharine Mende from De.EnCon GmbH for engineering assistance, a number of staff members and students from OOWV who were involved in operating the pilot plant and other members of the consortium for critical discussions. We are further very grateful to the German Ministry of Education and Research (BMBF) that financed this study as part of the MULTI-ReUse (Modular Water Treatment and Monitoring for Wastewater Reuse, 2016–2019) project under grant no. 02WAV1403A ('Future-oriented Technologies and Concepts to Increase Water Availability by Water Reuse and Desalination').

DATA AVAILABILITY STATEMENT

All relevant data are included in the paper or its Supplementary Information.

REFERENCES

Albergamo, V., Blankert, B., Cornelissen, E. R., Hofs, B., Knibbe, W.-J., van der Meer, W. & De Voogt, P. 2019 Removal of polar organic micropollutants by pilot-scale reverse osmosis drinking water treatment. *Water Research* **148**, 535–545.

Becker, D., Maaßen, S., Krömer, K., Nahrstedt, A., Schramm, E., Wencki, K., Zimmermann, B. & Zimmermann &, M. 2019 MULTI-ReUse: Zweckgebundene wasserwiederverwendung für industrie und weitere sektoren. *Korrespondenz Abwasser* **66**, 464–471.

Buysschaert, B., Vermijs, L., Naka, A., Boon, N. & De Gusseme, B. 2018 Online flow cytometric monitoring of microbial water quality in a full-scale water treatment plant. *npj Clean Water* **1**, 16.

Camper, A. K., LeChevallier, M. W., Broadway, S. C. & McFeters, G. A. 1985 Growth and persistence of pathogens on granular activated carbon. *Applied and Environmental Microbiology* **50**, 1178–1382.

Drewes, J. & Horstmeyer, N. 2015 Recent developments in potable water reuse. In: *The Handbook of Environmental Chemistry* (D. Barceló & A. G. Kostianoy, eds). Springer, Berlin-Heidelberg. Available from: https://link.springer.com/content/pdf/bfm%3A978-3-319-12907-5%2F1.pdf.

Drewes, J., Schramm, E., Ebert, B., Mohr, M., Beckett, M., Krömer, K. & Christina Jungfer, C. 2019 Potenziale und Strategien zur Überwindung von Hemmnissen für die Implementierung von Wasserwiederverwendungsansätzen in Deutschland. *KA Korrespondenz Abwasser, Abfall* **66**, 995–1003.

Egli, T. 2010 How to live at very low substrate concentration. *Water Research* **44** (17), 4826–4837.

Faour-Klingbeil, D. & Todd, E. C. D. 2018 The impact of climate change on raw and untreated wastewater use for agriculture, especially in arid regions: a review. *Foodborne Pathogens and Disease* **15**, 61–72.

Farhat, N., Hammes, F., Prest, E. & Vrouwenvelder, J. 2018 A uniform bacterial growth potential assay for different water types. *Water Research* **142**, 227–235.

Gatza, E., Hammes, F. & Prest, E. 2013 *Assessing Water Quality with BD C6 Flow Cytometer*. White Paper: BD Biosciences.

Gillespie, S., Lipphaus, P., Green, J., Parsons, S., Weir, P., Juskowiak, K., Jefferson, B., Jarvis, P. & Nocker, A. 2014 Assessing microbiological water quality in drinking water distribution systems with disinfectant residual using flow cytometry. *Water Research* **65**, 224–234.

Hull, N. M., Ling, F., Pinto, A. J., Albertsen, M., Jang, H. G., Hong, P. Y., Konstantinidis, K. T., LeChevallier, M., Colwell, R. R. & Liu, W. T. 2019 Drinking water microbiome project: is it time? *Trends in Microbiology* **27** (8), 670–677.

Jjemba, P. K., Weinrich, L. A., Cheng, W., Giraldo, E. & Lechevallier, M. W. 2010 Regrowth of potential opportunistic pathogens and algae in reclaimed-water distribution systems. *Applied Environmental Microbiology* **76**, 4169–4178.

Klindworth, A., Pruesse, E., Schweer, T., Peplies, J., Quast, C., Horn, M. & Glöckner, F. O. 2013 Evaluation of general 16S ribosomal RNA gene PCR primers for classical and next-generation sequencing-based diversity studies. *Nucleic Acids Research* **41**, e1.

Kozich, J. J., Westcott, S. L., Baxter, N. T., Highlander, S. K. & Schloss, P. D. 2013 Development of a dual-index sequencing strategy and curation pipeline for analyzing amplicon sequence data on the MiSeq Illumina sequencing platform. *Applied Environmental Microbiology* **79**, 5112–5120.

Learbuch, K. L. G., Lut, M. C., Liu, G., Smidt, H. & van der Wielen, P. W. J. J. 2019 *Legionella* growth potential of drinking water produced by a reverse osmosis pilot plant. *Water Research* **157**, 55–63.

McMurdie, P. J. & Holmes, S. 2013 phyloseq: an R package for reproducible interactive analysis and graphics of microbiome census data. *PLoS ONE* **8** (4), e61217.

Nahrstedt, A., Gaba, A., Zimmermann, B., Jentzsch, T., Krömer, K., Tiemann, Y., Harsanyi, L., Buchta, P., Dölchow, U., Lipnizki, J., Mende, K., Koch, T. & Rohn, A. 2020 Reuse of municipal wastewater for different purposes based on a modular treatment concept. *Journal of Water Reuse & Desalination* **10** (4), 301–316. https://doi.org/10.2166/wrd.2020.040.

Oesterholt, F., Medema, G., van der Kooij, D. & Martijnse, G. 2007 Health risk assessment of non-potable domestic water supplies in the Netherlands. *Journal of Water Supply: Research and Technology – AQUA* **56**, 171–179.

Ormerod, K. J. 2016 Illuminating elimination: public perception and the production of potable water reuse. *WIRES Water* **3**, 537–547.

Park, S. K. & Hu, J. Y. 2010 Assessment of the extent of bacterial growth in reverse osmosis system for improving drinking water quality. *Journal of Environmental Science Health, Part A: Toxic/Hazardous Substances and Environmental Engineering* **45** (8), 968–977.

Pinto, A. J., Xi, C. & Raskin, I. 2012 Bacterial community structure in the drinking water microbiome is governed by filtration processes. *Environmental Science & Technology* **46**, 8851–8859.

Prest, E. I., El-Chakhtoura, J., Hammes, F., Saikaly, P. E., van Loosdrecht, M. C. M. & Vrouwenvelder, J. S. 2014 Combining flow cytometry and 16S rRNA gene pyrosequencing: a promising approach for drinking water monitoring and characterization. *Water Research* **63**, 179–189.

Prest, E. I., Hammes, F., van Loosdrecht, M. C. M. & Vrouwenvelder, J. S. 2016 Biological stability of drinking water: controlling factors, methods, and challenges. *Frontiers in Microbiology* **7**, 45.

Proctor, C. R., Besmer, M. D., Langenegger, T., Beck, K., Walser, J.-C., Ackermann, M., Bürgmann, H. & Hammes, F. 2018 Phylogenetic clustering of small low nucleic acid-content bacteria across diverse freshwater ecosystems. *ISME Journal* **12**, 1344–1359.

Schloss, P. D., Westcott, S. L., Ryabin, T., Hall, J. R., Hartmann, M., Hollister, E. B., Lesniewski, R. A., Oakley, B. B., Parks, D. H., Robinson, C. J., Sahl, J. W., Stres, B., Thallinger, G. G., van Horn, D. J. & Weber, C. F. 2009 Introducing mothur: open-source, platform-independent, community-supported software for describing and comparing microbial communities. *Applied Environmental Microbiology* **75**, 7537–7541.

Schmid, S. & Bogner, F. X. 2018 What Germany's university beginners think about water reuse. *Water* **10**, 731.

Tolker-Nielsen, T., Brinch, U. C., Ragas, P. C., Andersen, J. B., Jacobsen, C. S. & Molin, S. 2000 Development and dynamics of *Pseudomonas* sp. biofilms. *Journal of Bacteriology* **182**, 6482–6489.

van der Kooij, D. 1992 Assimilable organic carbon as an indicator of bacterial regrowth. *Journal American Water Works Association* **84** (2), 57–65.

Van Nevel, S., Koetzsch, S., Proctor, C. R., Besmer, M. D., Prest, E. I., Vrouwenvelder, J. S., Knezev, A., Boon, N. & Hammes, F. 2017 Flow cytometric bacterial cell counts challenge conventional heterotrophic plate counts for routine microbiological drinking water monitoring. *Water Research* **113**, 191–206.

Weinrich, L. A., Jjemba, P. K., Giraldo, E. & LeChevallier, M. W. 2010 Implications of organic carbon in the deterioration of water quality in reclaimed water distribution systems. *Water Research* **44**, 5367–5375.

Wen, G., Kötzsch, S., Vital, M., Egli, T. & Ma, J. 2015 Biomig – A method to evaluate the potential release of compounds from and the formation of biofilms on polymeric materials in contact with drinking water. *Environmental Science & Technology* **49** (19), 11659–11669.

Zhao, X., Hu, H., Liu, S., Jian, F., Shi, X., Li, M. & Xu, X. 2013 Improvement of the assimilable organic carbon (AOC) analytical method for reclaimed water. *Frontiers of Environmental Science & Engineering* **7**, 483–491.

First received 11 March 2020; accepted in revised form 20 June 2020. Available online 31 August 2020

In-situ fixed bed denitrification in sequential biofiltration: laboratory testing of solid substrates

Josefine Filter, Christin Bosinsky, Sefine Oksal Kilinc,
Aki Sebastian Ruhl IWA and Martin Jekel

ABSTRACT

High nitrate concentrations in wastewater treatment plant effluents and aquifers can challenge sequential biofiltration systems in preventing nitrite and gas formation in the sand bed, as well as to achieve the regulated limit value for nitrate in potable water reuse applications. This study investigates the introduction of electron donors in the form of organic fixed bed materials as an *in-situ* anoxic zone into sequential biofiltration systems. Laboratory batch and column tests with straw, soft wood, peat, polylactic acid (PLA), and polycaprolacton (PCL) revealed incomplete denitrification with a hydraulic retention time below 10 h, high organic carbon leaching, especially during the first three months, and gas accumulation within the filter bed. Therefore, *ex-situ* denitrification prior to oxic biofilters or in a defined side-stream treatment is recommended. No enhanced transformation of trace organic chemicals was observed under nitrate reducing conditions. Peat revealed a sorption potential for 5-methyl-benzotriazole, carbamazepine, benzotriazole, and metoprolol.

Key words | DOM leaching, filling cost, fixed bed denitrification, gas clogging, sequential biofiltration, trace organic chemicals

Josefine Filter (corresponding author)
Christin Bosinsky
Sefine Oksal Kilinc
Aki Sebastian Ruhl IWA
Martin Jekel
Chair of Water Quality Control,
Technische Universität Berlin,
KF4, Straße des 17. Juni 135, 10623 Berlin,
Germany
E-mail: *josefine.filter@tu-berlin.de*

Aki Sebastian Ruhl
German Environment Agency,
Section II 3.1, Schichauweg 58, 12307 Berlin,
Germany

INTRODUCTION

Surface water used for potable water reclamation can be influenced by upstream discharges of waste water treatment plant (WWTP) effluents (Karakurt *et al.* 2019). Besides this de facto potable reuse, (semi-)closed water cycles can lead to the accumulation of poorly degradable or even persistent trace organic chemicals (TOrCs) in the aquatic environment. Managed aquifer recharge (MAR) systems allow a brine-free removal of pathogens and dissolved organic compounds like various TOrCs with low operational costs. Sequential biofiltration (SBF) systems or sequential managed aquifer recharge technologies (SMART) represent an advancement of conventional biofilters and MAR systems. They are characterized by two infiltration steps with an intermediate aeration or introduction of oxygen. While in the first filtration step, easily biodegradable DOC is removed associated with the consumption of oxygen, beneficial carbon limiting and oxic conditions are established in the second filtration step after aeration for enhanced TOrC removal. Those SBF systems can be realized in different forms. A pilot scale vertical column set up was tested and described by Müller *et al.* (2017). A further technical pilot application consists of a horizontal plug-flow reactor filled with technical sand and an *in-situ* introduction of oxygen (Karakurt-Fischer *et al.* 2020). Since these systems are promoting oxic conditions, contaminants such as nitrate, carbamazepine, or iodinated x-ray contrast agents which are predominantly removed or dehalogenated under anoxic and anaerobic conditions (König *et al.* 2016; Redeker

doi: 10.2166/wrd.2020.005

et al. 2018), might not be removed. This can be critical in potable reuse with a lack of dilution by natural waters.

The EU water framework directive seeks a good status of all waters by 2027 (The Council of the European Union 2015). Therefore, the limit of discharged total nitrogen into waterbodies sensitive to eutrophication is set below 15 mg/L for WWTPs exceeding 10,000 population equivalents (European Council 1991). The parametric value for drinking water is currently set at 50 mg/L as nitrate (11 mg NO_3^--N/L) (The Council of the European Union 2015). Especially in groundwater, critical nitrate concentrations resulting from excessive agriculture or decentralized wastewater treatment are becoming more and more challenging for drinking water supply. Meeting the required threshold values is essential to ensure a sufficient water quality in terms of nitrate concentrations for further indirect potable reuse options. A population-based cohort study of Schullehner *et al.* (2018) found statistically significant increased risks of colorectal cancer already at nitrate drinking water levels above 3.87 mg/L, which questions the adequacy of the current drinking water standard in regards to chronic effects.

In order to prevent the release of nitrate into aquifers, this study investigates possible options to integrate an anoxic zone as *in-situ* treatment as for a horizontal flow-through reactor (Figure S1, supporting information). Such a technical SBF system is characterized by a short HRT and limited filling volume. It is designed to treat WWTP effluents prior to aquifer recharge but also an application with surface water might be considerable. There are several possibilities to introduce electron donors. Continuous dosage of easily degradable organic carbon is a well-known technique to induce denitrification. However, it promotes biofilm growth and gas clogging in sand beds that cannot be backwashed. Another option is the application of H_2 as inorganic electron donor, but due to its low water solubility of only 1.7 mg/L (at 20 °C and normal pressure of 1.013 bar) it seems inapplicable for removing nitrate concentrations exceeding 16 mg/L in a non-pressurized system (assuming a demand of 0.106 mg H_2 per mg NO_3^-). Furthermore, the safe handling of gaseous hydrogen would require additional technical efforts, e.g. membrane contactors or exhaust air treatment.

Organic fixed bed materials appear promising for promoting denitrification in sequential biofiltration systems, since the release of electron-rich organic compounds, biomass growth, and gas formation could be limited to a defined zone. Furthermore, this approach might decrease operational effort and costs compared to a nitrate removal with membrane technologies or ion exchangers. Solid phase denitrification is broadly reviewed by Wang & Chu (2016), who identified woodchips as the most popular and competitive natural plant-like used for denitrification. Further plant-based materials for denitrification have already been evaluated in laboratory as well as in pilot studies (Gibert *et al.* 2008; Warneke *et al.* 2011). Schaffer *et al.* (2015) tested compost in a laboratory set up as the denitrifying layer in SAT applications. Besides natural substrates, synthetic but biodegradable polymers such as polycaprolacton (PCL), polylactic acid (PLA), polyhydroxyalkanoate (PHA), and polybutylene succinate (PBS) were investigated as electron and carbon sources for fixed bed denitrification (Boley *et al.* 2000; Shen & Wang 2011; Chu & Wang 2013).

In this study, natural as well as synthetic materials were tested as electron donors for *in-situ* denitrification in a SBF system. Batch experiments were conducted for a first evaluation of the different materials within a month, as reported by Gibert *et al.* (2008). A parallel set of laboratory columns was installed to simulate an anoxic filtration zone in a SBF system. Although there are plenty of studies available on TOrC degradation under anoxic conditions, to the best of our knowledge, there is hardly any information available on TOrC removal during fixed bed denitrification. Therefore, besides monitoring nitrate removal and dissolved organic carbon (DOC) release, this study also investigates the removal of selected TOrCs.

MATERIALS AND METHODS

Materials

Straw, pine woodchips (soft wood), peat, PCL, and PLA were applied as fixed bed materials. Sand from a MAR site located at Lake Tegel in Berlin (Germany) was used as a reference filling material. Detailed specifications of the substrates are given in Table S1 in the supporting information.

Batch experiments

Batch tests were carried out according to the standard DIN EN ISO 7827. A prepared solution was purged with nitrogen in order to reduce the dissolved oxygen concentration below 1 mg/L. Amounts of 0.25, 0.50, and 1.00 g of the respective fixed bed materials were mixed with 100 mL of the solution. To provide sufficient microorganisms, 15 mg TSS/L of washed activated sludge from a municipal WWTP (with biological phosphor removal, denitrification and nitrification described by Streicher *et al.* (2016)) was added. In addition to the fixed bed materials, H_2 was also tested as electron donor. Here, the prepared batches were purged with H_2 for 10 minutes at the beginning and additional H_2 was introduced during each sampling. The bottles were shaken at 120 rpm in the dark for 30 days. Samples were taken at day 5, 10, 20, and 30 under anoxic conditions by purging nitrogen or respectively hydrogen with hollow needles into the closed batches. After 30 days of incubation under anoxic conditions, batches were unsealed and shaken for another 24 h to investigate if backwards transformation of TOrCs into the parent compound occurs under oxic conditions.

Long-term column experiments

Prior to the column experiments, all materials were exposed to WWTP effluent for 48 h for inoculation with microorganisms. All materials were packed into glass columns (35 mm inner diameter, 200 mm length) which were operated in the dark and upstream with a mean flow rate of 19.0 ± 0.6 mL/h (396 individual measurements ($n = 396$)) leading to hydraulic retention times of less than 10 hours (Table 1).

Table 1 | Operational details of the packed columns

Material	Reference (Sand)	Rye straw	Soft wood	Peat	PCL	PLA
Applied dry matter (g)	323	30	20	49	153	177
Packing density (kg/L)	1.68	0.16	0.16	0.26	0.79	0.92
HRT (h)	9.0	7.8	8.8	9.0	2.9	5.3

The experiment was performed for almost 200 days at a mean temperature of 22 ± 1.4 °C ($n = 64$). The influent consisted of Berlin tap water containing a non-biodegradable DOC of 5.5 ± 0.4 mg/L ($n = 55$) which was spiked with sodium nitrate to a mean nitrate concentration of 46.7 ± 9.9 mg/L ($n = 26$). Accordingly, the applied test water revealed a low ratio of bio-degradable DOC (BDOC) to nitrogen (N). Oxygen concentrations were adjusted below 2 mg/L by aeration with nitrogen. Furthermore, a defined volume of a TOrC stock solution containing 10 mg/L of acesulfame (artificial sweetener), benzotriazole (corrosion inhibitor and antifreeze) and 5-methyl-benzotriazole (transformation product of benzotriazole), diclofenac (analgesic), carbamazepine and gabapentin (antiepileptic), iopromide and iomeprol (x-ray contrast agents), formylaminoantipyrine (metabolite of analgetic metamizol), and sulfamethoxazole (antibiotic) was added to the influent to adjust TOrC concentrations of 1 μg/L each. Along the column, sampling ports were installed at 7.8 and 15.6 cm height. The effluent passed a gas capture bottle to quantify released nitrogen. A scheme of the experimental setup is presented in Figure 1.

Analyses

Samples were taken under anoxic conditions and filtrated with 0.45 μm membranes prior to further analyses. Oxygen was quantified with an OXY-4 mini fiber-optic oxygen meter (PreSens, Precision Sensing GmbH, Germany). The DOC was measured with a TOC Cube analyzer (Elementar, Germany) and further characterized with liquid chromatography and continuous organic carbon detection (LC-OCD) (Huber *et al.* 2011). The selected TOrCs were quantified by high performance liquid chromatography coupled with

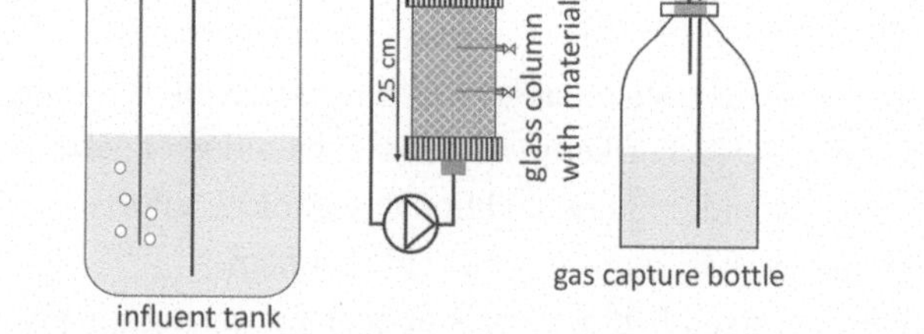

Figure 1 | Scheme of the column setup.

tandem mass spectrometry (HPLC-MS/MS), as described by Hellauer *et al.* (2017) in detail.

Data processing

The mean nitrate removal rate R_N (mg N/(L·d)) was calculated according to Equation (1) and the theoretical gas formation rate R_{N2} (mL/d) was calculated according to Equation (2) assuming a complete reduction of nitrate to molecular nitrogen:

$$R_N = \frac{Q}{BV} \cdot \Delta N \tag{1}$$

$$R_{N2} = \frac{\Delta NO_3^- \cdot Q}{\delta_{N2}} \tag{2}$$

with: $\Delta N = c_{NO_3\text{-}N;out} - c_{NO_3\text{-}N;out} - c_{NO_2\text{-}N;out}$

Q volume flow (L/d)

BV bed volume of the column (L)

$c_{NO_3\text{-}N;in}$ nitrate concentration in the column influent (mg/L)

$c_{NO_3\text{-}N;out}$ nitrate concentration in the column effluent (mg/L)

$c_{NO_2\text{-}N;out}$ nitrite nitrogen calculated as nitrate (mg/L)

δ_{N2} density of molecular nitrogen at 20 °C (1.16 mg/mL).

The Student's t-test was carried out with unpaired samples with differing variance. The probability for a similar behavior of two samples is expressed by the p-value. A p-value smaller than 0.05 is considered as significant difference between the samples. To examine if the observed removal in batch tests is significant, the dataset of relative removals was tested against 0 (= no removal).

RESULTS AND DISCUSSION

Nitrate removal

The batch tests showed a nitrate removal within 30 days for PCL, straw, soft wood, peat, and H_2. The nitrate removal rate constant k correlates with the DOC released from the materials (Figure S2, supporting information).

In the column experiment, a mean nitrogen removal R_N of 4–21 mg N/(L·d) was calculated according to Equation (1) for the tested substrates, except for the reference sand and PLA columns, which showed no nitrate removal (Figure S3). As presented in Figure 2, the highest removals were achieved with rye straw (30 g filter material) and PCL (153 g), which was already indicated by the batch tests.

The relative effluent concentrations of nitrate and nitrite reveal an incomplete denitrification to nitrite with PCL during the first four months of operation. The residual oxygen in the influent water as well as the short hydraulic retention time of only 3 h might have provoked incomplete denitrification and nitrite formation. Furthermore, biofilm seems to develop, since the total nitrogen removal increases with time. After day 125, an almost complete nitrogen removal is achieved resulting in a removal rate of 23 ± 5 mg N/(L·d) ($n = 9$). In contrast, Chu & Wang (2013) achieved a high removal rate of 182 mg N/(L·d) in a PCL reactor at 20 °C and 6–9 h HRT with spiked groundwater after 300 days of operation. They recommend high temperature and HRT during the initial phase. However, this could be difficult to implement in an *in-situ* application. Furthermore, they conducted microbial analyses of the biofilm and found denitrifying processes in deeper layers of the biofilm, requiring a certain biofilm thickness.

In the column experiments, straw achieved the highest average nitrate removal rate of 21 mg N/(L·d) during the first four months of operation. The rates were calculated only from measurements up to day 128, since air was accidentally pumped into the column due to a broken pumping hose. However, Cameron & Schipper (2010) reported comparable values for wheat straw during the first ten months after start up but a significantly lower average nitrate removal of 7.8 mg N/(L·d) between 10 and 23 months, indicating a decline in nitrate removal over time. This decline was also observed with straw in this study. During the first month nitrate removal rates reached 29 mg N/(L·d) and steadily decreased to 14 mg N/(L·d). Reasons might be the decreasing pore volume and HRT due to gas entrapment as well as the washout of readily degradable BDOC during the first month.

The observed average nitrate removal for soft wood of 5 mg N/(L·d) during six months of operation corresponds to the reported value of 4.9 g N/m^{3}d by Cameron & Schipper (2010). The high removal rates reported by Ghane *et al.* (2015) can be explained by the carbon turn-over of

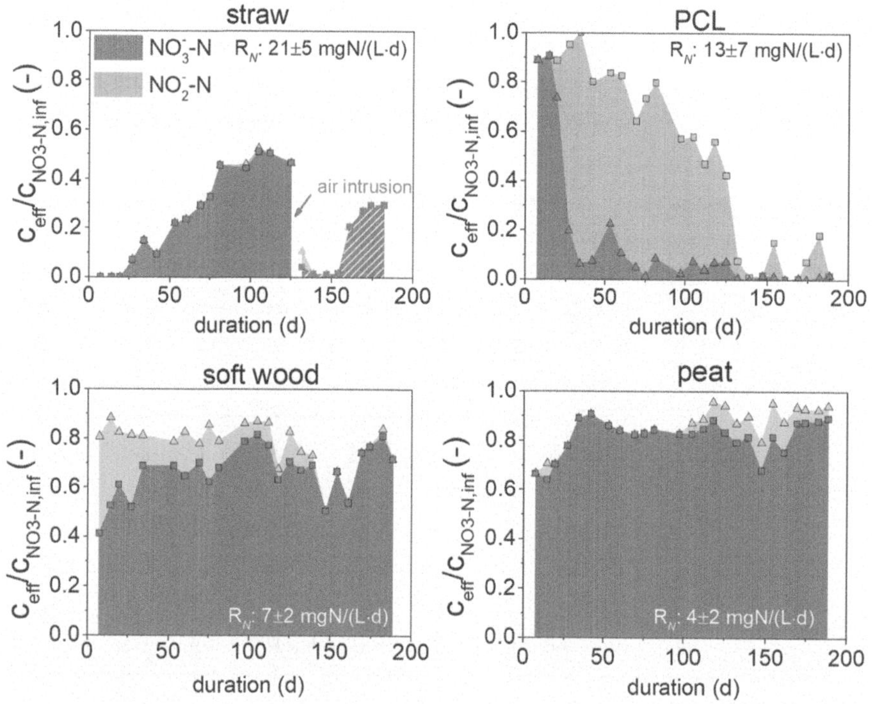

Figure 2 | Stacked effluent concentration of NO_3^--N and NO_2^--N relative to the NO_3^--N influent concentration. The mean volumetric nitrogen removal rate R_N with standard deviation was calculated between days 14 and 188 (only until day 128 for straw due to technical problems).

wastewater DOC. Therefore, soft wood seems to be a suitable fixed bed material for treating carbon-rich influents even under short HRT (Christianson *et al.* 2016). The results of Halaburka *et al.* (2017) suggest that denitrification in an aged woodchip bioreactor at constant temperature can effectively be modeled using even zero-order kinetics when nitrate concentrations are exceeding 2 mg N/L.

A direct comparison of the observed nitrate removal rates with literature data is not possible and always needs to be seen within the context of operational conditions, since HRTs are usually longer, temperatures differ, or the content of BDOC in the influent water is much higher.

Gas formation

The gas formation in the column experiments was quantified by measuring the weight loss of the replaced water in column and gas capture bottles. To transfer mass of replaced water, into a nitrogen equivalent volume, the mass was divided by 0.998 g/mL reflecting the density of water. The obtained gas formation rates correlate with the theoretical gas formation rates R_{N2} calculated from nitrogen removal, indicating that molecular nitrogen was the main gas formed during the experiment (Figure 3). Only a minor gas formation was observed for sand and PLA. In contrast, the gas retention capacity of the porous media was exceeded

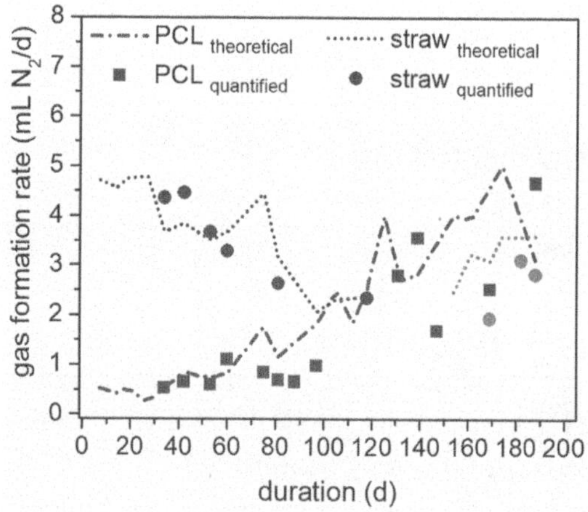

Figure 3 | Theoretical gas formation calculated from nitrogen removal compared to the gas formation quantified as replaced water in column and gas capture bottles for PCL and straw during the column experiment.

for straw, PCL, wood, and peat, as more than 70% of the formed gas was released from those columns into the gas capture bottles (Figure S4).

In a biofiltration system with horizontal or downstream flow conditions the gas retention capacity can be higher, since gas entrapments would not be 'flushed' out with the water stream as in the up-stream operated column tests. This might lead to a further decrease of the hydraulic conductivity, as reported by Cameron & Schipper (2010), especially for coarser grain size media.

DOC release

All materials that induced denitrification in the experiments also released DOC. Batch tests were able to indicate the amount of released DOC from the different substrates, especially for straw and peat (Figure 4(a)). A higher DOC release was observed for PCL and soft wood in the column tests compared to batch experiments.

In the column tests, especially PCL and straw showed an excessive release of DOC. While DOC release from straw mainly occurred within the first 10 days, DOC washout from PCL was observed for several weeks, as can be seen in Figure 4(b). LC-OCD analyses revealed a release of low molecular weight compounds from PCL (Figure S5) which were found to be biodegradable in batch tests. In contrast, Boley *et al.* (2000) reported only a slow DOC increase in the effluent of a PCL reactor. A high DOC release from wheat straw was also observed by Aslan & Türkman (2004). Although the DOC release from the PCL column decreased with time, denitrifying activity increased. This might be another indication for an incomplete biofilm development in the PCL column. In contrast to PCL, straw revealed a full nitrate removal during the first weeks, associated with the washout of DOC. Possibly, the DOC was more diverse and more easily accessible for microorganisms than the PCL polymers. Furthermore, straw as a natural material already might have contained microorganisms, which initiated denitrification after a short runtime.

Released DOC from peat was found to be mainly hardly biodegradable humic substances.

Removal of TOrCs

Data reported in the literature often show different observations for the removal of TOrCs under nitrate reducing conditions. Due to potential micro-milieus and insufficient resolution, it can be hard to distinguish between processes under micro-aerophilic and nitrate reducing conditions in field as well as in column studies; especially, since denitrifying bacteria are commonly facultative aerobes which can switch between aerobic and anoxic respiration (Madigan & Brock 2009).

Therefore, removal data from the batch experiments were evaluated to verify biotransformation under nitrate reducing conditions. To eliminate oxic transformation with residual oxygen, data from the first five days was excluded, as well as data when sulfate was reduced. The mean relative removal under nitrate reducing conditions, which can be

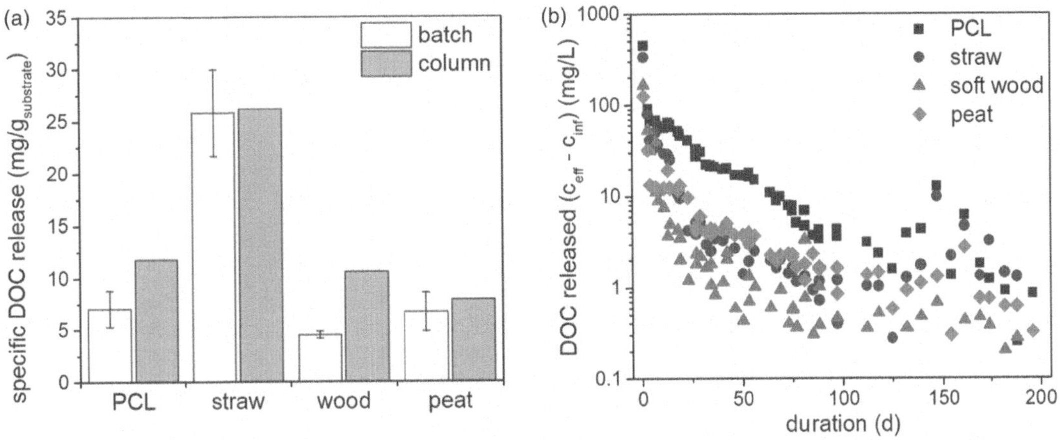

Figure 4 | (a) Comparison of specific DOC release from PCL, straw, soft wood, and peat in batch and column experiments; (b) DOC released during column experiments.

linked to biotransformation, is listed in Table 2. Acesulfame was removed by 44% in batch tests independently of the material under denitrifying conditions. The artificial sweetener is known to be transformed under micro-aerophilic and nitrate reducing conditions with sulfamic acid being the main transformation product (Castronovo *et al.* 2017). Gabapentin, as well, shows a significant removal of 11% under nitrate reducing conditions, which might be attributed to biotransformation. Schaper *et al.* (2019) already observed a removal of gabapentin in the absence of oxygen during their field study. Even though the Student's t-test also reveals a *p*-value below 0.05 for the removal of 5-methyl-benzotriazole, the biodegradability under nitrate reducing conditions found by Liu *et al.* (2011) cannot be clearly confirmed, due to its low mean removal of only 6%.

Column experiments also show a removal of 5-methylbenzotriazole in the nitrate reducing columns. However, we cannot exclude oxic transformation due to residual oxygen. As for all other compounds, no significant removal is observed for denitrifying conditions. Sulfamethoxazole is removed strictly under oxygen or sulfate reducing conditions (Figure S6), which concurs with the findings in sand column investigations of Baumgarten *et al.* (2011).

Comparing the different electron donors, especially peat, revealed a nitrate reduction and a high removal for some TOrCs in batch and column experiments (Figures 5 and S7). As for carbamazepine, which is considered persistent under anoxic conditions, the removal by peat can be attributed to sorption (0.13 ± 0.05 μg/g in batch experiments). Benzotriazole, 5-methyl-bezotriazole and metoprolol were removed with peat in batch experiments and especially during the initial phase in column experiments, that can be linked to adsorption (Figure S8). Due to sorption effects, peat batches were not taken into account for quantifying biotransformation in Table 2. Surprisingly, acesulfame was also well removed in the peat column. In contrast to carpamazepine or benzotriazole, the decreasing effluent concentrations during the column experiment indicate biotransformation as the main removal process.

Boley *et al.* (2006) reported a successful removal of a-endosulfan due to sorption on PCL. However, no removal of the investigated TOrCs linked to adsorption was observed with PCL in this study.

Denitrification with H_2 as electron donor did not show an enhanced removal of the investigated TOrCs in batch experiments (Figure S9).

Table 2 | Mean relative removal: $1 - c/c_0$ ($n = 9$) associated with biological transformation in batch tests under nitrate reducing conditions independent of the applied material (peat was excluded due to sorption effects)

TOrCs	Removal in batch tests	*p*-values (Student's t-test)	Literature
Acesulfame	0.44 ± 0.13	7.58×10^{-6}	+[a]
Gabapentin	0.11 ± 0.11	0.02	+[b]
5-Methyl-benzotriazole	0.06 ± 0.08	0.04	+[c]
4-Formylaminoantipyrine	0.05 ± 0.21	0.53	+[d]
Sulfamethoxazole	0.05 ± 0.18	0.44	+[e] +[d] −[f]
Metoprolol	0.04 ± 0.19	0.62	+[g]
Benzotriazole	0.05 ± 0.08	0.10	+[c]
Carbamazepine	0.05 ± 0.07	0.10	−[h] −[d]
Diclofenac	-0.02 ± 0.08	0.56	+[e] +[d]

Literature references which consider a TOrC as removable (+) or not removable (−) under nitrate reducing conditions.
[a]Castronovo *et al.* (2017).
[b]Schaper *et al.* (2019).
[c]Liu *et al.* (2011).
[d]Wiese *et al.* (2011).
[e]Banzhaf *et al.* (2012).
[f]Baumgarten *et al.* (2011).
[g]Wick *et al.* (2009).
[h]König *et al.* (2016).

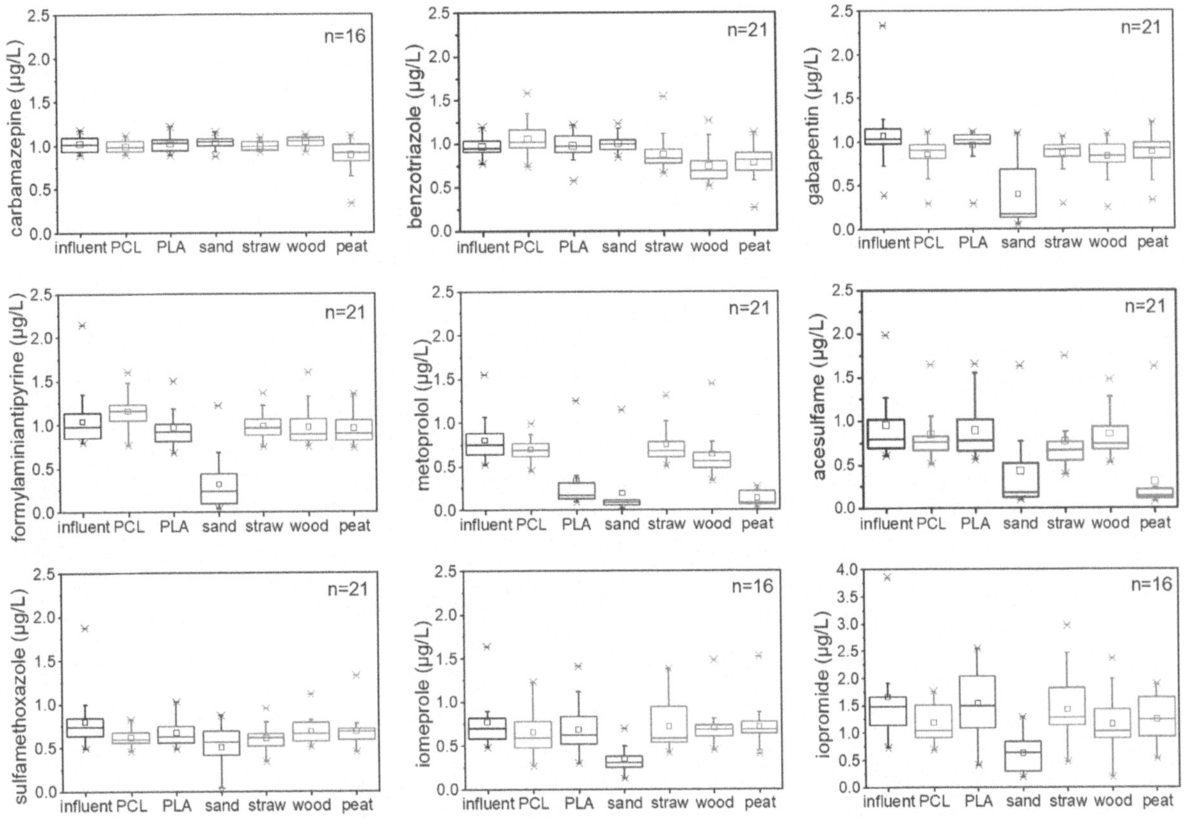

Figure 5 | Removal of TOrCs in column experiment with fixed bed materials. Please note, that only PCL, straw, soft wood and peat created anoxic conditions by significant nitrate removals. The columns filled with PLA and sand are still considered oxic, even though the influent concentration of dissolved oxygen was adjusted to less than 1 mg/L.

Highest TOrC removal was found in the oxic PLA and sand column. Generally, higher removal rates are found for oxic than for anoxic conditions (Greskowiak *et al.* 2017). The MAR sand removed most of the monitored TOrCs even under suboxic conditions below 1 mg/L. The sand might have had a more diverse microbiome, since it was used as filter material in soil aquifer treatment for years, whereas PLA was only inoculated with sewage sludge for 2 days. Furthermore, the HRT was almost double in the sand column compared to PLA, which might explain the better removal under suboxic conditions in MAR sand.

The availability of oxygen after anoxic conditions might cause a backwards transformation of already tranformed TOrCs. This might be the case after re-aeration in SBF systems. This effect was simulated in the end of the batch tests. After 30 days of anoxic conditions, the batch tests were opened and shaken for another 24 h. The introduction of oxygen significantly improved the general removal of TorCs, as shown in Figure S9 (Student's t-test $p = 0.02$). Only a slight increase of TOrC concentrations was observed in some batches after re-aeration, indicating no severe backwards transformation into the parent compound.

Material costs

An overview of the estimated material costs (Table 3) for one cubic meter of a fixed bed reactor illustrates the comparatively high costs for the synthetic materials PCL and PLA. Currently, they are more than 200 times higher than for natural filling materials. However, the specific costs related to nitrate removal can be lower, assuming that synthetic materials show a better long-term performance as electron donor as indicated in the column experiment with PCL.

For wood chips and wheat straw, Saliling *et al.* (2007) detected weight losses of 16.2% and 37.7% over 140 days while Chu & Wang (2013) calculated consumed PCL

Table 3 | Estimation of specific material costs per m³ filled reactor volume based on the applied packing density of the column experiments (Table 1)

Material	Reference (sand)	Rye straw	Soft wood	Peat	PCL	PLA
Estimated material costs (€/$m^3_{reactor}$)	34[a]	16[b]	78[c]	110[d]	3,958[e]	2,489[f]

References for estimated material costs: [a]THARWA SANDS (2019); [b]agrarheute (2019); [c]astsaege (2017); [d]HORNBACH (2019); [e]Boley *et al.* (2000); [f]alibaba (2020).

amounts that ranged between 1.6 and 3.7 g PCL/g NO_3^--N for their experiments.

CONCLUSIONS

An effective nitrate removal was achieved with PCL and straw at HRTs below 10 hours. Whereas the nitrogen removal with straw decreased over time, a complete removal was achieved with PCL after an adaptation time of 125 days. With a better adjustment of HRT, influent oxygen concentrations or even temperature, this time could be decreased and PCL might be a possible substrate for long-term fixed bed denitrification. Only a minor removal of TOrCs was observed under denitrifying conditions, except for peat, which partly removed benzotriazole, 5-methy-benzotriazole, metoprolol, carbamazepine, and acesulfame.

Especially during the initial phase, high DOC leaching from the materials and nitrite formation can occur, which would cause additional oxygen consumption in a subsequent oxic zone in an SBF system. *In-situ* fixed bed denitrification seems to be difficult to control, especially in a technical system with limited HRT or space. A better option might be the combination of denitrification with total suspended solids removal in the rapid sand filter prior to the technical SBF system (Figure S1). For this purpose, the rapid sand filter needs to include additional biofilm carriers for the denitrifiers and an optional dosage of easily degradable organic carbon. Treating just a part of the influent as a recirculated side-stream in a separated and controlled denitrification unit could be another possibility to achieve the required threshold values. This option allows for better control of the system conditions and the application of fixed bed materials as carbon and electron source can be reconsidered. By optimizing the HRT, the leaching of DOC and nitrite formation could be minimized.

ACKNOWLEDGEMENTS

Funding of this study and the open access publication by the German Federal Ministry of Education and Research within the project TrinkWavE (02WAV1404D) and the accompanying scientific project TransWavE (02WAV1400) as parts of the funding measure WavE is gratefully acknowledged.

SUPPLEMENTARY MATERIAL

The Supplementary Material for this paper is available online at https://dx.doi.org/10.2166/wrd.2020.005.

REFERENCES

agrarheute 2019 Available from: https://markt.agrarheute.com/futtermittel-3/stroh-19 (accessed 18 April 2019).

alibaba 2020 Available from: https://www.alibaba.com/showroom/pla-plastic-price.html (accessed 16 January 2020).

Aslan, S. & Türkman, A. 2004 Biological denitrification of drinking water using various natural organic solid substrates. *Water Science and Technology* **48** (11–12), 489–495.

astsaege 2017 Available from: https://astsaege.net/wie-viel-kosten-saegespaene/ (accessed 16 January 2020).

Banzhaf, S., Nödler, K., Licha, T., Krein, A. & Scheytt, T. 2012 Redox-sensitivity and mobility of selected pharmaceutical compounds in a low flow column experiment. *The Science of the Total Environment* **438**, 113–121.

Baumgarten, B., Jährig, J., Reemtsma, T. & Jekel, M. 2011 Long term laboratory column experiments to simulate bank filtration: factors controlling removal of sulfamethoxazole. *Water Research* **45** (1), 211–220.

Boley, A., Müller, W.-R. & Haider, G. 2000 Biodegradable polymers as solid substrate and biofilm carrier for denitrification in recirculated aquaculture systems. *Aquacultural Engineering* **22** (1–2), 75–85.

Boley, A., Unger, B., Müller, W.-R., Kuch, B. & Deger, A. 2006 Biological drinking water treatment for nitrate and pesticide (endosulfan) elimination. *Water Science and Technology: Water Supply* **6** (3), 123–127.

Cameron, S. G. & Schipper, L. A. 2010 Nitrate removal and hydraulic performance of organic carbon for use in denitrification beds. *Ecological Engineering* **36** (11), 1588–1595.

Castronovo, S., Wick, A., Scheurer, M., Nödler, K., Schulz, M. & Ternes, T. A. 2017 Biodegradation of the artificial sweetener acesulfame in biological wastewater treatment and sandfilters. *Water Research* **110**, 342–353.

Christianson, L. E., Lepine, C., Sharrer, K. L. & Summerfelt, S. T. 2016 Denitrifying bioreactor clogging potential during wastewater treatment. *Water Research* **105**, 147–156.

Chu, L. & Wang, J. 2013 Denitrification performance and biofilm characteristics using biodegradable polymers PCL as carriers and carbon source. *Chemosphere* **91** (9), 1310–1316.

DIN EN ISO 7827 *2013-03, Wasserbeschaffenheit- Bestimmung der "leichten", "vollständigen" aeroben biologischen Abbaubarkeit organischer Stoffe in einem wässrigen Medium- Verfahren mittels Analyse des gelösten organischen Kohlenstoffs (DOC) (ISO_7827:2010); Deutsche Fassung EN_ISO_7827:2012*: *Water quality* [ndash] *Determination of the "easy", "suitable" biodegradability of organic substances in an aqueous medium-method based on the analysis of dissolved organic carbon (DOC).* Beuth Verlag GmbH, Berlin. doi:10.31030/1939501.

European Council 1991 *Council Directive 91/271/EEC of 21 May 1991 concerning urban waste-water treatment*: *UWWT Directive*. Available from: https://eur-lex.europa.eu/legal-content/EN/TXT/PDF/?uri=CELEX:31991L0271&from=EN (accessed 11 May 2020).

Ghane, E., Fausey, N. R. & Brown, L. C. 2015 Modeling nitrate removal in a denitrification bed. *Water Research* **71**, 294–305.

Gibert, O., Pomierny, S., Rowe, I. & Kalin, R. M. 2008 Selection of organic substrates as potential reactive materials for use in a denitrification permeable reactive barrier (PRB). *Bioresource Technology* **99** (16), 7587–7596.

Greskowiak, J., Hamann, E., Burke, V. & Massmann, G. 2017 The uncertainty of biodegradation rate constants of emerging organic compounds in soil and groundwater – A compilation of literature values for 82 substances. *Water Research* **126**, 122–133.

Halaburka, B. J., LeFevre, G. H. & Luthy, R. G. 2017 Evaluation of mechanistic models for nitrate removal in woodchip bioreactors. *Environmental Science & Technology* **51** (9), 5156–5164.

Hellauer, K., Mergel, D., Ruhl, A., Filter, J., Hübner, U., Jekel, M. & Drewes, J. 2017 Advancing sequential managed aquifer recharge technology (SMART) using different intermediate oxidation processes. *Water* **9** (3), 221.

HORNBACH 2019 Available from: https://www.hornbach.de/shop/Floratorf-Floragard-150-L/3843684/artikel.html (accessed 09 April 2019).

Huber, S. A., Balz, A., Abert, M. & Pronk, W. 2011 Characterisation of aquatic humic and non-humic matter with size-exclusion chromatography–organic carbon detection–organic nitrogen detection (LC-OCD-OND). *Water Research* **45** (2), 879–885.

Karakurt, S., Schmid, L., Hübner, U. & Drewes, J. E. 2019 Dynamics of wastewater effluent contributions in streams and impacts on drinking water supply via riverbank filtration in Germany-A national reconnaissance. *Environmental Science & Technology* **53** (11), 6154–6161.

Karakurt-Fischer, S., Sanz-Prat, A., Greskowiak, J., Ergh, M., Gerdes, H., Massmann, G., Ederer, J., Regnery, J., Hübner, U. & Drewes, J. E. 2020 Developing a novel biofiltration treatment system by coupling high-rate infiltration trench technology with a plug-flow porous-media bioreactor. *Science of the Total Environment* **722**, 137890.

König, A., Weidauer, C., Seiwert, B., Reemtsma, T., Unger, T. & Jekel, M. 2016 Reductive transformation of carbamazepine by abiotic and biotic processes. *Water Research* **101**, 272–280.

Liu, Y.-S., Ying, G.-G., Shareef, A. & Kookana, R. S. 2011 Biodegradation of three selected benzotriazoles under aerobic and anaerobic conditions. *Water Research* **45** (16), 5005–5014.

Madigan, M. T. & Brock, T. D. 2009 *Brock Biology of Microorganisms*, 12th edn. Pearson/Benjamin Cummings, San Francisco, CA, USA.

Müller, J., Drewes, J. E. & Hübner, U. 2017 Sequential biofiltration – A novel approach for enhanced biological removal of trace organic chemicals from wastewater treatment plant effluent. *Water Research* **127**, 127–138.

Redeker, M., Wick, A., Meermann, B. & Ternes, T. A. 2018 Anaerobic transformation of the iodinated x-ray contrast medium iopromide, its aerobic transformation products, and transfer to further iodinated x-ray contrast media. *Environmental Science and Technology* **52** (15), 8309–8320.

Saliling, W. J. B., Westerman, P. W. & Losordo, T. M. 2007 Wood chips and wheat straw as alternative biofilter media for denitrification reactors treating aquaculture and other wastewaters with high nitrate concentrations. *Aquacultural Engineering* **37** (3), 222–233.

Schaffer, M., Kröger, K. F., Nödler, K., Ayora, C., Carrera, J., Hernández, M. & Licha, T. 2015 Influence of a compost layer on the attenuation of 28 selected organic micropollutants under realistic soil aquifer treatment conditions: insights from a large scale column experiment. *Water Research* **74**, 110–121.

Schaper, J. L., Posselt, M., Bouchez, C., Jaeger, A., Nuetzmann, G., Putschew, A., Singer, G. & Lewandowski, J. 2019 Fate of trace organic compounds in the hyporheic zone: influence of retardation, the benthic biolayer, and organic carbon. *Environmental Science & Technology* **53** (8), 4224–4234.

Schullehner, J., Hansen, B., Thygesen, M., Pedersen, C. B. & Sigsgaard, T. 2018 Nitrate in drinking water and colorectal cancer risk: a nationwide population-based cohort study. *International Journal of Cancer* **143** (1), 73–79.

Shen, Z. & Wang, J. 2011 Biological denitrification using cross-linked starch/PCL blends as solid carbon source and biofilm carrier. *Bioresource Technology* **102** (19), 8835–8838.

Streicher, J., Ruhl, A. S., Gnirß, R. & Jekel, M. 2016 Where to dose powdered activated carbon in a wastewater treatment plant for organic micro-pollutant removal. *Chemosphere* **156**, 88–94.

THARWA SANDS 2019 Available from: http://www.tharwasand.com.au/prices-for-sand-gravel-pebbles-rocks-topsoil-canberra/ (accessed 9 April 2019).

The Council of the European Union. 2015 *Council Directive 98/83/EC of 3 November 1998 on the Quality of Water Intended for Human Consumption: Drinking Water Directive.*

Wang, J. & Chu, L. 2016 Biological nitrate removal from water and wastewater by solid-phase denitrification process. *Biotechnology Advances* **34** (6), 1103–1112.

Warneke, S., Schipper, L. A., Matiasek, M. G., Scow, K. M., Cameron, S., Bruesewitz, D. A. & McDonald, I. R. 2011 Nitrate removal, communities of denitrifiers and adverse effects in different carbon substrates for use in denitrification beds. *Water Research* **45** (17), 5463–5475.

Wick, A., Fink, G., Joss, A., Siegrist, H. & Ternes, T. A. 2009 Fate of beta blockers and psycho-active drugs in conventional wastewater treatment. *Water Research* **43** (4), 1060–1074.

Wiese, B., Massmann, G., Jekel, M., Heberer, T., Dünnbier, U., Orlikowski, D. & Grützmacher, G. 2011 Removal kinetics of organic compounds and sum parameters under field conditions for managed aquifer recharge. *Water Research* **45** (16), 4939–4950.

First received 17 January 2020; accepted in revised form 20 March 2020. Available online 22 May 2020

Application of a decision support tool for industrial and agricultural water reuse solutions in international case studies

Kristina Wencki, Verena Thöne, Dennis Becker, Kerstin Krömer, Isabelle Sattig, Gunnar Lischeid and Martin Zimmermann

ABSTRACT

Treated wastewater is expected to constitute an essential part of the urban water cycle as an additional water resource in water-scarce or densely populated regions in the future. As decisions on the implementation of water recycling measures should always consider local conditions, the project 'MULTI-ReUse: Modular treatment and monitoring for wastewater reuse' has developed a comprehensive sustainability assessment tool, designed to support decision-makers in examining the technical feasibility, economic viability, ecological compatibility and social acceptance of alternative service water supply solutions at local level. This article describes the structure of this sustainability assessment tool and its underlying multi-criteria assessment approach based on 23 evaluation criteria. Already in the development phase, the tool was tested in a German and a Namibian case study. Both case studies are presented with a special focus on the technologies used and the results of the analysis with the sustainability assessment tool. Case study testing proved that the tool is applicable in various environmental and societal settings with widely differing climatic conditions, limited resource availability, for varying feed water qualities and water quality requirements. The comprehensive, straightforward assessment approach enabled the local users to identify the most sustainable supply system or strategy for their decision case.

Key words | assessment framework, multi-criteria decision analysis, practical implementation, water management, water reuse

Kristina Wencki (corresponding author)
Verena Thöne
IWW Rheinisch-Westfälisches Institut für Wasserforschung gemeinnützige GmbH,
Moritzstr. 26, 45476 Mülheim an der Ruhr,
Germany
E-mail: *k.wencki@iww-online.de*

Dennis Becker
DECHEMA e.V.,
Theodor-Heuss-Allee 25, 60486 Frankfurt am Main,
Germany

Kerstin Krömer
Isabelle Sattig
Oldenburgisch-Ostfriesischer Wasserverband,
Georgstraße 4, 26919 Brake,
Germany

Gunnar Lischeid
Leibniz-Zentrum für Agrarlandschaftsforschung (ZALF) e.V.,
Eberswalder Straße 84, 15374 Müncheberg,
Germany

Martin Zimmermann
Institute for Social-Ecological Research Ltd (ISOE),
Hamburger Allee 45, 60486 Frankfurt am Main,
Germany

HIGHLIGHTS

- The paper presents a multi-criteria assessment approach based on 23 evaluation criteria that can be used to assess the technical feasibility, economic viability, ecological compatibility and social acceptance of alternative service water supply solutions at local level.
- Assessment results from two different case studies comparing different solutions for industrial and agricultural water reuse are presented.

doi: 10.2166/wrd.2020.127

INTRODUCTION

In many countries, the effluent of wastewater treatment plants (WWTPs) is discharged into receiving waters without further use. In the long run, it is expected that treated wastewater will, as an additional water resource in water-scarce or densely populated regions, constitute an essential part of the urban water cycle, also in Germany. Although widespread use of water reuse technologies is currently hampered by insufficient public acceptance or regulatory obstacles, analyses and forecasts generally attest Germany – just as many other European countries – a high water reuse potential across all sectors of about 144 million m^3 per year that could be reached by as early as 2025 (Wintgens *et al.* 2006). The concepts of 'circular economy' and 'water smart societies' reinforce the importance of water reuse in European countries and beyond the near future as the use of wastewater and the recovery of byproducts can open up new business opportunities and help to cover the costs of new, innovative and adapted plants so that energy, nutrients, metals and other byproducts can be recovered (Water Europe 2016).

Currently, global annual freshwater withdrawal amounts to 4,000 km^3 per year and about 75% of the water is used for agricultural production (WWAP 2019), wherein an overwhelming proportion is used for irrigation of arable fields. The need for irrigation is the highest in water-scarce regions. Consequently, in these regions, a large fraction of treated or untreated wastewater is used for irrigation. For example, in Israel, 86% of wastewater is reused providing about 50% of the national irrigation water demand (Tal 2016). To meet the urgent need for additional irrigation water in countries of the European Union (EU), the European Commission adopted new rules to stimulate and facilitate water reuse in the EU for agricultural irrigation in May 2020 (European Commission 2020). So far, irrigation has played only a marginal role in Germany, comprising about 2.2% of the agricultural area (Statistisches Bundesamt 2014). Due to very rigid legal regulations, treated wastewater is used in a few exceptional cases only. However, facing a series of dry years and a predicted increase of dry seasons in the next decades due to climate change, there is increasing pressure on these water resources.

In contrast, water recycling in industry and commerce already plays a major role in several sectors in Germany. The process industry is a major water user and an important solution provider for innovative products, technologies and services that enable more sustainable water management. Technologies and concepts for water reuse promote the economical use of water resources, make companies independent from freshwater resources and thus bring clear locational advantages, both in Germany and internationally (competition for water use, quantitative and qualitative water shortage/stress). Some process industries, which require large quantities of water, have been recycling or reusing water for a long time. In the paper industry, loop separation and counterflow control as well as extensive mechanical circuit water purification are state of the art (Lyko 2014). The steel industry also circulates large quantities of water, especially for cooling and gas scrubbing, e.g. circuits between 500 and 5,000 m^3/h in blast furnaces or between 500 and 3,000 m^3/h in steelworks (Track & Kozariszczuk 2018). The aim is to reduce freshwater and wastewater costs, but also to recover resources or energy, for example.

The aim of the transdisciplinary research project 'MULTI-ReUse: Modular treatment and monitoring for wastewater reuse' was the development of advanced modular treatment process technologies to provide service water that is ideally suitable for industrial and agricultural purposes which do not require water of a quality suitable for drinking. Process technologies, for instance new reverse osmosis (RO) membranes and ultrafiltration (UF), were developed to process different qualities and quantities of treated wastewater to provide water that is fit for purpose and suitable to substitute the use of other drinking water resources at an economically competitive level (Schramm & Zimmermann 2018).

In order to examine advantages and disadvantages of the service water supply using MULTI-ReUse technologies with the current water supply concept, a comprehensive sustainability assessment tool was developed and applied in national and international case studies. In the subsequent sections, the methodological approach and the structure of this sustainability assessment tool are described. Furthermore, the applicability of the tool to assess the advantages of water reuse solutions compared with existing supply systems in an industrial and an agricultural context is demonstrated in a German and a Namibian case study.

MATERIALS AND METHODS

Multi-criteria evaluation approaches

In order to select a scientifically sound and practicable assessment methodology for the multi-criteria decision support tool, widely used multi-criteria analysis (MCA) approaches and procedures for the inclusion of quantitative and qualitative data were examined for their applicability and suitability for the comparative assessment of different water reuse solutions in national and international contexts. A focused literature review on methods for the sustainability assessment of technologies and measures in the context of water reuse was performed, considering selected peer-reviewed articles found in the Web of Science or Scopus databases. In total, 47 publications from the years 1998 to 2017 were recorded in the literature database and analyzed with regard to the evaluation method used, the evaluation object and the case study region. From these literature sources, it was possible to differentiate between 25 evaluation procedures, which consider at least one or, in some cases, also several evaluation dimensions in the evaluation process.

As examples of processes that only consider one of the presented assessment dimensions, mainly environmentally oriented assessment methods such as ecological risk analysis (El Heloui *et al.* 2016; Zaibel *et al.* 2016), ecological indicators (Carr *et al.* 2010; Uddin *et al.* 2014; Müller & Cornel 2017), life cycle assessment (Holloway *et al.* 2016; Kavvada *et al.* 2016), material flow analysis (Chen *et al.* 2012) or scenario-based hydrological modeling (Nies *et al.* 2016) could be identified. In the case of multidimensional methods, a distinction can be made between the methods usually declared as MCA, which are typically carried out in the form of a utility analysis, and the MCA approaches based on sequencing methods such as the analytical hierarchy process (Aydiner *et al.* 2016) or linear programming (Chu *et al.* 2004). The strongly economically oriented cost-benefit analysis (Haruvy 1998; Hernández et al. 2006; Chen *et al.* 2015) represents a third form of MCA that, together with the utility value analysis, is one of the most frequently used methods in the publications examined.

For the adaptation within a decision support tool, a large part of these procedures had to be excluded due to the one-dimensionality of the assessment, an undue degree of complexity of the assessments (e.g. goal, aspiration or reference level models and outranking methods) or the expectation of insufficient data availability for further use within the intended evaluation tool.

Based on these considerations, value measurement models were found to be best suitable for the assessment purposes of the MULTI-ReUse tool. Value measurement models are based on the idea that a numerical use value is calculated for each criterion. Afterwards, all the use values of the single criteria are ultimately synthesized to a global score using criteria weightings. The weighting and aggregation of the partial results can take place in a variety of ways and, if necessary, also lead to cost-benefit and cost-effectiveness considerations.

The final MCA approach developed within the project is based on 23 easy to handle and comprehensive evaluation criteria in four dimensions covering environmental impacts (e.g. net energy consumption, space requirement and volume of residual substances), social factors (e.g. compliance with national strategies, increasing environmental awareness and health risks), technical feasibility (e.g. anticipated expense for implementation, flexibility, adaptability and expandability) and economical viability (annual costs, potential for innovation leadership and competitiveness) of water supply alternatives (Table 1).

All criteria should be assessed individually for the particular case of application in order to identify advantages and disadvantages of the considered water supply variants. Results are presented in the form of normalized figures and graphic illustrations. In order to facilitate decision-making, the criteria assessments can be further combined to an overall assessment of each alternative based on the preference set by the decision-makers. The final result is a ranking of the individual alternatives that are assessed.

Based on this evaluation approach, sustainability assessments for two case studies were carried out based on different data sources. For the German case study, technical

Table 1 | Overview of evaluation criteria considered within sustainability assessment covering four dimensions

ID	Criterion	Possible indicator(s) and further information	Unit
ENV1	Space requirement	Total space required by the processing plants (including all space requirements associated with water supply, e.g. pipeline network for water distribution, and all areas unusable for other purposes due to the plant)	m^2 or m^2/m^3
ENV2	Energy consumption	Specific net energy consumption per m^3 produced water (including all energy consumption for water treatment and water distribution, measurement, control and regulation technology, and lighting and ventilation of the company buildings)	kWh/m^3
ENV3	Consumption of process and operating resources	Specific consumption of treatment materials and disinfectants per m^3 of water produced (if necessary, accounted separately for individual substance classes)	g/m^3
ENV4	Accumulation of residues	Specific residues use per m^3 produced water (including all residues removed from the system and properly disposed)	ml/m^3
ENV5	Additionally available water volume	Volume of water additionally available (if necessary subdivided into different water quality levels for different types of use)	m^3/a
ENV6	Water quality	Reduction of the load for selected parameters that are critical for the intended use (only considering parameters relevant for the individual case)	5-point scale from 'no reduction' to 'complete reduction'
ENV7	Ecosystem services	Contribution to maintaining or increasing (supporting, provisioning, regulating and cultural) ecosystem services in the area under consideration and adjacent ecosystems	5-point scale from 'negative' to 'positive'
SOC1	Personnel requirements	Number of personnel required for operation and maintenance (taking into account all phases of system planning, construction and commissioning, operation, administration and monitoring)	5-point scale from 'very high' to 'very low'
SOC2	Accordance with guiding principles	Consistency with national water and sanitation strategies (depending on the location of the system and the intended type of use)	5-point scale from 'large gap' to 'good correlation'
SOC3	Strengthening environmental awareness	Potential to sensitize stakeholders to issues of sustainable resource use (e.g. motivation to conscious and sustainable water and energy use)	5-point scale from 'no potential' to 'high potential'
SOC4	Acceptance	Stakeholder's acceptance for the intended water reuse as well as for the product produced by recycled water usage	5-point scale from 'no acceptance' to 'high acceptance'
SOC5	Health protection	Health protection of the user in contact with pathogenic microorganisms from the reuse water	5-point scale from 'unsatisfactory' to 'very good'
TEC1	Integrability in existing infrastructure	Technical integrability of the process chain into the existing infrastructure (incl. times required for planning, construction and commissioning)	5-point scale from 'impossible' to 'very good'
TEC2	Flexibility, adaptability, expandability	Potential for structural adaptations (e.g. modularity, scalability) with regard to changing boundary conditions	5-point scale from 'no potential' to 'very high potential'
TEC3	Operational safety	Control and review of accident prevention measures as well as occupational health and safety	5-point scale from 'defective' to 'very good'

(continued)

Table 1 | continued

ID	Criterion	Possible indicator(s) and further information	Unit
TEC4	Technical complexity	Minimum educational qualification required for operation (including monitoring, control) and maintenance of the system	5-point scale from 'no qualification' to 'with work experience and further education'
TEC5	Maintenance	Local availability of spare parts for repair services	5-point scale from 'not available' to 'very good'
TEC6	Process stability	Mean time to failure (MTTF)	years/failure
TEC7	Tolerance of disturbance against external perturbations	Tolerance of the systems against natural risks and hazards, operating error, manipulation and vandalism	5-point scale from 'inadequate' to 'high'
ECO1	Overall costs	Specific annual costs (total)	EUR ct/m^3
ECO2	Market position, competition	Potential for regional innovation leadership through the expansion of know-how and image improvements	5-point scale from 'no potential' to 'very high potential'
ECO3	Added local economic value	Positive economic effects that can be assigned to the water reuse measure	EUR/m^3
ECO4	Opportunity costs	Follow-up costs of a system failure with disruption or interruption of water supply	EUR/d

ENV, environmental impacts; SOC, social factors; TEC, technical feasibility; and ECO, economical viability.

and environmental data were taken directly from the pilot plant, which was operated as part of the project. Further information on economic and social aspects was provided by the case study owner, the local water company based on in-house cost calculations and expert guess. As the Namibian case study consists of a comparison of modified technical systems investigated in previous and current projects, named 'EPona: Water Reuse in Northern Namibia' and 'CuveWaters: Sustainable Water Management in Namibia', unpublished data were provided and modified by the project coordinators to match the MULTI-ReUse project's needs. Additional information was derived from structured interviews and expert estimates as part of the studies in the projects named above.

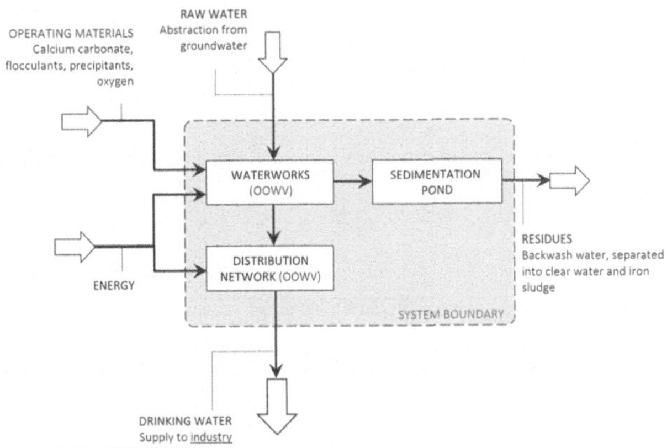

Figure 1 | Process diagram for water supply option 1-A in Nordenham, Germany.

Case study Nordenham, Germany

In North Germany, the water company Oldenburgisch-Ostfriesischer Wasserverband (OOWV) ensures water supply for municipal and industrial customers. In a pilot plant, OOWV already demonstrated water reclamation and reuse from the WWTP Nordenham with improved process technologies such as UF and RO (Figure 1). In the medium term, OOWV intends to substitute drinking water consumption in an industrial park in Nordenham by reuse water with low electric conductivity and low chlorine contents from a separate pipeline network. The demand for industrial service water is expected to increase by 800,000 m^3 per year. By using the sustainability assessment tool developed within the MULTI-ReUse project, OOWV aimed at performing a sound cost-effectiveness consideration for the planned water recycling project (option 1-B, Figure 2) by comparing it to the current water supply

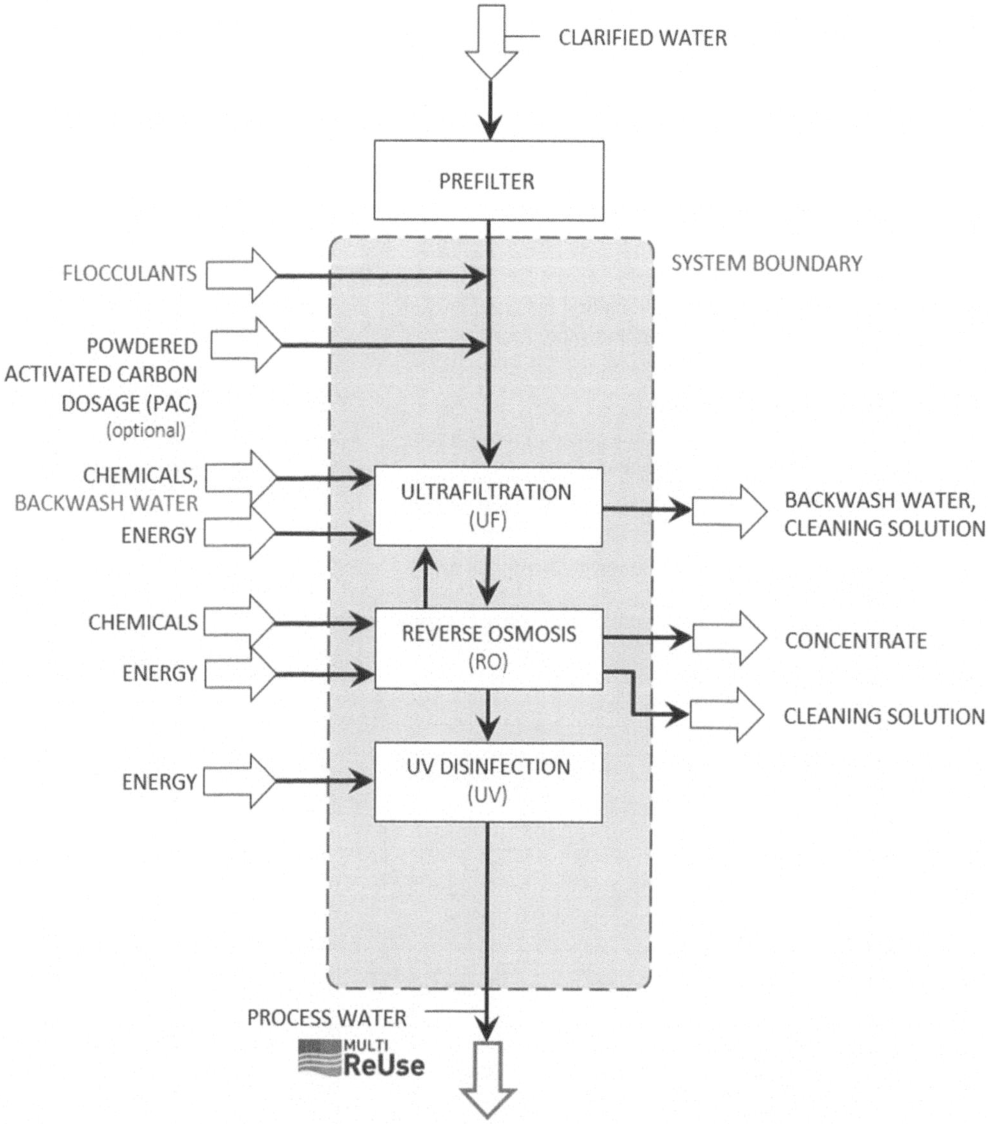

Figure 2 | Process diagram for water supply option 1-B using the MULTI-ReUse process chain in Nordenham, Germany.

system (option 1-A, Figure 1). In order to ensure the selection of the most sustainable solution to cope with the local challenges identified for today as well as in long-term planning, the assessment is performed from a current point of view (year 2020) and with a view to the future (year 2030).

In both temporal scenarios, option 1-A covers the service water supply for industrial customers in the City of Nordenham with drinking water from the municipal waterworks Großenkneten, Nethen and Sandlermöns. Thus, investment and operating costs for the provision of drinking water include plant-specific costs for groundwater extraction, intermediate storage und water pumping (including storage pumping stations).

Option 1-B describes an alternative supply utilizing reclaimed process water. In this option, investment costs for land, a separate non-potable water pipeline network, storage tanks as well as buildings including the plant engineering, plant and pump houses must be taken into consideration. Operating costs, such as energy costs for pumping (for water supply to the treatment plant, RO or UF and process water supply to the customer), are particularly relevant in this context. Cleaning chemicals, precipitants (ferrous or aluminum salts), disinfectants and personnel costs are further examples of expenditures that need to be assessed.

Case study Outapi, Namibia

In the northern Namibian town of Outapi, a water reuse system for domestic wastewater of approximately 1,000 inhabitants has been operating successfully for more than 6 years. The system feeds a drip irrigation system with reuse water for the production of vegetables and fruit for human consumption. Due to the semi-arid climate, the reuse of water is a viable alternative for the cultivation of plants in the region. However, there are several alternatives of reuse systems available. Based on the developed sustainability assessment approach, the potential positive and negative impacts of three different water use options were assessed. Option 2-A is the above-mentioned system, which consists of an up-flow anaerobic sludge blanket (UASB), rotating biological contactors (RBC), and a micro-screen and ultraviolet (UV) irradiation for disinfection before the irrigation water is stored in a pond (Figure 3). In option 2-B, the micro-screen and UV irradiation of option 2-A is replaced by an UF as tested in the MULTI-ReUse project (Figure 4). The remaining system of this hypothetical option is identical to the first one. Option 2-C uses existing wastewater ponds that are already used for a large part of Outapi's wastewater disposal (Figure 5). These ponds are supplemented by a pre-treatment with either UASB reactors or a micro-screen, a stone filter for post-treatment after the ponds and an additional UF membrane for irrigation water production.

RESULTS

Tool description

Based on the considerations on different multi-criteria evaluation approaches and the selection of a suitable

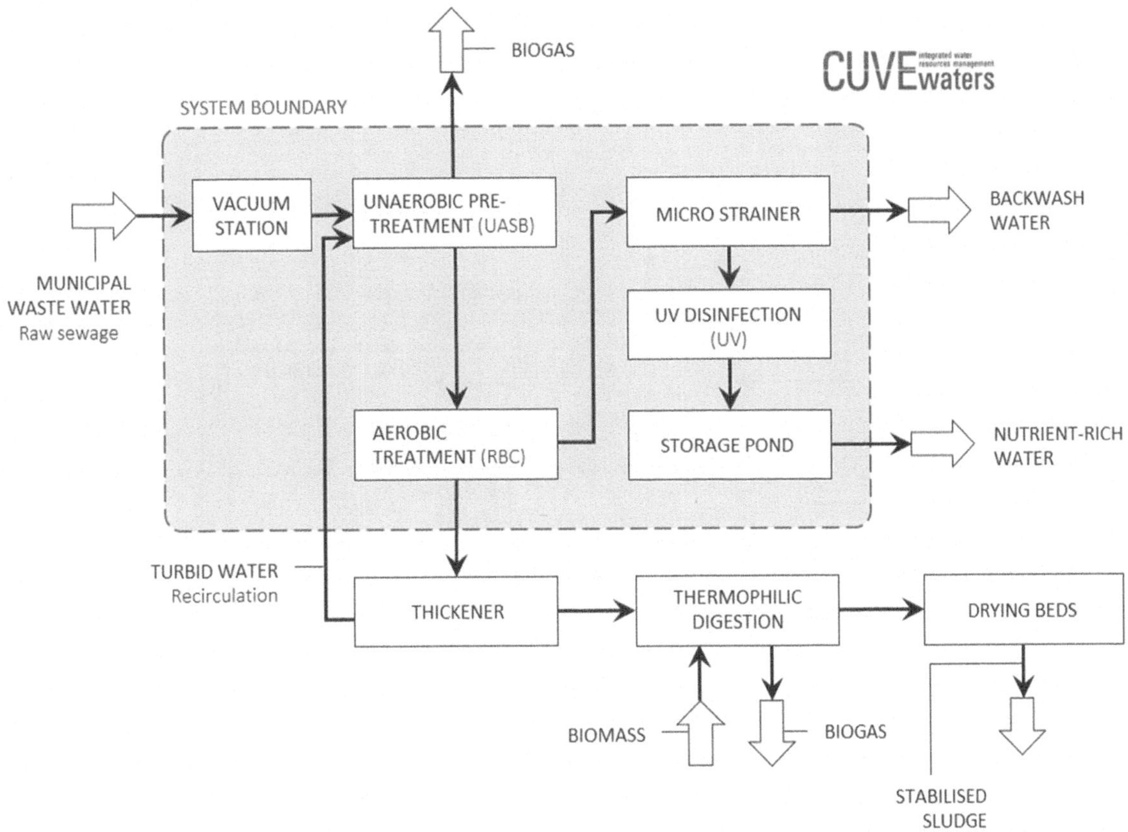

Figure 3 | Process diagram for water supply option 2-A in Outapi, Namibia.

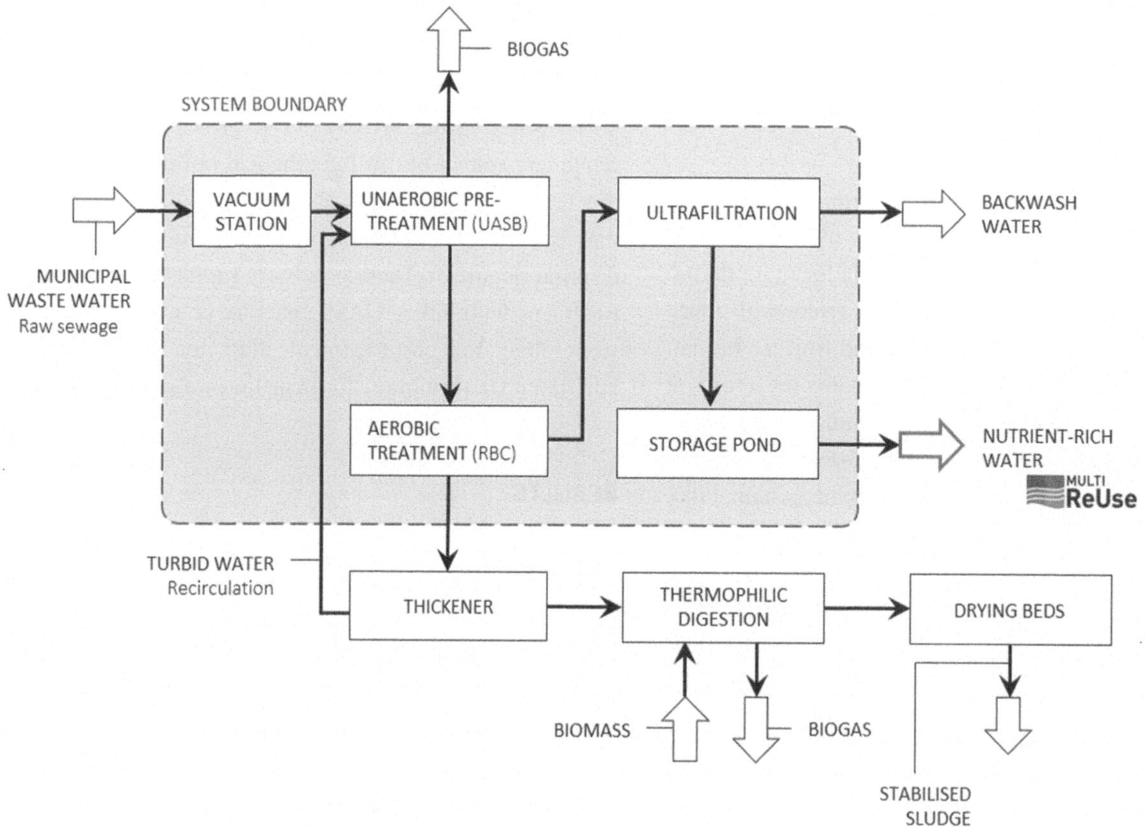

Figure 4 | Process diagram for water reuse option 2-B using the MULTI-ReUse process chain in Outapi, Namibia.

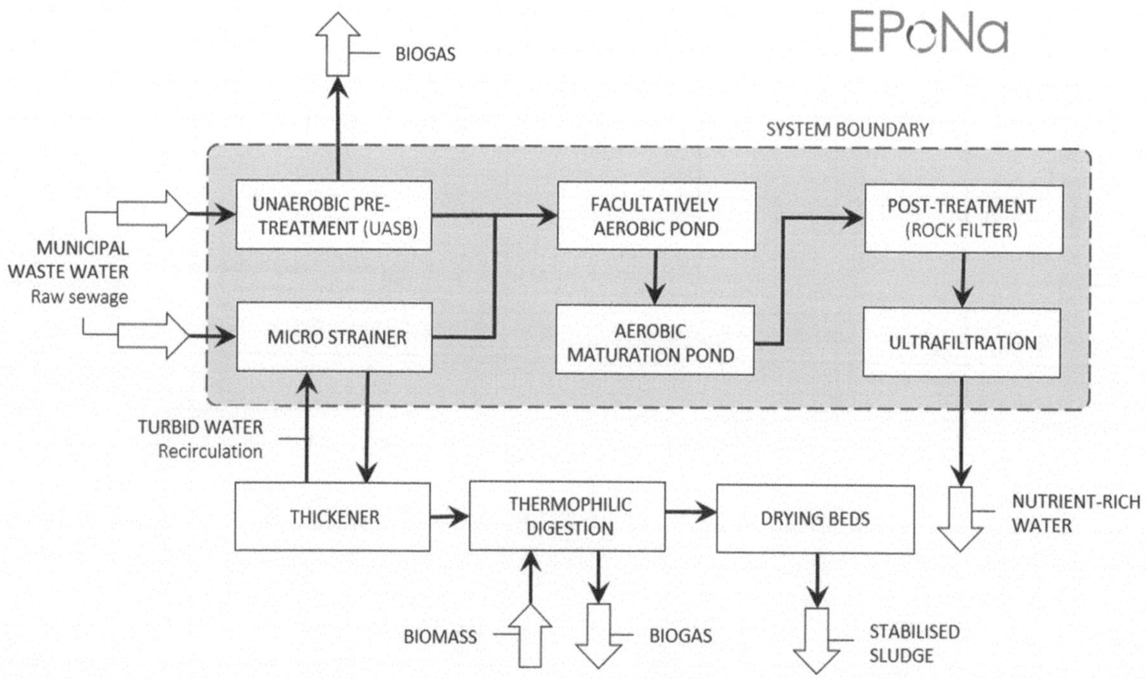

Figure 5 | Process diagram for water reuse option 2-C in Outapi, Namibia.

assessment approach described above, a multi-criteria decision tool has been developed and tested for various case studies. It is intended to be used by water treatment plant (WTP) and WWTP managers, municipal decision-makers, operators of industrial plants and consulting engineers in an initial phase of the planning process. It allows the decision-makers to evaluate and compare the sustainability of conventional water supply and innovative process chains for water reclamation and reuse for service water supply considering the case-specific environmental and societal settings.

The tool is suitable for comparing a minimum of two and up to five water supply variants in parallel. On the basis of a checklist, termination criteria are verified at the beginning of the sustainability evaluation. The termination criteria include core requirements, such as water demand, social acceptance of water reuse and disposal options for residues. The total of 23 evaluation criteria recorded apply to the categories of environment, social issues, technology and economics. Criteria that are to be assessed qualitatively can be entered using 5-level Likert-type scales. Based on the input data, the tool creates an integrated assessment of the water supply variants to be compared. To provide the user with an overview of the data entered, input data are presented in the form of separate bar charts per indicator. The diagrams can be used to check the data entered.

The centerpiece of the calculations is the MCA approach: In the context of a utility value analysis, in a first step, utility values between 0 and 1 are calculated for each criterion and each of the variants, whereas a higher utility value is always indicating a higher consistency of the solution with the respective target. Scaling and standardization are performed depending on the case-specific minimum or maximum values for the indicators which are derived from the input data inserted for the different options and regulatory thresholds – where applicable. With the purpose of simplification of the assessment in the initial phase of planning that the tool is designed for, input values between minima and maxima are determined by linear interpolation, accepting inaccuracies in the scale of values, at least for some of the criteria proposed.

The total utility value of a variant is calculated by aggregating the partial values, taking into account the specification of different weighting scenarios, and presented

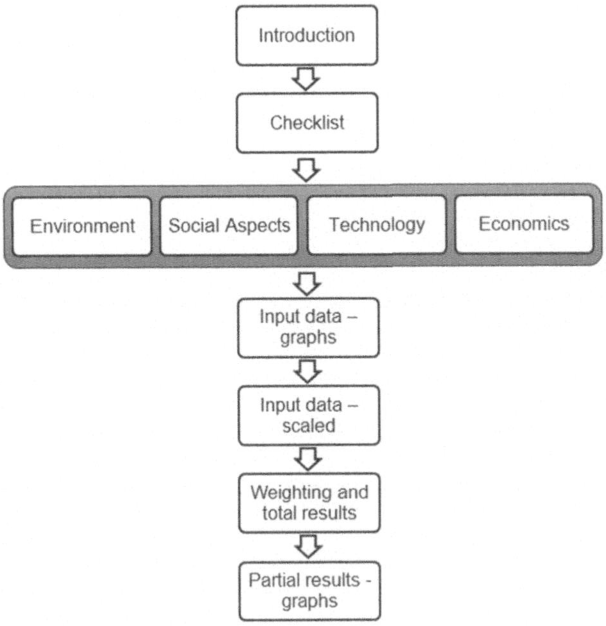

Figure 6 | Process scheme of the MULTI-ReUse sustainability assessment tool.

in the form of a summary table. The tool creates a ranking list, which takes into account the user's preference in addition to a set of nine predefined weighting scenarios, and spider graphs, which allow for a direct comparison of individual variants.

A process scheme of the MULTI-ReUse sustainability assessment tool is depicted in Figure 6.

Assessment results – case study Nordenham, Germany

The application of the decision support tool for the assessment of the northern German case study confirms that currently the existing water supply system is preferable from a technical, social and economic perspective (Figure 7). For example, the specific energy requirement for the provision of process water from water reuse (option 2-B) is approximately 1 kWh/m^3, whereas the total specific energy demand for drinking water supply (option 1-A) is only 0.7 kWh/m^3 on average (Sattig 2019). The calculated specific costs for the process water supply are thus about 10% higher than the costs for a process water supply from drinking water resources.

However, as an increase in water abstraction or an installation of additional extraction wells is restricted by

No.	Weighting scenario	Environment	Social aspects	Technology	Economy	Option 1-A *Drinking Water (2020)*	Option 1-B *MULTI ReUse (2020)*	Option 1-A *Drinking Water (2030)*	Option 1-B *MULTI ReUse (2030)*
1	User's choice								
2	Example from **MULTI ReUse**	26%	13%	**38%**	23%	0,697 1	0,592 2	0,632 1	0,619 2
3	Focus **Environment**	**40%**	20%	20%	20%	0,674 1	0,590 2	0,612 2	0,618 1
4	Focus **Social aspects**	20%	**40%**	20%	20%	0,748 1	0,668 2	0,692 2	0,715 1
5	Focus Technology	20%	20%	**40%**	20%	0,725 1	0,619 2	0,669 1	0,653 2
6	Focus Economy	20%	20%	20%	**40%**	0,693 1	0,611 2	0,611 2	0,638 1
7	Focus **Environment + Social aspects**	**30%**	**30%**	20%	20%	0,711 1	0,629 2	0,652 2	0,666 1
8	Focus **Environment + Technology**	**30%**	20%	**30%**	20%	0,662 1	0,605 2	0,641 1	0,635 2
9	Focus **Technology + Economy**	20%	20%	**30%**	**30%**	0,709 1	0,615 2	0,628 2	0,646 1
10	Equal Weighting	25%	25%	25%	25%	0,710 1	0,622 2	0,640 2	0,656 1

Figure 7 | Assessment result for case study in Nordenham (options 1-A 'drinking water', 1-B 'MULTI-ReUse' for a current point of view (2020) and with a view to the future (2030)) from the sustainability assessment tool with total utility values and ranking positions of the options (right columns).

current water rights for this location, water reuse should still be considered as an alternative to drinking water supply in the future. It is expected that water reuse will reduce the pressure on scarce regional groundwater resources in the long run by meeting the increased industrial water needs with appropriate water quality at acceptable prices. Other contributions to the protection of ecosystems, that can already be recognized today, include a reduced land requirement, a potential improvement in the discharge quality with regard to the annual loads (especially regarding the discharge of trace organic contaminants) when dosing the powdered activated carbon and a reduction in the entry of pathogenic germs into the surface water by membrane installations (Figure 8). Significantly lower chlorine and salt contents of the reuse water also enable multiple water recirculation within industrial processes, so that the

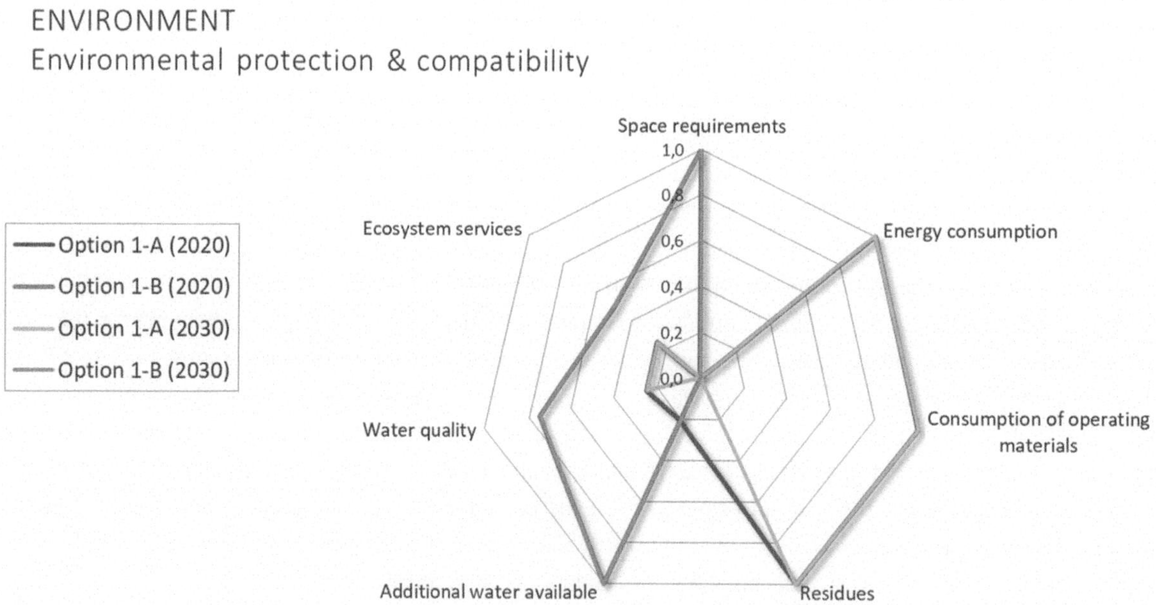

Figure 8 | Spider diagram of the assessment result for the case study Nordenham with regard to environmental criteria (scale: normalized non-dimensional utility value between 0 (minimum) and 1 (maximum)).

consumption of water resources for production can even be reduced. In addition, option 1-B also allows greater flexibility in terms of quantity and quality requirements for customers, and it makes a greater contribution to local value creation and increases environmental awareness. Furthermore, the use of the MULTI-ReUse approach (option 1-B) leads to a minimized use of cleaning chemicals (e.g. flocculants and precipitants) in the various process steps.

Assessment results – case study Outapi, Namibia

The sustainability assessment of the three options in the Namibian case study shows that option 2-A is the most sustainable system for water reuse, followed by option 2-B and option 2-C (Figure 9). This is mainly due to the good environmental performance of option 2-A. In all weighting scenarios, option 2-C ranks last as it seems to combine all the disadvantages of the other two options. Option 2-A also has the highest utility value when greater weight is given to individual assessment dimensions, thus focusing on either environmental, social, technical or economic criteria (Figure 9). If both environmental and social assessment criteria are given greater weight, option 2-A remains the most sustainable option, followed by option 2-B and finally option 2-C. The ranking does not change even if technical and economic aspects or ecological and technical aspects are given priority.

In terms of environmental criteria, option 2-A is the most sustainable option as it is more water and energy efficient, produces less residues and requires fewer chemicals for operation (Figure 10). Option 2-A is followed by option 2-B, *inter alia*, because of its lower specific space requirements.

Within the social dimension, options 2-A and 2-B achieved equal scores. Both options perform better than option 2-C in terms of human resource requirements and consistency with national water and sanitation strategies. In terms of technical criteria, all three options considered have their strengths. In particular, option 2-C scores better than the other two options in terms of integration into the existing infrastructure and availability of spare parts. Options 2-A and 2-B are more adaptable than option 2-C.

Apart from this, option 2-A has slight advantages compared with the other two options in terms of the technical complexity of the plant and option 2-B in terms of process stability. When focusing on economic aspects, option 2-A proves to be the most sustainable option due to its ability to innovation leadership and its lower opportunity costs (follow-up costs due to a system failure). In contrast, option 2-C has the comparatively lowest specific annual

No.	Weighting scenario	Environment	Social aspects	Technology	Economy	Option 2-A *UASB-RBC-UV-pond*	Option 2-B *MULTI-ReUse*	Option 2-C *UASB-ponds-UF*
1	User's choice							
2	Example from **MULTI ReUse**	26%	13%	**38%**	23%	0,558 1	0,490 2	0,310 3
3	Focus Environment	**40%**	20%	20%	20%	0,629 1	0,494 2	0,272 3
4	Focus Social aspects	20%	**40%**	20%	20%	0,540 1	0,476 2	0,320 3
5	Focus Technology	20%	20%	**40%**	20%	0,529 1	0,486 2	0,322 3
6	Focus Economy	20%	20%	20%	**40%**	0,553 1	0,473 2	0,473 2
7	Focus **Environment + Social asp**	**30%**	**30%**	20%	20%	0,584 1	0,485 2	0,296 3
8	Focus **Environment + Technolo**	**30%**	20%	**30%**	20%	0,550 1	0,490 2	0,297 3
9	Focus **Technology + Economy**	20%	20%	**30%**	**30%**	0,541 1	0,479 2	0,330 3
10	Equal Weighting	25%	25%	25%	25%	0,563 1	0,482 2	0,313 3

Figure 9 | Assessment results for the case study in Outapi (options 2-A 'UASB-RBC-UV-pond', 2-B 'MULTI-ReUse' and 2-C 'USAB-ponds-UF') from the sustainability assessment tool with total utility values and ranking positions of the options (right columns).

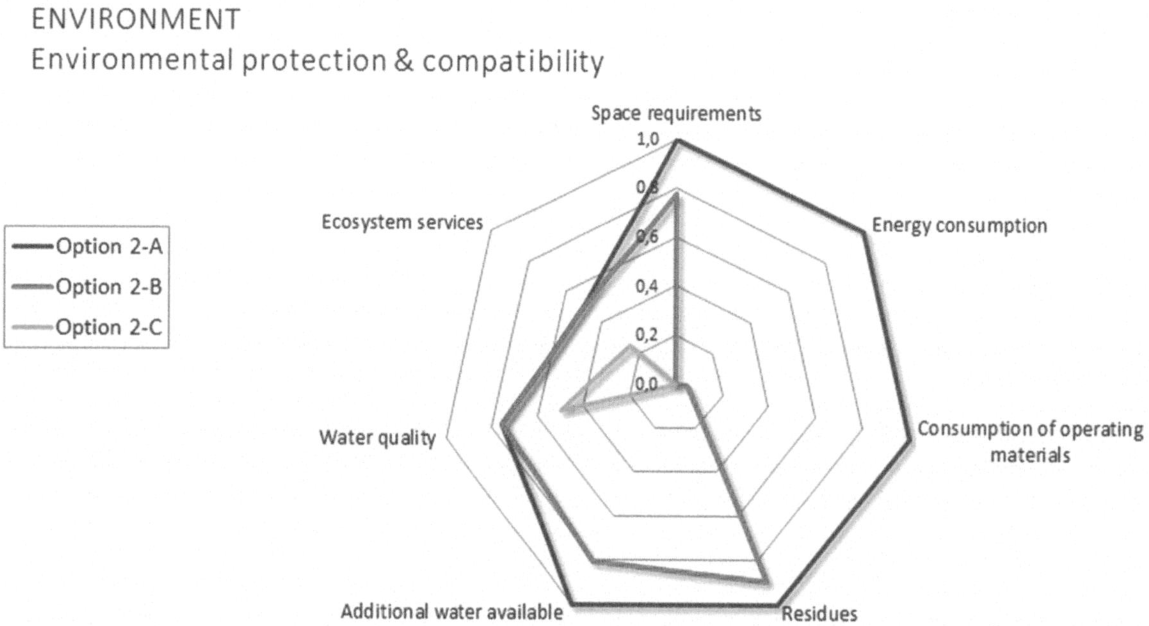

Figure 10 | Spider diagram of the assessment results for the case study Outapi with regard to environmental criteria (scale: normalized non-dimensional utility value between 0 (minimum) and 1 (maximum)).

costs. All three options considered show a comparable contribution to local value creation.

DISCUSSION

One of the key objectives of the MULTI-ReUse project was to develop a sustainability assessment tool to evaluate different water recycling solutions against the current system configuration in order to identify the most sustainable water supply system for the future. This represents a major challenge since current water supply systems regularly developed historically, are economically written off and optimized in a way that they can operate efficiently in large scale implementation. In contrast, innovative water reuse technologies are usually making use of different water treatment technologies and are currently implemented on much smaller scales. Thus, it is a really challenging task to find common system boundaries and comparable water supply systems. The MULTI-ReUse assessment tool helps the user to identify the advantages and disadvantages of different system configurations. However, a mandatory requirement for its application is the existence of basic concepts for water reuse which have to be developed and made comparable by the tool user individually.

Another challenge in meeting this objective was to identify a suitable multi-criteria assessment methodology that is comprehensive enough to deal with the diverging objectives attached to such decision cases and at the same time offers the highest level of flexibility to be applicable in various contexts. As described, the development of the MULTI-ReUse decision support tool was based on a focused review of existing multi-criteria approaches for water reuse technologies and a structured compilation and review of criteria lists, including experts from different disciplines served to ensure a common understanding of terminology and consideration of contradicting stakeholders' viewpoints of water reuse technology implementation. However, it must be stated that the MCA approach used allows only for a rather superficial assessment of the considered solution. Nevertheless, the practice partners involved in the development process of the MULTI-ReUse tool, attached great importance to define such rather simple but holistic approach as this enables them to provide key information relevant in an early planning phase of water reuse projects, as a more detailed assessment of the solutions for implementation planning has to be carried out in-house using specific modeling and calculation approach. Thus, creating a consistent approach for detailed assessments that is useful for all types of users is very difficult to find and might not be appropriate.

The limitations of the MCA approach used in the tool are principally based on data requirements and data availability at such an early phase of the planning process which might force the user to incorporate expert judgements in the weighting of individual indicators required for aggregation along the evaluation and in the evaluation of (especially qualitative) indicators themselves, which could potentially result in uncertainty and inconsistency if the assumptions made are not adequately documented. This is, however, a common issue of MCA approaches. Depending on available data for a given case study, this uncertainty can be reduced through the higher employment of measured or modeled data, keeping the use of assumptions to a minimum. Furthermore, social science approaches can be used to reduce the impact of subjective rating within the MCA approach.

A subsequent challenge was to transfer this assessment approach into a tool, providing transparent and applicable decision support to its different user groups. The application of the decision support tool for the sustainability assessment of the two case studies provided its developers with key insights about advantages as well as limitations that were necessary to modify the prototype of the tool and transform it into the practice-oriented tool. The involvement of hypothetical end users in the evaluation of the application tests has helped to reveal lack of understanding, gaps and inconsistencies.

As the implementation of water reuse technologies is occasionally hampered by risk concerns and lack of social acceptance, these factors should be considered in decision-making in an early phase of planning. Further development requirements for the tool exist, in particular with regard to a stronger incorporation of risk management approaches requiring additional data. However, by focusing on local-scale evaluations of comparable technical solutions, the tool provides accessible, lean and yield results that are directly relatable and actionable for stakeholders and decision-makers.

CONCLUSION

The application of the developed decision support tool in two international case studies proved that the tool is applicable in various environmental and societal settings with widely differing climatic conditions, limited resource availability, for varying feed water qualities and water quality requirements. Due to its user-friendly design, a transparent valuation approach as well as the clear and comprehensible presentation of results, the local users became more aware of the strengths and weaknesses of the considered option and were able to identify the most sustainable supply system or strategy for their decision case. However, due to its simplified semi-quantitative approach, the assessment tool is designed and more suitable for the application in an early planning phase. Thus, the decision support tool can be a good starting point to foster in-house and local discussion on the implementation of water reuse solutions. As the approach requires information input from various sources, it is well suited for collaborative decision-making. In order to make it also suitable as a decision-support tool in the following implementation phases, future research should focus on connecting models and tools to the MULTI-ReUse decision support tool that allow for a more detailed and reliable assessment of preselected water supply options.

ACKNOWLEDGEMENTS

The project MULTI-ReUse (Modular Water Treatment and Monitoring for Wastewater Reuse, 2016–2019) has received funding from the German Federal Ministry of Education and Research as part of the funding measure 'Future-oriented Technologies and Concepts to Increase Water Availability by Water Reuse and Desalination (WavE)', https://water-multi-reuse.org/en/.

DATA AVAILABILITY STATEMENT

All relevant data are included in the paper or its Supplementary Information.

REFERENCES

Aydiner, C., Sen, U., Koseoglu-Imer, D. & Can, E. 2016 Hierarchical prioritization of innovative treatment systems for sustainable dairy wastewater management. *Journal of Cleaner Production* **112**, 4605–4617.

Carr, G., Nortcliff, S. & Potter, R. B. 2010 Water reuse for irrigated agriculture in Jordan: challenges of soil sustainability and the role of management strategies. *Philosophical Transactions: Mathematical, Physical and Engineering Sciences* **368**, 5315–5321.

Chen, Z., Ngo, H. H. & Guo, W. 2012 A critical review on sustainability assessment of recycled schemes. *Science of the Total Environment* **426**, 13–31.

Chen, R., Wang, X. & Liu, Y. 2015 Water reuse system in Xi'an Municipality of China. *Journal of Water Reuse and Desalination* **5** (3), 407–418.

Chu, J., Chen, J., Wang, C. & Fu, P. 2004 Wastewater reuse potential analysis: implications for China's water resources management. *Water Research* **38** (11), 2746–2756.

El Heloui, M., Mimouni, R. & Hamadi, F. 2016 Impact of treated wastewater on groundwater quality in the region of Tiznit (Morocco). *Journal of Water Reuse and Desalination* **6**, 454–463.

European Commission 2020 *Regulation (EU) 2020/741 of the European Parliament and of the Council of 25 May 2020 on Minimum Requirements for Water Reuse.*

Haruvy, N. 1998 Wastewater reuse – regional and economic considerations. *Resources, Conservation and Recycling* **23**, 57–66.

Hernández, F., Urkiaga, A., Fuentes, L. D. L., Bis, B., Chiru, E., Balazs, B. & Wintgens, T. 2006 Feasibility studies for water reuse projects: an economical approach. *Desalination* **187–188** (1–3), 253–261.

Holloway, R., Miller-Robbie, L., Patel, M., Stokes, J., Munakata Marr, J., Dadakis, J. & Cath, T. 2016 Life-cycle assessment of two potable water reuse technologies: MF/RO/UV-AOP treatment and hybrid osmotic membrane bioreactors. *Journal of Membrane Science* **507**, 165–178.

Kavvada, O., Horvath, A., Stokes-Draut, J., Hendrickson, T., Eisenstein, W. & Nelson, K. 2016 Assessing location and scale of urban non-potable water reuse systems for life-cycle energy consumption and greenhouse gas emissions. *Environmental Science & Technology* **50** (24), 13184–13194.

Lyko, H. 2014 Wassermanagement in der Papierindustrie mit und ohne Membran. *F & S Filtrieren und Separien* **28** (4), 206–211 (in German).

Müller, K. & Cornel, P. 2017 Setting water quality criteria for agricultural water reuse purposes. *Journal of Water Reuse and Desalination* **7** (2), 121–135.

Nies, L., Jafvert, C. & Wiener, J. 2016 The assessment of water use and reuse through reported data: a US case study. *Science of the Total Environment* **539**, 70–77.

Sattig, I. 2019 *Technische und wirtschaftliche Optimierung der Wasserrückgewinnung aus Kläranlagenabläufen.* Master Thesis, Hochschule Bremen.

Schramm, E. & Zimmermann, M. 2018 *Das MULTI-ReUse-Verfahren: Häusliches Betriebswasser aus Siedlungsabwasser, fbr – wasserspiegel, No. 2/18 (in German).*

Statistisches Bundesamt 2014 *Statistisches Jahrbuch Deutschland und Internationales 2014.* Statistisches Bundesamt, Wiesbaden, Herausgeber (in German).

Tal, A. 2016 Rethinking the sustainability of Israel's irrigation practices in the drylands. *Water Research* **90**, 387–394.

Track, T. & Kozariszczuk, M. 2018 Industriewasser 4.0 – Digitalisierung im industriellen Wassermanagement. Presentation at ProcessNet-Jahrestagung 33. DECHEMA-Jahrestagung der Biotechnologen, 10–13 September 2018, Eurokongress Aachen (in German).

Uddin, S. M. N., Li, Z., Mang, H.-P., Schüßler, A., Ulbrich, T., Huba, E. M., Rheinstein, E. & Lapegue, J. 2014 Opportunities and challenges for greywater treatment and reuse in Mongolia: lessons learnt from piloted systems. *Journal of Water Reuse and Desalination* **4** (3), 182–193. https://doi.org/10.2166/wrd.2014.008.

Water Europe 2016 *The Value of Water: Towards a Future-Proof Model for a European Water-Smart Society.* Brussels.

Wintgens, T., Bixio, D., Thoeye, C., Jeffrey, P., Hochstrat, R. & Meli, T. 2006 *Reclamation and Reuse of Municipal Wastewater in Europe – Current Status and Future Perspectives Analysed by the AQUAREC Research Project.* AQUAREC Policy Brief.

WWAP (UNESCO World Water Assessment Programme) 2019 *The United Nations World Water Development Report 2019: Leaving No One Behind.* UNESCO, Paris.

Zaibel, I., Zilberg, D., Groisman, L. & Arnon, S. 2016 Impact of treated wastewater reuse and floods on water quality and fish health within a water reservoir in an arid climate. *Science of the Total Environment* **559**, 268–281.

First received 2 April 2020; accepted in revised form 29 September 2020. Available online 23 October 2020

Economic evaluation of different treatment options for water reuse in industrial parks using modular cost functions

Jens Hilbig IWA, Birte Boysen, Philipp Wolfsdorf and Karl-Ulrich Rudolph IWA

ABSTRACT

Industrial parks (IPs) play a significant role in the context of economic growth as well as urban and regional development strategies. They rely on the availability of factors of production and an enabling environment, which also includes the legal framework and economic conditions. The availability of water is essential for the operation and expansion of IPs. Sustainable, efficient and reliable water supply is crucial for IPs and the companies located in the IPs. Reusing wastewater to reduce the amount of drinking or process water necessary for production requires economically viable treatment processes. From an economic point of view, it is important to compare the costs of different treatment trains and to ensure that technical solutions generate an economic benefit for the operators of the IPs. Based on data from municipal wastewater treatment, the authors derive cost functions for individual treatment processes and develop tools for a modular economic assessment of water reuse in IPs.

Key words | calculation model, cost functions, economic evaluation, industrial wastewater treatment, water reuse

Jens Hilbig IWA (corresponding author)
Birte Boysen
Philipp Wolfsdorf
Karl-Ulrich Rudolph IWA
IEEM gGmbH – Institute of Environmental Engineering and Management at the Witten/Herdecke University,
Alfred-Herrhausen-Straße 44, 58455 Witten, Germany
E-mail: *hilbig@uni-wh-ieem.de*

Karl-Ulrich Rudolph
Faculty of Management and Economics, Witten/Herdecke University,
Alfred-Herrhausen-Str. 50, 58455 Witten, Germany

HIGHLIGHTS

- Comparison of costs of different wastewater treatment trains.
- Based on municipal wastewater treatment, a meta-analysis has been conducted to derive cost functions of different treatment technologies.
- Development of a calculation model that can serve as a strategic decision-making tool.
- The objective is a sustainable and economically feasible treatment and recycling of wastewater streams in industrial parks.

INTRODUCTION

Industrial parks (IPs) play a significant role in the context of economic growth and urban and regional development strategies. They rely on the availability of factors of production and an enabling environment, which also includes the legal framework and economic conditions. The availability of water is essential for the operation and expansion of IPs. The concentration of industrial production, on one hand, a major challenge with regard to increasing water demand, (regional) water shortages and other environmental impacts. In northern and western China, for example, economic development is limited by water shortages that are conflicting

doi: 10.2166/wrd.2020.032

with the establishment or expansion of industrial production sites (Bauer *et al.* 2019). On the other hand, IPs open up a wide range of options for saving resources, energy and costs.

Water reuse in IPs is an important measure to reduce the water demand of IPs and to mitigate the depletion of available freshwater resources in the catchments around the industrial areas. The implementation of sustainable reuse concepts is essential to overcome possible future water shortages that would cause huge problems for industries, economies and societies. Based on studies of several IPs in Germany, China and Vietnam, a sustainable water reuse concept for IPs is being developed under the frameworks of the joint research project Water Reuse in Industrial Parks (WaReIp). The main goal of the WaReIp project is to increase water reuse for general infrastructure purposes, in-plant reuse and reuse within specific production processes is not considered.

Sustainable, efficient and reliable water supply is crucial for IPs and the companies located in the IPs. Reusing wastewater to reduce the necessary amount of drinking or process water requires economically viable treatment processes. The pollution of industrial wastewater is often more complex compared with municipal wastewater and requires a more elaborate treatment. Wastewater treatment plants (WWTPs) in IPs often provide several treatment trains for different wastewater inputs. Nonetheless, if sophisticated treatment technologies are already in place (for example, in order to comply with discharge qualities), it sometimes requires only minor adjustments to make the water suitable for infrastructure reuse purposes (Rudolph *et al.* 2014; Zaffaronia *et al.* 2016). Closing the water circuit within IPs (Rudolph 2011) should be considered before tapping new external water resources.

From an economic point of view, it is important to compare the costs of different treatment trains and to ensure that technical solutions generate an economic benefit for the operators of IPs, not limited to a specific kind or size. This will create an incentive to invest in innovative wastewater treatment concepts and technologies. Furthermore, the economic evaluation of different treatment options for water reuse described in this paper is part of a multi-criteria analysis consisting of economic, environmental, technical and other criteria that is conducted under the WaReIp project (Wasser Abwasser Technik 2020).

Among others, a method is being developed to support the decision-making process of industrial users in order to identify and select measures, processes or changes of use for industrial water, wastewater and water reuse. The objective is a sustainable and economically feasible treatment and recycling of wastewater streams in IPs by reducing the demand from natural resources and providing reuse water for several infrastructure purposes, for example sanitation (toilet flushing), irrigation and street cleaning. Within WaReIp, an infrastructure reuse factor of up to 25% has been calculated, which is defined as reuse-flows from the central WWTP (Bauer *et al.* 2019).

The economic benefit of WaReIp depends both on the costs of the existing water supply and wastewater treatment systems and on external factors. These factors, such as legal requirements, effluent standards, water shortages or demand-related behaviour, are important additional drivers of water reuse. In contrast, water reuse systems tend to be more expensive than 'traditional' water supply systems if the price of the existing water supply is subsidised or not calculated on a full cost recovery basis. To compare the cost of the entire water cycle of IPs, the required investment and annual costs of water treatment plants can be calculated based on the cost functions.

The objective of this paper is to show an approach of how to estimate the cost of wastewater treatment in an IP with municipal investment cost functions and a developed MS-based calculation model.

Using cost functions for possible reuse projects has been done before for municipal wastewater. For example in a project funded by the European Union and the Italian Ministry of Education, Universities and Scientific Research (Sipala *et al.* 2003), the costs of different municipal wastewater treatment and reuse scenarios were investigated. With the help of an online tool, non-experts can also explore different reuse options, the legal framework, process combinations and the associated costs. The cost functions for individual procedures and procedure combinations are based on Italian and international literature research. Functions are given for primary, secondary treatment, filtration, nitrification/denitrification and filtration, nitrification/denitrification plus phosphate decomposition and filtration, coagulation flocculation, carbon adsorption as well as reverse osmosis (see Table 1). They are specified for WWTPs either smaller

Table 1 | Cost functions of individual treatment technologies (Sipala *et al.* 2003)

Treatment alternatives	***X*** **< 30,000 p.e.**	***X*** **> 30,000 p.e.**
Primary treatment	$Y = 0.317–9 \times 10^{-6} \times X$	$Y = 0.132–5 \times 10^{-7} \times X$
Secondary treatment	$Y = 0.474–7 \times 10^{-6} \times X$	$Y = 0.309–4 \times 10^{-7} \times X$
Filtration	$Y = 0.507–7 \times 10^{-6} \times X$	$Y = 0.342–4 \times 10^{-7} \times X$
Nitrification/denitrification + filtration	$Y = 0.559–8 \times 10^{-6} \times X$	$Y = 0.369–5 \times 10^{-7} \times X$
Nitrification/denitrification + phosphate removal + filtration	$Y = 0.602–8 \times 10^{-6} \times X$	$Y = 0.393–5 \times 10^{-7} \times X$
Coagulation–flocculation	$Y = 0.939–2 \times 10^{-5} \times X$	$Y = 0.471–5 \times 10^{-7} \times X$
Carbon adsorption	$Y = 1.132–1 \times 10^{-5} \times X$	$Y = 0.730–5 \times 10^{-7} \times X$
Reverse osmosis	$Y = 1.503–2 \times 10^{-5} \times X$	$Y = 0.907–5 \times 10^{-7} \times X$

Y indicates the unit cost in €/m^3.
X indicates the number of population equivalent (p.e.).
For *X* < 1,000 p.e., a constant cost is assumed equal to that obtained for 1,000 p.e.
For *X* > 200,000 p.e., a constant cost is assumed equal to that obtained for 200,000 p.e.

or larger than 30,000 inhabitants. However, no distinction is made between investments and operating costs. The specific total costs may also serve to validate the calculated combined data of investment and operational expenditures but are not used for the average WaReIp cost functions.

Another example of the use of cost functions when differentiation between investment and operational expenditure is not needed, the results of the Spanish project NOVEDAR-Consolider can be adduced (Hernandez-Sancho *et al.* 2011). Within the project in 2010, cost data from 341 WWTPs were evaluated. Depending on the treatment technology, the cost functions can include not only the capacity but also the age of the treatment plant, the degradation rate of settleable solids, chemical oxygen demand (COD), biological oxygen demand (BOD), nitrogen and phosphorus, as shown in Table 2. Cost functions for a total of seven treatment stages are given: extended aeration (EA); activated sludge without nutrient removal (AS); activated sludge with nutrient removal (NR); bacterial beds (BB); peat beds (PB); biodisk (BD) and tertiary treatment (TT) (Hernandez-Sancho *et al.* 2011). The Spanish cost functions are not used to derive the calculated cost functions but for the purpose of the WaReIp project, and they can be used for verification if the corresponding degradation rates are known.

In this publication, cost functions are further examined and broken down in order to adjust them for industry and country-specific circumstances in the next step.

METHODICAL APPROACH

Industrial wastewater treatment requires industry-specific technical process stages, for example pre-flotation in the

Table 2 | Cost functions of individual treatment technologies (according to Hernandez-Sancho *et al.* (2011))

Technology	**Cost functions**	R^2
EA (extended aeration)	$C = 169.4844\ V^{04540} e^{(0.0009A+06086SS)}$	0.6133
AS (activated Sludge without nutrient removal)	$C = 2.1165\ V^{07128} e^{(0.0174A+1.5122SS+0.0372BOD)}$	0.6849
NR (activated Sludge with nutrient removal)	$C = 2.518 V^{07153} e^{(0.518+1455COD+0258N+0243P)}$	0.7301
BB (bacterial beds)	$C = 17.3617\ V^{0.5771} e^{(0.1006A+0.6932COD)}$	0.9862
PB (peat beds)	$C = 1{,}510.8400\ V^{0.2596} e^{(0.0171SS)}$	0.5240
BD (biodisk)	$C = 28.9522\ V^{0.4493} e^{(2.3771SS)}$	0.8058
TT (tertiary treatment)	$C = 3.7732\ ^{V0.7223} e^{(0.6721COD+0.0.1958N+0.7603P)}$	0.9029

food industry. In Germany, requirements for 55 different branches are given in the wastewater ordinance (Abwasserverordnung 2020). Furthermore, industrial WWTPs are often enclosed and automated and therefore vary from municipal WWTPs. The ideal treatment process depends not only on the source of wastewater but on the reuse purposes as well. Wherever the reuse water is produced for agricultural purpose nutrients (phosphorous and nitrogen) are valuable and there is no advantage in removing them before reuse. When the reuse water is purposed, for example, for electronic production processes or potable reuse nutrients have to be removed. To cover as many possibilities as possible, a broad spectrum of treatment processes is required. However, a literature review showed that most of the data and cost functions published for treatment options are based on municipal wastewater treatment. Interviews with companies and IP operators in Germany, China and Vietnam led to similar results: either detailed data on individual treatment steps are not collected in IPs or companies are quite reluctant to publish process details and cost functions, as they are concerned to disclose commercially sensitive information.

To bypass this problem, data of municipal wastewater treatment have been collected in a meta-analysis using several sources. The data collected date back to the 1990s and include different studies, research projects, technical guidelines or public reference data mainly from Germany. The data have been converted into Euros to a common reference year (2017) taking into account inflation rates as well as exchange rates for data from other countries than Germany. By this, an average cost function is derived for each different treatment processes. Despite the limitation of the underlying data (e.g. long observation period, different size of treatment plants in the individual studies and heterogeneous study design and number of plants evaluated), the resulting cost functions provide sufficient information for further use. Under the WaReIp project, the cost functions are used to show general trends and allow for a comparison of different treatment steps as well as a model calculation of capital (CAPEX) and operational expenditure (OPEX)) as well as annual costs and costs per cubic metre of different treatment trains. This calculation is conducted in a specially developed MS office calculation model to allow transparency and easy adaptability. Unlike the described cost functions within the WaReIp project, municipal CAPEX and OPEX data are collected and calculated separately in an approach to adapt input parameters individually according to industry- or country-specific conditions and then calculate the cost per cubic metre or annual cost accordingly.

DERIVATION OF COST FUNCTIONS

By setting up cost functions for individual treatment steps, possible process chains and reuse of partially treated wastewater within the IP can be economically evaluated in a modular fashion and with varying volumes. In the following paragraphs, some of the cost functions incorporated are described.

Cost functions and data for individual process steps are available for municipal sewage treatment plants from various sources (Bohn 1993; Türk *et al.* 2013; Horstmeyer *et al.* 2014). The cost functions of treatment technologies published in several different studies include the usual process steps for municipal WWTPs: screen, grit chamber, primary and secondary clarification, phosphate precipitation/flocculation, activated sludge process, UV disinfection, ozonation, sludge thickening, mechanical sludge dewatering and digestion/stabilisation.

Publications explaining the cost functions of several treatment stages mostly date back to a time when sewage treatment plants were increasingly built in Germany, such as Bohn (1993), Schoenenberg (1996) as well as Günthert & Reicherter (2001). These publications use data from existing WWTPs as examples to derive cost functions. Some of these functions are divided into civil works and mechanical equipment. The derived functions are generally linear or degressive. Economies of scale are clearly identifiable, as can be seen in Figure 1 showing investments of activated sludge.

The Hessisches Ministerium für Umwelt, ländlichen Raum und Verbraucherschutz (Hessian Ministry for the Environment, Rural Areas and Consumer Protection) has published cost functions which are to be used as reference values for the investment of new WWTPs (Hessisches Ministerium für Umwelt ländlichen Raum und Verbraucherschutz 2006). In the published graphs, economies of scale are apparent with limitations as shown in Figure 2.

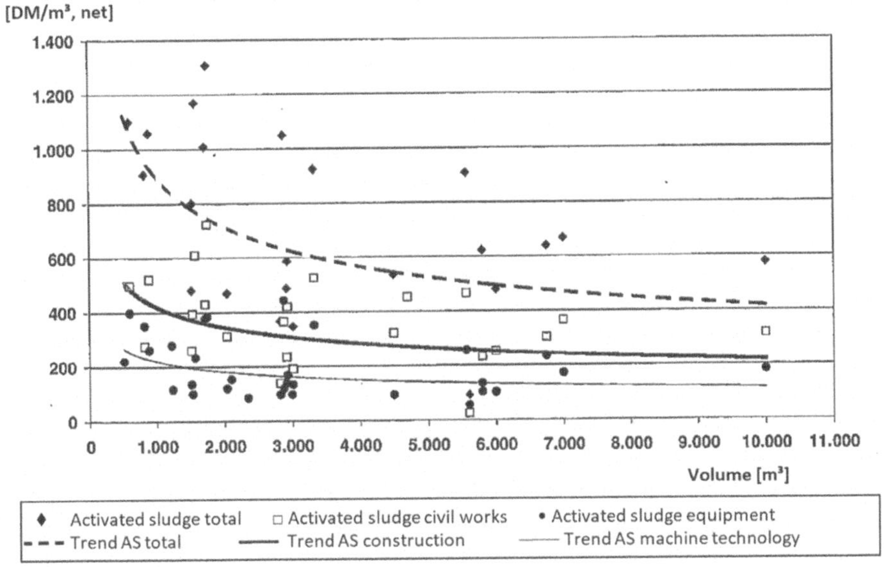

Figure 1 | Cost functions investment of activated sludge (Günthert & Reicherter 2001).

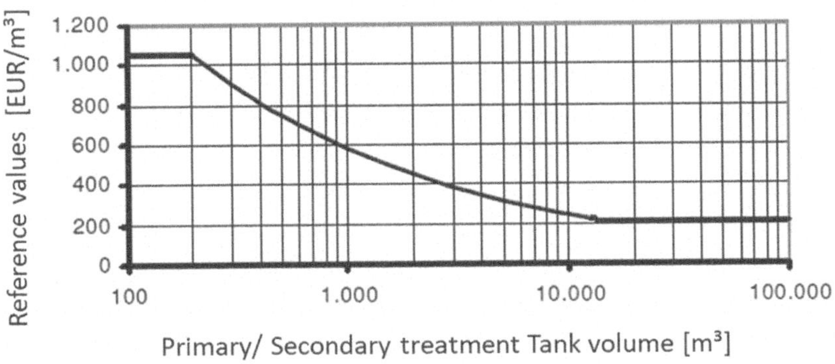

Figure 2 | Cost function primary/secondary treatment tanks (Hessisches Ministerium für Umwelt ländlichen Raum und Verbraucherschutz 2006).

Horstmeyer *et al.* (2014) set out various cost functions of individual treatment steps. However, due to a partial lack of data basis or too wide a spread, it is not possible to calculate the costs for all treatment steps. Not only the cost functions determined by Horstmeyer *et al.* but also the data points are included in the underlying database of this study and used to develop cost functions.

In its study 'Micropollutants from municipal wastewater', the Swiss Federal Office for the Environment refers to data from Hunziker-Betatch from a previous study in 2008 (Hunziker-Betatech 2008; Abegglen & Siegrist 2012). The cost analysis includes the investment and operating costs for two different dosages for ozonation and activated carbon adsorption at six sites in Switzerland (Hunziker-Betatech 2008). The specific extension costs of UV systems are described by Müller *et al.* (2009). A cost function can be derived from the examples given.

In the study 'Economic benefit of upgrading municipal wastewater treatment plants to eliminate organic trace substances, pharmaceuticals, industrial chemicals, bacteriologically relevant germs and viruses' by Türk *et al.* (2013), operating and investment costs of activated carbon treatment and ozonation are compared. The cost functions of the investment are based on constructed plants. The investments consist of civil works, equipment, electronic instrumentation and control (E/I&C) and additional costs

(Türk *et al.* 2013). Some of these treatment plants are also part of the data basis of Hunziker-Betatech (2008), and the dual use of the data in the cost functions slightly influences the arithmetic mean.

Metzger *et al.* (2014) evaluate in their study extensions of WWTPs by adsorption stages from Baden-Württemberg. For the six plants considered, the capital costs and operating costs are given with further subdivisions. A study commissioned by the German Federal Environment Agency on an area-wide coverage of the fourth treatment stage in Germany shows the total costs for ozonation and treatment with activated carbon. Among others, data from Hunziker-Betatech (2008) and Metzger *et al.* (2014) are used (Hillenbrand *et al.* 2015, 2016).

Palmowski & Pinnekamp (2018) use the costs of adding activated carbon to an existing filter with data published by Bornemann *et al.* (2015) to derive a cost function for the investment of silos and dosing stations. In a study by Helmreich *et al.* (2017), the costs of phosphate precipitation are analysed in relation to the size of the WWTP. For municipal WWTPs between 1,000 and 25,000 inhabitants, a distinction is made between investments, precipitant costs, costs for sludge disposal and costs for labour and maintenance. The specific costs for additional phosphorus removed are also given (Helmreich *et al.* 2017). As a result of the recent discourse on the fourth treatment stage and the resulting additional costs, there are more and more publications on the possible process steps of ozonation, activated carbon and disinfection with UV (Hillenbrand *et al.* 2016; Jekel *et al.* 2016; Palmowski & Pinnekamp 2018).

For other treatment processes like Membrane Bioreactors, a sufficient database from Germany is not given. Pinnekamp (2007), for example, published an invest cost function for German MBR based on six plants. The availability of International data is better (DeCarolis *et al.* 2004, 2007). To avoid the issue of different cost levels from different countries as much as possible, it is refrained from combining and publishing this data. As described above, there is no sufficient data from industrial WWTPs for the individual process steps. Therefore, the published municipal data and internal IEEM data from former projects are used. To cover a broad spectrum of possible process chains, a rather large number of process steps is used: screen, grit chamber, primary and secondary clarification, phosphate precipitation/flocculation, anaerobic reactors, activated sludge, UV disinfection, ozonation, activated carbon adsorption, sludge thickening, mechanical sludge dewatering, digestion/stabilisation, sandfiltration and chlorination.

In order to be able to compare the above-described cost functions of different references, the functions must be converted to one price level and one unit. Cost data from other countries are converted (via price index of the respective country) to a 2017 level and then converted into Euros. The data are mostly based on German WWTPs with exceptions, for example, from Hunziker-Betatech (2008) and the data for chlorination (US EPA 1999; Das 2002). For compatibility, the results are converted to €/m^3/d and €/m^3. The average daily runtime of machines was used from Schoenenberg (1996), and it was assumed that the treatment plant runs 365 days a year. As most of the functions were given as net costs, it is assumed when no information of the value-added tax (VAT) is available those are net as well.

By calculating the arithmetic function of the individual cost functions from literature and introducing a trend line, an average cost function (potential (average)) of the municipal WWTPs can be derived for individual process steps as shown in Figure 3. This also covers treatment sizes that are not considered in all of the underlying publications of the database. Despite the limitations of the underlying database, it is possible to determine cost functions that allow for an estimation of the costs for individual treatment processes.

The determination of an average net cost function is carried out for the possible treatment technologies listed in Table 3.

Cost functions for investment are as far as available from the literature and useful broken down to civil works, equipment and if available E/I&C (see Figure 4) to later calculate depreciations and adapt the sections separately to country-specific conditions. In case a specific breakdown is not available, a percentage allocation is made.

OPEX is calculated separately. There are a large number of publications on operational costs for individual treatment technologies, such as activated carbon adsorption or ozonation (Hunziker-Betatech 2008; Türk *et al.* 2013; Metzger *et al.* 2014). For other process steps, the operating costs can be estimated individually via publications on energy, consumable and labour demand as well as maintenance requirements.

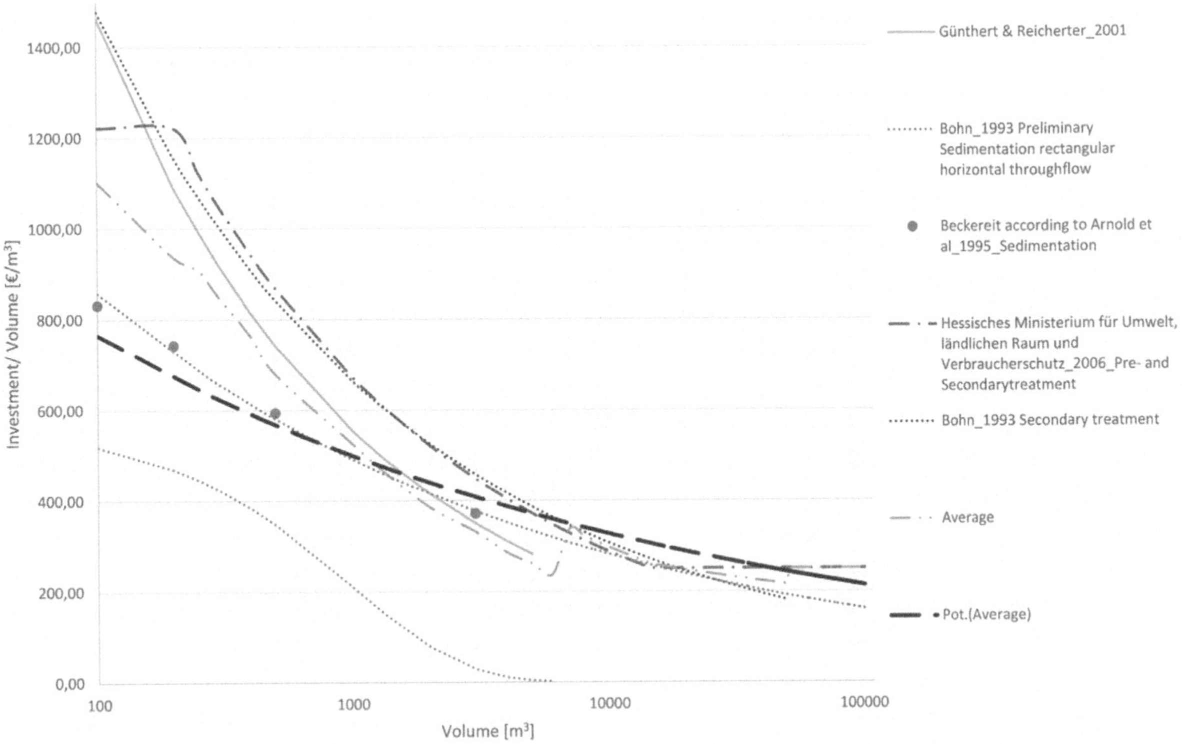

Figure 3 | Determined cost functions primary/secondary treatment.

Table 3 | Determined cost functions

Process	x	y	Average cost function investment	Range for x
Screen	m^3/d	€/m^3/d	$y = 680.78x^{-0.297}$	100–50,000
Grit chamber	m^3/d	€/m^3/d	$y = 2215.1x^{-0.392}$	100–20,000
Pre-/secondary treatment	m^3	€/m^3	$y = 1791.5x^{-0.185}$	100–100,000
Activated sludge	m^3	€/m^3	$y = 1682.7x^{-0.148}$	100–70,000
Phosphate precipitation/flocculation	m^3	€/m^3	$y = 11.950x^{-0.574}$	800–20,000
UV disinfection	m^3/d	€/m^3/d	$y = 1209.2x^{-0.328}$	100–100,000
Ozonation	m^3/d	€/m^3/d	$y = 348.02x^{-0.069}$	100–100,000
Activated carbon adsorption	m^3/d	€/m^3/d	$y = 29.458x^{-0.533}$	700–100,000
Sandfiltration	m^3/d	€/m^3/d	$y = 2045.3x^{-0.273}$	3,000–100,000
Chlorination	m^3/d	€/m^3/d	$y = 3.416x^{-0.422}$	3,000–100,000
Digestion/stabilisation	m^3	€/m^3	$y = 8192.8x^{-0.322}$	400–50,000
Sludge thickening	m^3	€/m^3	$y = 2220.8x^{-0.226}$	10–10,000
Mechanical sludge dewatering	m^3/d	€/m^3/d	$y = 39.745x^{-0.155}$	30–14,000

The energy demand of individual process steps depending on the size of the plant is also frequently discussed, and energy-intensive processes are the focus of these publications (Schoenenberg 1996; Bolle & Pinnekamp 2011). The energy demand is often described as a linear function depending on the volume of wastewater.

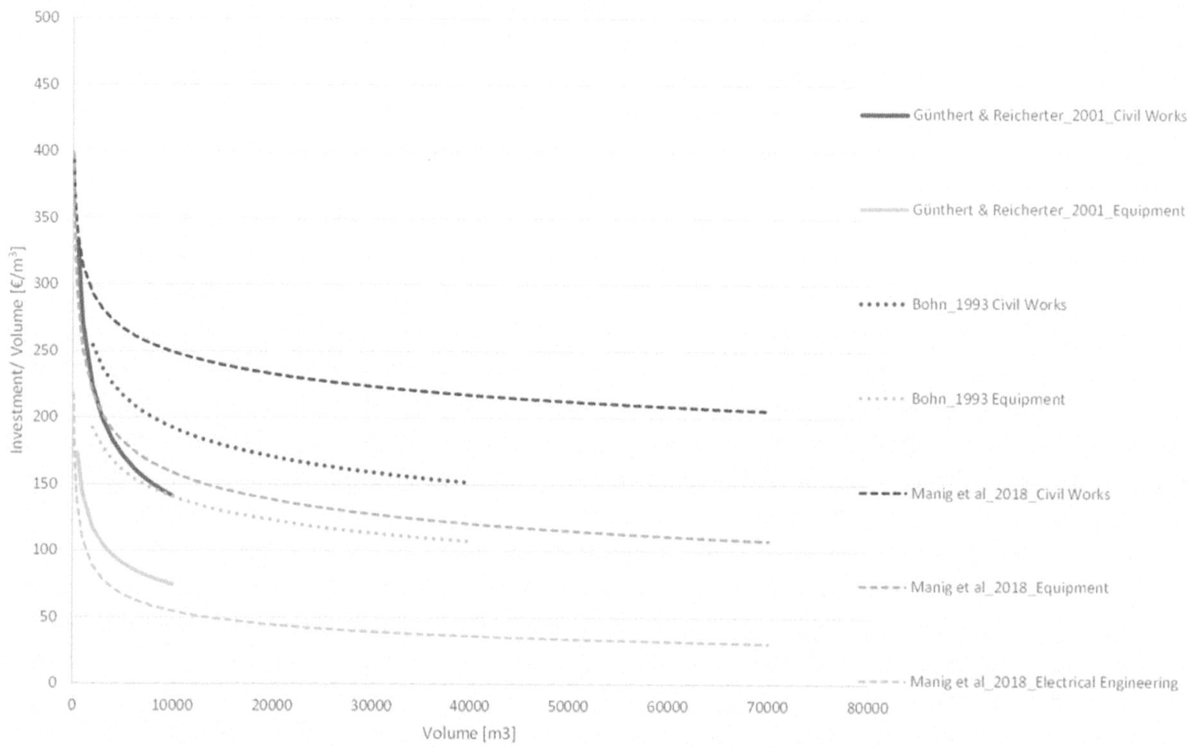

Figure 4 | Literature examples of the breakdown of the activated sludge cost function.

In the leaflet 'DWA M-271' published by the German Association for Water, Wastewater and Waste nomograms of the labour requirements in relation to the size of the plant are given, for example, the biological treatment in Figure 5 (DWA 2017). These can be multiplied by an hourly wage definition of the labour costs. In this case, the wage should not only include one person but an engineer (e.g. 15%), a foreman (e.g. 25%) and technical staff (e.g. 60%).

The average cost of maintenance is presented, for example, for activated carbon and ozonation in Hunziker-Betatech (2008) and Türk *et al.* (2013). In cases where no further information regarding maintenance is available percentages of the investment cost can be used for civil works 0.5%, for equipment 2.5% and E/I&C 1% (Jekel *et al.* 2016). The operational expenses like energy and consumable demand are based on the input data from the module library developed under the WaReIp project and therefore no overall operating expenses cost functions are assembled.

The provided data or customised data can be entered into the calculation model to estimate overall OPEX, annual and cost per cubic metre.

CALCULATION MODEL

In the next step, the derived cost functions and the OPEX data are incorporated into a specifically developed MS Excel-based calculation model that can serve as a strategic decision-making tool when weighing potential water reuse investment alternatives for IPs. Therefore, the model can calculate up to three different investment alternatives in parallel and compare the resultant costs per cubic metre of produced reuse water over time in order to support the decision-making process.

Guidelines and descriptions of how to calculate the cost of water treatment are widely available (EPA 2014; Ruiz-Rosa *et al.* 2016; Papapetrou *et al.* 2017). MS Excel has been chosen as a widespread and easy accessible tool. The developed calculation model in MS Excel is therefore ideal since input data, as well as formulas, can be altered.

Considering the available data basis for the derived cost functions, the model provides the best results for IPs for various sizes. With the derivation of cost functions from multiple sources with different sizes, an enlarged range

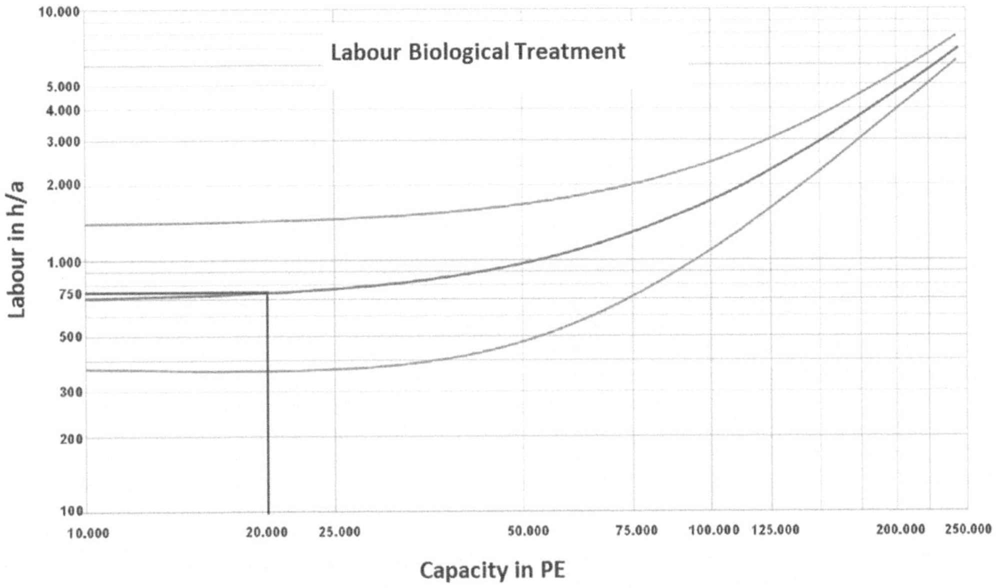

Figure 5 | Nomogram of labour requirements for biological treatment (based on DWA M-271).

can be covered. For example, primary/secondary treatment from 100 m^3 tanks up to 100,000 m^3.

The model was designed in a modular way to allow comparability of different compositions of the technologies and process chains, including sand filtration, UV disinfection, chlorination, and the required distribution network. The main modules are basic set-up, investment expenditures and operative expenditures. The basic set-up module allows definition of several variables that are valid for all three investment alternatives throughout the model, e.g. inflation rate, used currency, VAT rate or operative days per year of the plant. Furthermore, for each investment alternative individually, design capacities, as well as inflow and outflow volumes of each process step, are defined here.

The investment module includes the derived cost functions for investment costs in order to calculate investment amounts and depreciation. Investments are calculated separately for each process step and the required distribution network. For further comparability and more accurate depreciation periods, each technology is (where applicable) separated into investment expenditures for general costs/fees/reserves, environmental works, civil works, equipment and E/I&C. Necessary reinvestments depending on the depreciation periods are automatically taken into account by underlying formulas.

The operating expenditure module similarly differentiates between the processes and distribution network on the one hand and between labour, maintenance, energy and consumable expenditures on the other hand. Where available, the derived cost functions for operative expenditures are included; otherwise, alternative calculations have been applied (e.g. maintenance as a percentage of investment costs). Default values and prices can be used or individually adjusted and therefore calculated with current and specific prices.

The focus of the calculation model is to determine both the total costs and the costs per m^3 of reuse water per scenario and to make them comparable in order to support the decision-making process from an economic point of view. Due to the modular structure, each individual module can also be compared with each other and its influence on the overall cost can be assessed. The model can either be used with the initially derived cost functions, be adapted to a particular company or industry needs or be transferred to meet the conditions of a specific country and industry.

Alternatively, the model can also be used to analyse sensitivities for one investment scenario by comparing it with itself under slightly altered assumptions. In addition, it can theoretically be used as a standalone tool for project calculation, if actual data for all aspects of investment are available and the cost functions are not needed.

CONCLUSION AND OUTLOOK

A literature review, as well as interviews with companies and IP operators in Germany, China and Vietnam, led to the result that either detailed data on individual treatment steps are not collected in IPs or companies are quite reluctant to publish process details and costs. With the approach of using average municipal investment cost functions and divided data on operational expenses and adapting them to industry and country requirements, estimations of industrial wastewater treatment are possible. In a first step, cost functions and costs from municipal plants were collected in a meta-analysis. The cost functions are used to show general trends and allow for a comparison of different treatment steps as well as a model calculation of CAPEX and OPEX as well as annual cost and cost per cubic metre in the developed calculation model.

The cost functions and the calculation model are subject to certain limitations. First, the accuracy of the calculation model's results is dependent on the quality of the derived cost functions. The cost functions as described date back to the 1990s and are not comprehensive, especially with the focus on processes used in the industrial sector. Additional data on similar applicable projects are desirable to further improve the cost functions, as data are from studies with heterogeneous designs and vital information like the inclusion of VAT is often missing. By including data from the United States of America and Switzerland, the cost level is not limited to Germany. Additionally, the model required a generalist approach in design and neglects detailed customisation options for individual calculations in favour of comparability between investment alternatives and transferability of the model to similar conditions. The determined investment cost functions can be used for rough estimates and first impressions of possible investments but need to be adapted to industry and country specifics. Furthermore, the cost degression in the investment is clearly visible. With the ability to adapt the modular calculation model to specific requirements, it is broadly applicable, for example, by IP planners or operators.

To validate the database and model outputs, further expert interviews are currently ongoing. The main objective is to transfer the data from municipal WWTPs to IP requirements and to assess the transferability to other countries, e.g. water-scarce regions. It is recognisable that investments tend to be higher in industrial WWTPs compared with municipal plants due to a high degree of specialisation and automation. The investment in automation technology may be even higher when the WWTP is connected to the existing control and monitoring system of the IP. On the other hand, labour requirements tend to be lower, but salaries tend to be higher compared with municipal WWTPs. A plausible explanation besides the automation is a more efficiently operated plant in IPs.

At this point, the methodology including the cost functions and the calculation model is tested to evaluate the WaReIp Model Industrial Park Reuse Plant which is described in Boysen *et al.* (2020). For this purpose, the German cost functions need to be transferred to Chinese conditions and circumstances, to match the Model IP that is located in China. To achieve this, a data query was started and the first inputs for the Chinese model were incorporated. First results show that, for the majority of the indicators, the installation of the water reuse plant seems to be beneficial for all examined reuse options. From an economic point of view, reuse options with larger volumes of water treated will be preferred since the more water is treated, the cost per cubic metre of reuse water is reduced due to the economies of scale.

Even in cases where water reuse is compared with 'traditional' water supply systems not favourable from purely an economic point of view, it can provide a viable and efficient option to increase water availability for IPs, especially in water-scarce regions. The most promising strategy is to use reuse water for applications that require low water qualities and little additional treatment efforts, e.g. some infrastructural reuse applications (Bauer *et al.* 2019).

ACKNOWLEDGEMENT

The authors would like to express their gratitude to the BMBF for the financial funding of the WaReIp project (grant no. 02WAV1409), as well as to the experts and partners in Germany, Vietnam and China for their cooperation and contribution and to the anonymous reviewers for their valuable comments.

DATA AVAILABILITY STATEMENT

All relevant data are included in the paper or its Supplementary Information.

REFERENCES

Abegglen, C. & Siegrist, H. 2012 *Mikroverunreinigungen aus kommunalem Abwasser (Micropollutants in Municipal Wastewater)*. Available from: http://www.bafu.admin.ch/uw-1214-d (accessed 31 March 2020)

Abwasserverordnung 2020 Verordnung über Anforderungen an das Einleiten von Abwasser in Gewässer (AbwV) (Ordinance on requirements for the discharge of wastewater into waters (Wastewater Ordinance). Waste Water Ordinance in the version promulgated on 17 June 2004 (BGBl. I p. 1108, 2625), last amended by Article 1 of the Ordinance of 16 June 2020 (BGBl. I p. 1287).

Bauer, S., Dell, A., Behnisch, J., Linke, H. J. & Wagner, M. 2019 Sustainability requirements of wastewater management concepts for new industrial park developments in water-stressed regions. In *Book of Abstracts: IWA Water Reuse 2019. 12th IWA International Conference on Water Reclamation and Reuse*, 16–20 June 2019, Overcoming Water Stress, Berlin, Germany, pp. 274–280.

Bohn, T. 1993 *Wirtschaftlichkeit und Kostenplanung von kommunalen Abwasserreinigungsanlagen (Economic Efficiency and Cost Planning of Municipal Wastewater Treatment Plants)*. Expert-Verlag (Schriftenreihe des Institutes für Baubetriebslehre der Universität Stuttgart, 34), Ehningen, Germany.

Bolle, F. & Pinnekamp, J. 2011 *Energieeinsparung bei Membranbelebungsanlagen – Phase I (Energy Savings of Membrane Bioreactors – Phase I)*. Available from: https://www.lanuv.nrw.de/fileadmin/tx_mmkresearchprojects/Abschlussbericht_EnReMem_Phase_1.pdf (accessed 18 February 2020).

Bornemann, C., Alt, K., Böhm, F., Hachenberg, M., Kolisch, G., Nahrstedt, A. & Taudien, Y. 2015 *Technische Erprobung des Aktivkohleeinsatzes zur Elimination von Spurenstoffen in Verbindung mit vorhandenen Filteranlagen 'FilterAK+', Abschlussbericht (Technical Testing of the Activated Carbon Application for the Elimination of Trace Substances in Connection with Existing Filter Systems, Final Report)*. Available from: https://www.lanuv.nrw.de/fileadmin/tx_mmkresearchprojects/151117_Abschlussbericht_FilterAK_.pdf (accessed 27 March 2020)

Boysen, B., Cristóbal, J., Hilbig, J., Güldemund, A., Schebek, L. & Rudolph, K.-U. 2020 Economic and environmental assessment of water reuse in industrial parks. *Water Reuse and Desalination* **10** (4), 475–489. doi: 10.2166/wrd.2020.034.

Das, T. K. 2002 Evaluating the life cycle environmental performance of chlorine disinfection and ultraviolet technologies. *Clean Technologies and Environmental Policy* **4** (1), 32–43. doi:10.1007/s10098-002-0139-x.

DeCarolis, J., Adham, S., Grounds, J., Pearce, B. & Wasserman, L. 2004 *Proceedings of the Water Environment Federation* (10) 401-411. doi:10.2175/193864704784131383

DeCarolis, J., Adham, S., Pearce, W. R., Hirani, Z., Lacy, S. & Stephenson, R. 2007 Cost trends of MBR systems for municipal wastewater treatment. *Proceedings of the Water Environment Federation* **2007** (15), 3407-3418. doi:10.2175/193864707787973734.

Deutsche Vereinigung für Wasserwirtschaft, Abwasser und Abfall e.V. 2017 *Personalbedarf für den Betrieb kommunaler Kläranlagen (Staff Requirements for the Operation of Municipal Sewage Treatment Plants)*. DWA-Regelwerk A 271, Hennef, Germany.

Günthert, F. W. & Reicherter, E. 2001 *Investitionskosten der Abwasserentsorgung (Investment Costs for Sewage Disposal)*. Oldenbourg-Industrieverlag, München, Germany.

Helmreich, B., Huber, M., Muntau, M., Athanasiadis, K. & Steinle, E. 2017 *Analyse einer möglichst weitestgehenden Phosphorelimination bei kommunalen Kläranlagen (Analysis of an Extensive Phosphorus Elimination in Municipal Wastewater Treatment Plants)*. Available from: https://www.bgu.tum.de/fileadmin/w00blj/sww/Publikationen/Endbericht-20170919_LfU-P-Elimination.pdf (accessed 25 March 2020)

Hernandez-Sancho, F., Molinos-Senante, M. & Sala-Garrido, R. 2011 Cost modelling for wastewater treatment processes. *Desalination* **268** (1–3), 1–5. doi:10.1016/j.desal.2010.09.042.

Hessisches Ministerium für Umwelt, ländlichen Raum und Verbraucherschutz 2006 *Verordnung zur Änderung der Verordnung über Zuweisungen zum Bau von Abwasseranlagen (Hessian Ordinance on Funding Regulations for the Construction of Wastewater Systems)*. Available from: https://umwelt.hessen.de/sites/default/files/media/hmuelv/arbeitsexemplar_-_verordnung_zur_aenderung_der_verordnung_ueber_zuweisungen_zum_bau_von_abwasseranlagen.pdf (accessed 22 March 2020)

Hillenbrand, T., Tettenborn, F., Menger-Krug, E., Marscheider-Weidemann, F., Fuchs, S., Toshovski, S., Kittlaus, S., Metzger, S., Tjoeng, I., Wermter, P., Kersting, M. & Abegglen, C. 2015 *Maßnahmen zur Verminderung des Eintrages von Mikroschadstoffen in die Gewässer (Measures to Reduce Micropollutant Emissions to Water)*. Available from: https://www.umweltbundesamt.de/sites/default/files/medien/378/publikationen/texte_85_2014_massnahmen_zur_verminderung_des_eintrages_von_mikroschadstoffen_in_die_gewaesser_0.pdf (accessed 21 March 2020)

Hillenbrand, T., Tettenborn, F., Fuchs, S., Toshovski, S., Metzger, S., Tjoeng, I., Wermter, P., Kerstin, M., Hecht, D., Werbeck, N., Wunderlin, P., Geiger, Y., Wöhler, L., Zörb, F., Palm, N., Tettinger, S., Karl, H., Ebben, T. & Meier, A. 2016 *Maßnahmen zur Verminderung des Eintrages von Mikroschadstoffen in die*

Gewässer – Phase 2 (Measures to Reduce Micropollutant Emissions to Water – Phase 2). Available from: https://www.umweltbundesamt.de/sites/default/files/medien/378/publikationen/texte_85_2014_massnahmen_zur_verminderung_des_eintrages_von_mikroschadstoffen_in_die_gewaesser_0.pdf (accessed 21 March 2020)

Horstmeyer, N., Rapp-Fiegle, S., Helmreich, B. & Drewes, J. E. 2014 *Kosten der Abwasserbehandlung. Finanzierung, Kostenstrukturen und Kostenkenndaten der Bereiche Kanal, Sonderbauwerke und Kläranlagen (Costs of Waste Water Treatment. Financing, Cost Structures and Cost Data of Sewers, Special Buildings and Sewage Treatment Plants)*. DIV Dt. Industrieverlag (Edition GWF Wasser, Abwasser), München, Germany.

Hunziker-Betatech 2008 *Maßnahmen in ARA zur weitergehenden Elimination von Mikroverunreinigungen – Kostenstudie (Measures to Further Eliminate Micropollutants in WWTPs – Cost Study)*. Winterthur, Switzerland. Available from: https://www.micropoll.ch/fileadmin/user_upload/Redaktion/Dokumente/01_Berichte/04_Kosten/2008_BAFU_Mikroverunreinigungen_Kostenstudie.pdf (accessed 21 March 2020)

Jekel, M., Altmann, J., Ruhl, A. S., Sperlich, A., Schaller, J. & Gnirß, R. 2016 *Integration der Spurenstoffentfernung in Technologieansätze der 4. Reinigungsstufe bei Klärwerken. Abschlussbericht IST4R (Integration of Micropollutant Removal in Technology Approaches for the Fourth Stage of Wastewater Treatment, Final Report)*. Universitätsverlag der TU Berlin, Berlin, Germany.

Metzger, S., Tjoeng, I., Rößler, A., Schwenter, G. & Rölle, R. 2014 Kosten der Pulveraktivkohleanwendung zur Spurenstoffelimination am Beispiel ausgeführter und in Bau befindlicher Anlagen (Costs of the application of powdered activated carbon to eliminate micropollutants). *KA – Korrespondenz Abwasser, Abfall* **61** (11), 1029–1037.

Müller, K., Bleisteiner, S., Pirchner, A., Gnirß, R. & Hübner, M. 2009 Desinfektion von biologisch gereinigtem abwasser (Disinfection of biologically treated wastewater). *KA – Korrespondenz Abwasser, Abfall* **56** (6), 593–599.

Palmowski, L. & Pinnekamp, J. 2018 *Abschlussbericht: Entwicklung und Integration innovativer Kläranlagentechnologie für den Transformationsprozess in Richtung Technikwende 'E-Klär'. (Towards the Energy-Optimised Wastewater Treatment Plant of the Future – Development and Model-Based Integration of Innovative Treatment Technologies for Transformation Processes. Final Report of the Research Project)*. RWTH Aachen, Aachen, Germany.

Papapetrou, M., Cipollina, A., La Commare, U., Micale, G., Zaragoza, G. & Kosmadakis, G. 2017 Assessment of methodologies and data used to calculate desalination costs. *Desalination* **419**, 8–19. doi:10.1016/j.desal.2017.05.038.

Pinnekamp, J. 2007 *Abschlussbericht: Optimierung der Gestaltung und des Betriebes von Membrananlagen in der kommunalen Abwasserreinigung*. (Optimisation of the design and operation of membrane plants in municipal wastewater treatment. Final Report of the Research Project). RWTH Aachen, Aachen, Germany.

Rudolph, K.-U. 2011 Case 3: The Contribution of Water Technology to Job Creation and Development of Enterprises. UNW-DPC's Capacity Pool, Quarterly magazine from the UN-Water Decade Programme on Capacity Development, Issue No. 13, 2011, S. 31–33.

Rudolph, K.-U., Kreuter, S., DuyCam, B. & Nguyen, v. L. 2014 Abwasserkonzept für Industriezonen in Vietnam (Wastewater concept for industrial zones in Vietnam). *Wasserwirtschaft Wassertechnik* **9**, 8–13. Special Industrie + Wasser.

Ruiz-Rosa, I., García-Rodríguez, F. J. & Mendoza-Jiménez, I. 2016 Development and application of a cost management model for wastewater treatment and reuse processes. *Journal of Cleaner Production* **113**, 299–310.

Schoenenberg, H. 1996 *Betriebstechnik, Kosten und Rechtsgrundlagen der Abwasserreinigung (Operating Technology, Costs and Legal Regulations of Wastewater Treatment)*, 4th edn. Ernst & Sohn, Berlin, Germany.

Sipala, S., Mancini, G. & Vagliasindi, F. G. A. 2003 Development of a web-based tool for the calculation of costs of different wastewater treatment and reuse scenarios. *Water Science and Technology: Water Supply* **3** (4), 89–96. doi:10.2166/ws.2003.0049.

Türk, J., Dazio, M., Dinkel, F., Ebben, T., Hassani, V. & Herbst, H. 2013 *Abschlussbericht zum Forschungsvorhaben 'Volkswirtschaftlicher Nutzen der Ertüchtigung kommunaler Kläranlagen zur Elimination von organischen Spurenstoffen, Arzneimitteln, Industriechemikalien, bakteriologisch relevanten Keimen und Viren (TP 9)' (Economic Benefit of the Refurbishment of Municipal WWTP to Eliminate Micropollutants, Final Report)*. Available from: https://www.lanuv.nrw.de/fileadmin/forschung/wasser/klaeranlage_abwasser/TP%209%20Abschlussbericht_Langfassung_final_140305.pdf (accessed 21 March 2020)

United States Environmental Protection Agency 1999 *Wastewater Technology Fact Sheet: Chlorine Disinfection*. Available from: https://www3.epa.gov/npdes/pubs/chlo.pdf (accessed 25 March 2020)

US EPA 2014 *Work Breakdown Structure-Based Cost Models for Drinking Water Treatment Technologies*. https://nepis.epa.gov/Exe/ZyPURL.cgi?Dockey=P100MYKY.TXT (accessed 29 October 2020)

Wasser Abwasser Technik 2020 Über WaReIp – Water Reuse in Industrieparks (About WaReIp – Water Reuse in industrial parks). *Wasser Abwasser Technik* **1/2020**, 49.

Zaffaronia, C., Daiggerb, G., Nicolc, P. & Leed, T. W. 2016 Wastewater treatment challenges faced by the petrochemical and refinery industry, and opportunities for water reuse. *Water Practice & Technology* **11** (1), 104–117. doi:10.2166/wpt.2016.012.

First received 4 April 2020; accepted in revised form 29 September 2020. Available online 22 October 2020

Impact assessment of water and nutrient reuse in hydroponic systems using Bayesian Belief Networks

Martin Zimmermann and Michaela Fischer

ABSTRACT

Water-saving agricultural practices can reduce negative environmental impacts in water-scarce regions all over the world. This study deals with an innovation that combines hydroponic crop production and municipal wastewater reuse for irrigation purposes. The research question was what impacts such hydroponic water reuse systems have on product confidence, economic viability, groundwater recharge, biodiversity and landscape quality. It should also be clarified under which conditions and with which measures these systems can be sustainable. To answer these questions, a number of generic hydroponic water reuse systems were modeled and assessed using a Bayesian Belief Network that included both numerical values and expert knowledge. The hydroponic water reuse systems with the most positive overall impacts are small-scale food production systems (tomatoes) equipped with lighting and heating whose products are marked with a quality label or with a label for regional products. The systems are located in a former industrial area. In addition, a wetland system and landscape integration are implemented as landscaping measures. Hydroponic systems can be operated economically viable, their products have a high level of product confidence and their ecological impacts can be positive. No tradeoffs have to be accepted between economic, social and ecological goals.

Key words | biodiversity, economic viability, landscaping, modeling, product confidence, social acceptance

Martin Zimmermann (corresponding author)
Michaela Fischer
ISOE – Institute for Social-Ecological Research,
Hamburger Allee 45, 60486 Frankfurt am Main,
Germany
E-mail: *zimmermann@isoe.de*

HIGHLIGHTS

- The study shows that hydroponic water reuse systems can be operated economically viable, that their products have a high level of product confidence and that their ecological effects can be positive if appropriate landscaping measures are taken.
- Landscaping as well as acceptance measures should be carried out to accompany hydroponic crop production to improve its social and ecological impacts.

INTRODUCTION

Agriculture accounts for approximately. 70% of withdrawals of global freshwater resources and is thus by far the largest consumer of water compared with other sectors. In arid and semi-arid areas, the withdrawals can even be as high as 85% (Chmielewski 2011). About 40% of the agricultural food produced is irrigated (Chmielewski 2011). Agriculture is dependent on irrigation not only in arid and semi-arid zones. Even in intensively cultivated agricultural areas in Germany, agriculture is confronted with problems of water scarcity. In total, 451,800 ha (or 1.3%) of agricultural land in Germany are irrigated (Statistisches Bundesamt 2017). In the German strategy for adaptation to climate change, it was found that

doi: 10.2166/wrd.2020.026

regional water use conflicts could arise with regard to surface and groundwater during dry periods (BMU 2008). Therefore, new water-saving cultivation methods and additional water resources are of interest in Germany in order to avoid conflicts of use, but also for the protection of groundwater. Intensification of agriculture can reduce land use and therefore reduce negative environmental impacts (Ellis *et al.* 2013).

One option for the water-efficient cultivation of crops is hydroponic greenhouse production. In hydroponics, plants are grown without soil. Instead, mineral nutrient solutions dissolved in water are used (dos Santos *et al.* 2013). Crops are cultivated in such a way that only their roots are hanging in the nutrient solution. The roots can be physically supported by an inert medium such as mineral wool, volcanic stones or other substrates. One possibility of operation is that the nutrient solution continuously flows past the roots (continuous-flow solution culture). This particularly facilitates the adjustment of the nutrient concentration (Rockel 1997). The nutrients used in hydroponic systems can come from various sources, e.g. fish excrement, chemical fertilizers or nutrient solutions (Jones 2004). Crops that are usually grown hydroponically include tomatoes and lettuce, but also ornamental plants. The main advantage of hydroponics is the significantly lower water consumption compared with conventional agriculture (Zhang *et al.* 2018). In particular, water loss through evapotranspiration is low in a closed greenhouse. This enables the cultivation of food even in extremely dry areas.

One possible resource for feeding hydroponic systems is the reuse of wastewater. In water reuse, wastewater is treated with technical means, so that it can be reused for other purposes (e.g. agricultural irrigation). The use of recycled water instead of fresh water for corresponding purposes can be a water-efficient measure. Water reuse can thus be part of sustainable water management. This can reduce water scarcity and alleviate pressure on groundwater and other water resources (Andersson *et al.* 2016). In addition, not only can water be saved when reusing wastewater, but the nutrients contained in the wastewater can also be used to fertilize the plants.

Although studies have already been carried out on hydroponic systems in which reused wastewater is used, no large-scale implementation has yet been carried out. The use of treated wastewater as a nutrient solution has so far been little investigated (Magwaza *et al.* 2020). In the HypoWave project, a new concept was tested, which deals with the connection of treated municipal wastewater and plant production (Bliedung *et al.* 2019). The key question was to what extent municipal wastewater has to be treated in order to be able to use it in hydroponic systems with as little additional nutrient input as possible. A pilot plant with a modular set-up of different wastewater treatment processes was implemented in Wolfsburg Hattorf, Germany, on the site of the municipal wastewater treatment plant (WWTP) (Bliedung *et al.* 2019). In addition to the option of using the biologically treated (activated sludge including nitrification and denitrification) WWTP runoff, a number of alternative treatment processes have been tested. These processes include an anaerobic expanded granular sludge bed reactor and an aerobic sequencing batch reactor for nitrification. In this way, a large part of the nitrate remains in the treated water. In addition, a biological activated carbon biofilter for the removal of trace substances and, if necessary, ozonation to eliminate pathogens was used. Depending on the process combination, water of different quality could be produced for the hydroponic system, which among other things resulted in different nutrient levels.

The hydroponic system of the pilot plant consisted of several parallel rows of pipes in which lettuce plants were cultivated. The pipes were fed with different water qualities. In addition, lettuce was grown in one pipe in a conventional Hoagland solution with a 50% concentration for comparison. The hydroponic system was designed as a mixture of nutrient film technique and deep water culture.

In addition to the above-mentioned advantages of hydroponic water reuse systems in terms of water and nutrient efficiency, a number of impacts are also unclear, such as questions about the social acceptance of the produced crops, as well as ecological consequences and economic viability. Accordingly, the key research questions are: (1) What are the social, economic and ecological impacts of selected hydroponic water reuse systems? (2) Under which conditions are hydroponic water reuse system socially, economically and ecologically sustainable? To answer these questions, a number of generic hydroponic water reuse systems were modeled and assessed using a software-based modeling process that can make use of both numerical values and expert knowledge.

METHODS AND MATERIALS

General methodology and state of the art

The assessment of social, ecological and economic impacts of hydroponic water reuse systems was carried out using Bayesian Belief Networks (BBNs). BBNs are statistical multivariate models based on the theorem of Thomas Bayes (1740) for calculating conditional probabilities. BBNs combine a qualitative, graphical representation of a system with a quantitative, probabilistic evaluation of the interactions between the variables of such a system (Castelletti & Soncini-Sessa 2007; Barton *et al.* 2012). BBNs are particularly suitable for depicting complex systems in a comprehensible way (Molina *et al.* 2013), can process qualitative and quantitative data (Chen & Pollino 2012) and allow the inclusion of stakeholder and expert knowledge (Bergmann *et al.* 2010). BBNs consist of a directed acyclical graph containing variables and nodes. These are connected by directional arrows that represent cause–effect relationships between the nodes. Conditional probability tables (CPTs) assigned to each node reflect the conditional probabilities for the values of a node as a function of the values of the parent nodes (Cain 2001). A disadvantage of conventional BBN is that feedbacks cannot be mapped. Furthermore, spatial and temporal dynamics are difficult to represent (Uusitalo 2007).

BBNs are used as versatile tools for the investigation of interactions between variables of a system and as decision support systems for the development of management strategies (Kuikka *et al.* 1999; Cain 2001; Castelletti & Soncini-Sessa 2007; Jensen & Nielsen 2007; Pagano *et al.* 2013; Reichert *et al.* 2015), in the fields of water resource management and ecology (Cain 2001; Bromley 2005; Marcot *et al.* 2006; Castelletti & Soncini-Sessa 2007; Pollino *et al.* 2007; Aguilera *et al.* 2011) and on environmental impact assessment and risk assessment (Voie *et al.* 2010; Liu *et al.* 2012; Tighe *et al.* 2013; Money *et al.* 2014; Brandmayr *et al.* 2015).

The structure of the BBN (i.e. system variables and their interrelations) was pre-developed together with external experts and project partners. For this purpose, a total of six expert interviews with researchers from different disciplines (e.g. civil and environmental engineering, plant science, ecology and landscape architecture) were carried out. Five social, economic and ecological impact variables were jointly chosen, namely product confidence, economic viability, groundwater recharge, biodiversity and landscape quality. The data required to model the BBN were collected through qualitative or semi-quantitative expert surveys in the case of product confidence, biodiversity and landscape quality (Table 1). Other impacts, namely economic viability and groundwater recharge, were calculated or assessed based on literature values and assumptions (Table 1). The relations among system variables (nodes) were expressed in the form of a CPT in the case of probabilistic values and in the form of a function table in the case of deterministic values (Supplementary Material, Appendix A). The BBN was finally evaluated using the software Netica. The BBN eventually shows which social, ecological and economic impacts the experts attributed to various hydroponic water reuse systems and selected measures.

Calculation of the economic viability

The economic viability was determined for a total of 16 different generic hydroponic water reuse systems. These systems differed in terms of product (food vs. non-food crops), size (small-scale vs. large-scale), production process (with or without lighting and heating) and the water regime applied (open vs. closed mode of operation). Furthermore, landscaping measures (i.e. wetland systems and landscape integration) were considered in the cost calculation.

Table 1 | Methods used to assess specific impacts of hydroponic water reuse systems

Impact dimension	Social impacts	Economic impacts	Ecological impacts		
Impacts	Product confidence	Economic viability	Groundwater recharge	Biodiversity	Landscape quality
Method	Expert survey	Calculation	Qualitative estimation	Expert survey	
Type of parameter	Probabilistic	Deterministic	Deterministic	Probabilistic	

Tomatoes were selected as food crops and Chrysanthemum as non-food crop or ornamental plants. The size of the hydroponic systems corresponded either to the water quantity of 1,000 (small-scale) or 10,000 (large-scale) population equivalents. The greenhouses were either equipped with lighting and heating for improved growing conditions or not. Finally, an open and a closed water regime were distinguished. In the first case, the water flows continuously through the hydroponic system which makes it nutrient-efficient (e.g. regarding micro- and macronutrients). In the second case, the water is circulated in the system which makes it water-efficient. Fertilizer, however, has to be added regularly. The water is circulated until a certain salinity threshold (e.g. 8 mmol NaCl) is reached. It can then be treated by reverse osmosis for further reuse.

Based on these preconditions, investments and operating costs (e.g. labor costs, energy costs, overhead and additional water treatment) of the 16 systems were determined. A depreciation period of 10 years has been applied. Capital costs have been calculated on the basis of an interest rate of 3%. Literature values for expenses and revenues were taken from the KWIN model (WUR 2018). Additional costs for the water reuse treatment steps and the landscaping measures were added. Finally, the economic viability was measured based on the annual rate of return which represents the profit expressed as the proportion of the investments and costs.

The market price for tomatoes was set at 1.86 EUR with lighting and heating and 1.74 EUR for tomatoes without lighting and heating as suggested by the KWIN model for European conditions (WUR 2018). The market price for Chrysanthemum Santini with lighting and heating was adjusted due to recent market fluctuations from 0.38 EUR to 0.28 EUR per stem. For Chrysanthemum Santini without lighting and heating, the price remained as suggested by the model at 0.17 EUR per stem. Wages were set at 12.00 EUR per hour for workers, 18.75 EUR for trained staff and 20.00 EUR for operators. For the landscaping measures, a land value (lease) of 800 EUR per ha and year was applied according to a related case study (Mohr *et al.* 2019). Additional costs for construction and operation of the landscaping measures, however, were not considered due to the diversity of possible options (e.g. wetland systems, hedges and trees).

Expert surveys and estimations of social and ecological impacts

A total of 12 experts from research and practice were surveyed using questionnaires regarding product confidence, biodiversity and landscape quality, covering different disciplines and sectors. Even if the number of 12 experts cannot be representative, their positions and experience mean that they are multipliers who can provide a well-founded appraisal of the impacts surveyed.

Consumer acceptance of the food and non-food products made using the hydroponic water reuse system was considered an important factor. The decision of food and flower retailers to add products to their range depends heavily on how they assess consumer acceptance. Therefore, the product confidence of corresponding retailers was chosen as an indicator of consumer acceptance. Additionally, measures to increase product confidence were taken into account. In addition to taking no action, these measures comprise products with a quality label (e.g. products are harmless to health, sustainable or without genetic engineering), products that are labeled as regionally produced as well as branded products (e.g. Bonduelle and Ardo).

The product confidence was determined by using standardized expert surveys. In particular, grocery retailers from organic supermarkets and farmers' markets as well as florists were interviewed. The interviewees were asked to rate the product confidence on a 5-level Likert scale from 1 to 5, with 1 meaning that the respondent would include the product in their range without hesitation and 5 that the respondent would under no circumstances do so. Two separate questionnaires for two separate groups of interviewees were used for vegetables (tomatoes) and ornamental plants (Chrysanthemum). One question, for example, was: 'Would you include foods (e.g. tomatoes) produced with the described hydroponic system in the assortment of your market if they are labelled as regional products?' The responses were converted to probabilities by first calculating the mean and then performing a linear interpolation where a mean of 1 corresponded to a product confidence of 100% and a mean of 5 to a product confidence of 0%.

The implementation of hydroponic greenhouses can have adverse ecological impacts, for instance increased land consumption or impervious surfaces. Landscaping measures to mitigate these effects can be the creation of wetland systems around the greenhouses and the integration of greenhouses into the landscape by using hedges, trees or extensive farming elements such as short rotation forestry. Both measures can also be combined and might have positive impacts on biodiversity and landscape quality.

Similar to the case of product confidence, the impacts of landscaping measures (i.e. wetland systems and landscape integration) on biodiversity and landscape quality were determined by means of standardized expert surveys. In particular, experts from science and practice (e.g. ecologists and landscape architects) were interviewed. The interviewees were asked to rate the likelihood of positive impacts of the landscaping measures on biodiversity and landscape quality on a scale from 0 to 100% where 100% represents the strongest positive effect. Two questionnaires were presented to the interviewees – one for the impacts on biodiversity and one for the impacts on landscape quality. It was assumed that wetland systems would expand the area of the hydroponic systems (including the greenhouses) by 20%. Measures for landscape integration would expand the area by 40% and both measures combined by 60%. These values were derived from the proposal to keep more than 50% of the area free from intensive use in order to preserve biodiversity in the long term (Ellis 2019). Furthermore, the questionnaires also took into account whether the hydroponic systems were implemented in former industrial or agricultural areas and whether they were small-scale (approximately 3–6 ha or 1,000 population equivalents) or large-scale systems (approximately 27–58 ha or 10,000 population equivalents). One question, for example, was: 'What is the probability that greenhouses built on a former agricultural area with a size of 100 ha and a continuous green area with wetland systems of an additional 20 ha have a positive and/or negative impact on biodiversity?'

Impacts on the local groundwater recharge were qualitatively estimated on the basis of land consumption and impervious surface of the hydroponic systems as well as the capacity of the considered measures to support rainwater infiltration. Five qualitative levels of groundwater recharge were distinguished, namely highly reduced, reduced, neutral, improved and highly improved. Different rates of groundwater recharge were assessed on the basis of two different initial situations, namely former agricultural areas (pervious) and former industrial areas (impervious). The implementation of various landscaping measures (i.e. wetland systems and landscape integration) was then considered for these areas. The area used for landscaping measures is assumed to be permeable.

RESULTS AND DISCUSSION

Structure of the BBN

The expert discussions resulted in a BBN that consists of five system variables describing properties of hydroponic water reuse systems, two measures relating to product confidence and landscaping as well as the pre-defined five social, economic and ecological impact variables. The system variables each have two instances, namely food production systems (tomatoes) vs. systems for the production of ornamental plants (Chrysanthemum), small-scale (1,000 population equivalents) vs. large-scale systems (10,000 population equivalents), systems with lighting and heating vs. systems without lighting and heating, systems with a closed water regime (water-optimized) vs. systems with an open one (nutrient-optimized), and finally systems implemented in a former industrial area vs. systems in a former agricultural area. Furthermore, the BBN includes two measures which represent possible interventions to avoid or mitigate undesirable impacts. The measures are aimed at social and ecological impacts and consist of the four above-mentioned instances each.

The expert discussions also revealed how the system variables, measures and impacts (i.e. product confidence, economic viability, groundwater recharge, biodiversity and landscape quality) are interrelated which eventually leads to the BBN (Figure 1 and Supplementary Material, Appendix C). A total of 512 combinations ($2^5 \times 4^2$) of the system variables and measures are possible. Figures with examples of the model's behavior are attached in Supplementary Material, Appendix B.

Selected hydroponic water reuse systems and their impacts

The hydroponic water reuse system with the most positive overall impacts is a small-scale food production system (3–6 ha, tomatoes) equipped with lighting and heating

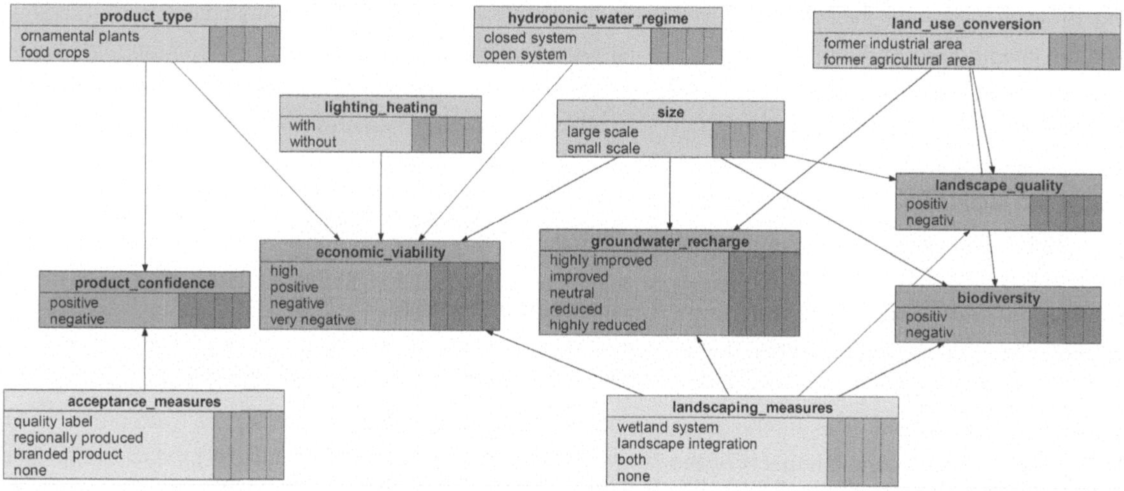

Figure 1 | BBN of hydroponic water reuse systems consisting of system variables (blue), measures (green) and impacts (red). Please refer to the online version of this paper to see this figure in color: http://dx.doi.org/10.2166/wrd.2020.026.

whose products are marked with a quality label (Table 2). The system is located in a former industrial area, and both above-mentioned landscaping measures (wetland system and landscape integration) are implemented. With regard to the investigated impacts, it is irrelevant whether it is a system in which the water is circulated (closed system) or through which the water flows only once (open system). Such a system would achieve a very high product confidence (90%), a high economic viability (6% rate of return), a greatly improved groundwater recharge, a very positive impact on biodiversity (77%) and a very positive effect on the landscape quality (84%). A slight variation in this system in which the tomatoes are marked with a label for regional products would achieve a slightly lower product confidence (70%). Apart from that, the system achieves the same very positive results. Small-scale food production systems without lighting and heating, however, have a very poor economic viability (−12% rate of return).

Large-scale food production systems (tomatoes with quality label) in former industrial areas also perform fairly well but only without lighting and heating (Table 2). In this case, the economic viability is just positive (Table 2).

Table 2 | Hydroponic water reuse systems with most positive impacts

Systems	Measures	Product confidence	Economic viability	Groundwater recharge	Biodiversity	Landscape quality
Food production (tomatoes); small-scale; lighting and heating; open or closed system; former industrial area	Quality label; wetland system, landscape integration	90%	High	Greatly improved	77%	84%
Food production (tomatoes); small-scale; lighting and heating; open or closed system; former industrial area	Labeled as regional product; wetland system, landscape integration	70%	High	Greatly improved	77%	84%
Food production (tomatoes); large-scale; no lighting and heating; open or closed system; former industrial area	Quality label; wetland system, landscape integration	90%	Positive	Greatly improved	76%	80%
Food production (tomatoes); large-scale; no lighting and heating; open or closed system; former industrial area	Labeled as regional product; wetland system, landscape integration	70%	Positive	Greatly improved	76%	80%

Apart from this, the large-scale system scores slightly worse than the small-scale system with regard to the criteria biodiversity (76 instead of 77%) and landscape quality (80 instead of 84%) (Table 2). Large-scale food production systems with lighting and heating, however, do not appear to be economically viable (rate of return −5%, Table 5).

As in the case of small-scale systems, product confidence drops from 90 to 70% if the tomatoes are labeled as regionally produced. Food systems without consumer acceptance measures (regardless of whether they are large- or small-scale) lead to a loss of product confidence (56 instead of 90 or 70%). Branded products even lead to a product confidence of only 42%.

In the case of food production systems in former agricultural areas, the positive ecological effects are not so pronounced. Assuming that both landscaping measures are carried out, the groundwater recharge is neutral in small-scale systems with lighting and heating, the biodiversity is 55 instead of 77% and landscape quality 67 instead of 84%. For large-scale systems without lighting and heating, groundwater recharge is also neutral, the biodiversity is 53 instead of 76% and landscape quality 53 instead of 80%.

Hydroponic water reuse systems for the production of ornamental plants (in our case: Chrysanthemum) have a negative or very negative economic viability in all calculated scenarios (Table 5). However, these systems can still achieve positive social and ecological impacts. Ornamental plants (Chrysanthemum) show high product confidence if they are labeled with a quality label (100%) or as regionally produced (100%) and even if neither of these two consumer acceptance measures have been carried out (92%). The potential for very good ecological effects only exists if the system is implemented in a former industrial area and not on an agricultural one. Conclusions regarding the advantages and disadvantages of systems with an open or closed water regime cannot be drawn either for food production systems or for systems for the production of ornamental plants.

Specific impacts and their interdependencies

Social impacts

The highest product confidence is found in ornamental plants (Chrysanthemum) with (100%) and without (92%) consumer acceptance measures (quality label and/or labeled as regionally produced) as well as in food products (tomatoes) with quality labels (90%) followed by ones that are labeled as regionally produced (70%). In general, food retailers were positive about potential products from hydroponic water reuse systems. They said that they could imagine including the products in their range. The best measure to improve the product confidence was a quality label which can be compared with labels that certify sustainable production or GMO-free labels. Labeling as a regionally produced product also improved product confidence. Branded food products even had a lower product confidence (42%) than products without any consumer acceptance measure (56%). A sensitivity analysis of the factors influencing product confidence shows that the influence of acceptance measures is significantly greater than that of the product type.

Florists did not mention concerns with selling ornamental flowers produced with hydroponic water reuse systems. The use of labels (quality label or labeling as a regional product) was preferred over no label. However, it was also argued that labels are not required in case of ornamental plants since customers would not be concerned. They were said to prefer quality and price over other factors. In some cases, it was said that costumers would ask where the flowers were made. This might indicate an interest in regional products and their social or environmental impacts.

The results indicate that the production of flowers from hydroponic water reuse systems would be well accepted. Flowers labeled as regionally produced could even influence the consumer's purchase decision in a positive way if they are interested in the social and environmental sustainability of the production. Apart from this, the attitude of grocery retailers towards food crops from hydroponic systems was predominantly positive. Nevertheless, quality labels or labeling as a regional product might be required to increase product confidence. Big supermarket chains could not be reached by means of this study. Further research is necessary to understand how the big players will position themselves towards the circular economy and future developments in food production. To our knowledge, there are no other detailed studies on product confidence concerning products from hydroponic systems with water reuse.

Ecological impacts

Implementing both landscaping measures (i.e. wetland system and landscape integration) leads to the most positive impacts in terms of biodiversity and landscape quality. This is irrespective of whether the hydroponic system was set up in a former industrial or agricultural area or whether it is a small-scale or large-scale system (Table 3). Undertaking no landscaping measures in a former agricultural area leads to the most negative effects in this regard.

Hydroponic systems in former agricultural areas need at least landscape integration measures to achieve positive impacts on landscape quality. In terms of biodiversity, both landscaping measures are necessary to achieve positive impacts. In contrast, hydroponic systems in former industrial areas have positive effects on landscape quality even without any landscaping measures. Regarding biodiversity, at least measures of landscape integration are necessary to reach positive impacts.

The evaluation of impacts on biodiversity and landscape quality can only provide first indications for decisions about landscaping measures. Activities to improve biodiversity need to be focused on a larger spatial scale than the rather narrow system boundaries, especially with a view to mobile organisms such as birds (Söderström *et al.* 2001). From a legal point of view, nature conservation legislation in Germany, for instance, prohibits a deterioration of land to avoid negative effects on nature and landscape ('Verschlechterungsverbot'). According to the German Federal Nature Conservation Act, interventions which cannot be avoided need to be compensated. The landscaping measures considered here could play a role in connection with these compensations.

Hydroponic systems implemented in former industrial areas always show an improved or greatly improved groundwater recharge if landscaping measures are taken (Table 4). Systems implemented in former agricultural areas have at best a neutral effect on groundwater recharge if both landscaping measures are carried out. Without any measures or with only one of the considered measures, groundwater recharge is highly reduced or at least reduced.

For the assessment, it was assumed that former industrial areas are impervious. Therefore, there would be no further reduction of groundwater recharge in the case of the construction of a greenhouse. The surfaces of the landscaping measures mentioned above are regarded as pervious. Apart from surface infiltration, various construction measures can be taken to improve groundwater recharge such as infiltration basins or percolation trenches.

A sensitivity analysis of the influencing factors on all three ecological impact variables shows that the influence of landscaping measures and the former type of land use is significantly greater than that of the size of the greenhouses.

Table 3 | Impacts of landscaping measures on biodiversity and landscape quality

Size	Measures	Biodiversity (%)	Landscape quality (%)
Former agricultural area:			
Large-scale (27–58 ha)	No measures	0	0
	Wetland system	21	28
	Landscape integration	37	52
	Both measures	53	53
Small-scale (3–6 ha)	No measures	0	7
	Wetland system	34	35
	Landscape integration	41	58
	Both measures	55	67
Former industrial area:			
Large-scale (27–58 ha)	No measures	23	53
	Wetland system	47	63
	Landscape integration	62	70
	Both measures	76	80
Small-scale (3–6 ha)	No measures	26	57
	Wetland system	48	66
	Landscape integration	58	76
	Both measures	77	84

Table 4 | Impacts of landscaping measures on groundwater recharge

Measures	Groundwater recharge
Former agricultural area:	
No measures	Highly reduced
Wetland system	Reduced
Landscape integration	
Both measures	Neutral
Former industrial area:	
No measures	Neutral
Wetland system	Improved
Landscape integration	
Both measures	Highly improved

To our knowledge, there are no other detailed studies on the effects of hydroponic systems with water reuse on biodiversity, landscape quality or groundwater recharge. Studies dealing with the environmental impacts of hydroponic systems deal either with their general land requirements (Moore *et al.* 2020), water pollution (Tewolde *et al.* 2016), energy demand (Barbosa *et al.* 2015; Chen *et al.* 2020) or greenhouse gas emissions, respectively, their carbon footprint (Martinez-Mate *et al.* 2018; Vinci & Rapa 2019).

Economic impacts

As mentioned above, good economic viability (6% rate of return) is associated with small-scale food production systems with lighting and heating (Table 5). Large-scale food production systems without lighting and heating are at least barely economically viable. Other studies confirm that hydroponic systems can be economically viable under various conditions (Waldhauer & Soethoudt 2015; Becraft 2017; de Franca Xavier *et al.* 2018; Souza *et al.* 2019). Systems for ornamental plant production or small-scale food production systems without lighting and heating, however, always have a poor economic viability. A sensitivity analysis of the factors influencing the economic viability shows that the influence of the product type, the size of the greenhouses as well as lighting and heating is greater than that of the water regime and landscaping measures.

Additional water treatment costs are lowest for systems producing ornamental plants. Corresponding costs range between 0.2 and 0.5% of the production costs in systems with an open water regime and between 2.0 and 4.6% in closed ones. In food production systems, they range between 2.0 and 5.5% in open systems and between 2.4 and 6.5% in closed ones.

Market prices of the products and energy costs have a big influence on the profitability. The calculations presented in Table 5 are based on rather conservative assumptions. A further sensitivity analysis was made with different assumptions considering energy supply and market prices. If prices and energy production are used as initially suggested by the model, six scenarios are highly profitable (15–35%), further four scenarios are positive, two scenarios are negative, and four scenarios are very negative (−6 to −13%). Furthermore, a combined heat and power plant for the production of heat and electricity would result in a surplus and therefore an income from sales of electricity. The market prices for Chrysanthemum (with lighting and heating) were assumed to be 0.38 EUR according to the model.

Costs for disposal of residues from water treatment were not considered in the calculations. In systems with closed water regimes, reverse osmosis is suggested to recycle the greenhouse runoff. It would be possible to irrigate a short rotation plantation in the landscape park which is much less sensitive to salinity than crops in hydroponic systems. In this way, water treatment costs and the question of disposal of residue would not arise.

Table 5 | Annual rate of return and size of hydroponic water reuse systems

		Food production (tomatoes) Lighting and heating		Ornamental plant production (Chrysanthemum) Lighting and heating	
		with	**without**	**with**	**without**
Large-scale	Open system	27 ha −5%	27 ha 1%	33 ha −10%	33 ha −4%
	Closed system	58 ha −5%	58 ha 0.5%	51 ha −14%	51 ha −8%
Small-scale	Open system	3 ha 6%	3 ha −12%	3 ha −17%	3 ha −12%
	Closed system	6 ha 6%	6 ha −12%	5 ha −19%	5 ha −13%

CONCLUSIONS

The results of the study show that hydroponic water reuse systems can be operated in an economically viability way, that their products have a high level of product confidence and that their ecological effects can be positive if appropriate landscaping measures are undertaken. This applies in particular to the small-scale food production system described above. The results mean that no tradeoffs have to be accepted between economic goals on the one hand, as well as social and ecological goals on the other. In order to achieve this, however, the results suggest that landscaping as well as acceptance measures should be carried out to accompany hydroponic crop production. Land use through the construction of greenhouses should be compensated in particular by measures such as wetland systems and landscape integration in order to avoid or at least mitigate undesirable effects on the environment.

The BBN method allowed a rough, but nevertheless comprehensive and quick overview of the interdependencies of generic hydroponic systems. On the one hand, the modeling of generic systems has the advantage that the results can offer a general orientation and can be transferred to a specific case under certain conditions. On the other hand, the circumstances of a specific case can be such that the effects of the modeled systems might not apply. From a business perspective, however, such a model can serve as a decision support for companies (e.g. horticultural companies) or investors for a pre-feasibility study. The advantage of using a method such as the BBN is that quantitative assessments (e.g. cost–benefit analysis) are expanded to integrate qualitative expert knowledge. This means that impacts that are difficult to assess and to quantify can be included, which are often neglected in purely quantitative assessments.

Further investigations could address questions that remained open. One of these aspects is to clarify how an adequate quality management for hydroponic crop production including water reuse should be. In connection with this, the possibilities for the certification of the products with regard to consumer acceptance could be examined in more detail. Another aspect would be to include further assessment criteria such as measuring the carbon footprint in order to get a more complete picture of the impacts of hydroponic systems. A life cycle analysis could, in particular, provide insights into energy and material consumption for water reuse and hydroponic crop production in greenhouses. Unfortunately, this was not possible within the scope of the study. Nevertheless, the findings of this work may contribute to a successful take-off of the discussed innovation and the associated transformation processes in agriculture and water management.

ACKNOWLEDGEMENTS

The authors thank all project partners and participants who took part in the expert interviews and the survey. The research project 'HypoWave – New Pathways Towards Wastewater Re-Use in Agriculture' was funded by the German Federal Ministry of Education and Research (grant no. 02WAV1402).

DATA AVAILABILITY STATEMENT

All relevant data are included in the paper or its Supplementary Information.

REFERENCES

Aguilera, P. A., Fernàndez, A., Fernàndez, R., Rumí, R. & Salmerón, A. 2011 Bayesian networks in environmental modelling. *Environmental Modelling & Software* **26** (12), 1376–1388.

Andersson, K., Rosemarin, A., Lamizana, B., Kvarnström, E., McConville, J., Seidu, R., Dickin, S. & Trimmer, C. 2016 *Sanitation, Wastewater Management and Sustainability: From Waste Disposal to Resource Recovery.* United Nations Environment Programme and Stockholm Environment Institute, Stockholm.

Barbosa, G. L., Almeida Gadelha, F. D., Kublik, N., Proctor, A., Reichelm, L., Weissinger, E., Wohlleb, G. M. & Halden, R. U. 2015 Comparison of land, water, and energy requirements of lettuce grown using hydroponics vs. conventional agricultural methods. *International Journal of Environmental Research and Public Health* **12**, 6879–6891.

Barton, D. N., Kuikka, S., Uusitalo, L., Henriksen, H. J., Borsuk, M. & Linnell, J. D. C. 2012 Bayesian networks in environmental and resource management. *Integrated Environmental Assessment and Management* **8** (3), 418–429.

Becraft, A. 2017 Potential for home-use hydroponic systems to increase food security in Cape Eleuthera, Bahamas. *i-ACES* **3** (1), 12–18.

Bergmann, M., Jahn, T., Knobloch, T., Krohn, W., Pohl, C. & Schramm, E. 2010 *Methoden Transdisziplinärer Forschung. Ein Überblick mit Anwendungsbeispielen (Methods of Transdisciplinary Research. An Overview with Application Examples).* Campus Verlag, Frankfurt am Main.

Bliedung, A., Dockhorn, T., Ebert, B., Germer, J., Mohr, M., Schramm, E. & Winker, M. 2019 Hypowave – Ressourceneffiziente Nutzung von gereinigtem Abwasser in hydroponischen Systemen (HypoWave – resource-efficient use of treated wastewater in hydroponic systems). *Zentralblatt für Geologie und Paläontologie* **I**, 95–104.

Brandmayr, C., Kerber, H., Winker, M. & Schramm, E. 2015 Impact assessment of emission management strategies of the pharmaceutical Metformin and Metoprolol to the aquatic environment using Bayesian networks. *Science of the Total Environment* **532**, 605–616.

Bromley, J. 2005 *Guidelines for the use of Bayesian Networks as A Participatory Tool for Water Resource Management.* Wallingford, UK.

Bundesministerium für Umwelt, Naturschutz und nukleare Sicherheit (BMU) 2008 *Deutsche Anpassungsstrategie an den Klimawandel (German Adaptation Strategy to Climate Change).* Available from: https://www.bmu.de/fileadmin/bmu-import/files/pdfs/allgemein/application/pdf/das_gesamt_bf.pdf (accessed 24 March 2020)

Cain, J. 2001 *Planning Improvements in Natural Resources Management.* Center for Ecology & Hydrology, Wallingford.

Castelletti, A. & Soncini-Sessa, R. 2007 Bayesian networks and participatory modelling in water resource management. *Environmental Modelling & Software* **22**, 1075–1088.

Chen, S. H. & Pollino, C. A. 2012 Good practice in Bayesian network modelling. *Environmental Modelling & Software* **37**, 134–145.

Chen, P., Zhu, G., Kim, H.-J., Brown, P. B. & Huang, J.-Y. 2020 Comparative life cycle assessment of aquaponics and hydroponics in the Midwestern United States. *Journal of Cleaner Production* **275**, 122888.

Chmielewski, F.-M. 2011 Wassernutzung und Eingriffe des Menschen in den Wasserhaushalt. Wasserbedarf in der Landwirtschaft. In: *WARNSIGNAL KLIMA: Genug Wasser für alle?* 3rd edn (J. L. v. Lozán, H. Graßl, P. Hupfer & K. L. Schönwiese, eds). Wiss. Auswertungen, Hamburg.

de Franca Xavier, J., Vieira de Azevedo, C. A., de Q.A. Azevedo, M. R., Fernandes Monteiro Filho, A. & Rocha da Silva, C. 2018 Economic viability and profitability of lettuce in hydroponic system using different effluents. *Journal of Agricultural Science* **10** (7), 342–351.

dos Santos, J. D., Lopes da Silva, A. L., da Luz Costa, J., Scheidt, G. N., Novak, A. C., Sydney, E. B. & Soccol, C. R. 2013 Development of a vinasse nutritive solution for hydroponics. *Journal of Environmental Management* **114**, 8–12.

Ellis, E. 2019 To conserve nature in the anthropocene, half earth is not nearly enough. *One Earth* **1**, 163–167.

Ellis, E. C., Kaplan, J. O., Fuller, D. Q., Vavrus, S., Klein Goldewijk, K. & Verburg, P. H. 2013 Used planet: a global history. *Proceedings of the National Academy of Sciences of the United States of America* **110** (20), 7978–7985.

Jensen, F. V. & Nielsen, T. D. 2007 *Bayesian Networks and Decision Graphs.* Springer Science + Business Media, New York.

Jones Jr., J. B. 2004 *Hydroponics: A Practical Guide for the Soilless Grower,* 2nd edn. CRC Press, Boca Raton, London, New York, Washington, DC, pp. 153–166.

Kuikka, S., Hildén, M., Gislason, H., Hansson, S., Sparholt, H. & Varis, O. 1999 Modeling environmentally driven uncertainties in Baltic cod (*Gadus morhua*) management by Bayesian influence diagrams. *Canadian Journal of Fisheries and Aquatic Sciences* **56**, 629–641.

Liu, K. F.-R., Lu, C.-F., Chen, C.-W. & Shen, Y.-S. 2012 Applying Bayesian belief networks to health risk assessment. *Stochastic Environmental Research and Risk Assessment* **26**, 451–465.

Magwaza, S. T., Magwaza, L. S., Odindo, A. O. & Mdithswa, A. 2020 Hydroponic technology as decentralised system for domestic wastewater treatment and vegetable production in urban agriculture: a review. *Science of the Total Environment* **698**, 134154.

Marcot, B. G., Steventon, J. D., Sutherland, G. D. & McCann, R. K. 2006 Guidelines for developing and updating Bayesian belief networks applied to ecological modeling and conservation. *Canadian Journal of Forest Resources* **36** (12), 3064–3074.

Martinez-Mate, M. A., Martin-Gorriz, B., Martínez-Alvarez, V., Soto-García, M. & Maestre-Valero, J. F. 2018 Hydroponic system and desalinated wáter as an alternative farm-productive proposal in water scarcity areas: energy and greenhouse gas emissions analysis of lettuce production in southeast Spain. *Journal of Cleaner Production* **172** (20), 1298–1310.

Mohr, M., Koch, M., Schramm, E., Ebert, B., Germer, J. & Bürgow, G. 2019 Nutzung des Ablaufs einer Teichkläranlage zum Anbau von Gemüse im hydroponischen System im Landkreis Gifhorn. *Zentralblatt für Geologie und Paläontologie Teil I* **2019**, 131–138.

Molina, J.-L., Pulido-Velázquez, D., García-Aróstegui, J. L. & Pulido-Velázquez, M. 2013 Dynamic Bayesian networks as a decision support tool for assessing climate change impacts on highly stressed groundwater systems. *Journal of Hydrology* **479**, 113–129.

Money, E. S., Barton, L. E., Dawson, J., Reckhow, K. H. & Wiesner, M. R. 2014 Validation and sensitivity of the FINE Bayesian network for forecasting aquatic exposure to nano-silver. *Science of the Total Environment* **473–474**, 685–691.

Moore, E., Ward, J. & Lennard, W. 2020 End-of-pipe horticultural reuse of recirculating aquaculture system effluent: comparing the hydro-economics of two horticulture systems. *Water* **12**, 1409.

Pagano, A., Vurro, M., Giordano, R., Portoghese, I., Fratino, U., Berardi, F., Duro, A., Campopiano, F., Fiori, G., Vitucci, G. & Palmieri, V. 2013 A DSS risk management on drinking water infrastructures during extreme events. *Environmental Engineering and Management Journal* **12** (S11), 125–128.

Pollino, C. A., Woodberry, O., Nicholson, A., Korb, K. & Hart, B. T. 2007 Parametrisation and evaluation of a Bayesian network for use in an ecological risk assessment. *Environmental Modelling & Software* **22**, 1140–1152.

Reichert, P., Langhans, S. D., Lienert, J. & Schuwirth, N. 2015 The conceptual foundation of environmental decision support. *Journal of Environmental Management* **154**, 316–332.

Rockel, P. 1997 Growth and nitrate consumption of sunflowers in the rhizostat, a device for continuous nutrient supply to plants. *Journal of Plant Nutrition* **20** (10), 1431–1447.

Söderström, B., Svensson, B., Vessby, K. & Glimskär, A. 2001 Plants, insects, birds in semi-natural pastures in relation to local habitat and landscape factors. *Biodiversity and Conservation* **10** (11), 1839–1863.

Souza, S. V., Toesca Gimenes, R. M. & Binotto, E. 2019 Economic viability for developing hydroponic system in emerging countries: a differentiated risk adjustment proposal. *Land Use Policy* **83**, 357–369.

Statistisches Bundesamt (Destatis) 2017 *Land- und Forstwirtschaft, Fischerei. Bewässerung in landwirtschaftlichen Betrieben/ Agrarstrukturerhebung 2016 (Agriculture, Forestry and Fisheries. Irrigation on Agricultural Holdings/Agricultural Structure Survey 2016)*. Available from: https://www.destatis.de/DE/Themen/Branchen-Unternehmen/Landwirtschaft-Forstwirtschaft-Fischerei/Landwirtschaftliche-Betriebe/Publikationen/Downloads-Landwirtschaftliche-Betriebe/betriebe-bewaesserung-5411205169004.pdf?__blob=publicationFile (accessed 30 March 2020)

Tewolde, F. T., Takagaki, M., Oshio, T., Maruo, T., Kozai, T. & Kikuchi, Y. 2016 Environmental impact of tomato production under different hydroponic systems. *Acta Hortic* **1112**, 267–271.

Tighe, M., Pollino, C. A. & Wilson, S. C. 2013 Bayesian networks as a screening tool for exposure assessment. *Journal of Environmental Management* **123**, 68–76.

Uusitalo, L. 2007 Advantages and challenges of Bayesian networks in environmental modelling. *Ecological Modelling* **203** (3–4), 312–318.

Vinci, G. & Rapa, M. 2019 Hydroponic cultivation: life cycle assessment of substrate choice. *British Food Journal* **121** (8), 1801–1812.

Voie, O. A., Johnsen, A., Stromseng, A. & Longva, K. S. 2010 Environmental risk assessment of white phosphorus from the use of munitions – a probabilistic approach. *Science of the Total Environment* **408**, 1833–1841.

Waldhauer, N. & Soethoudt, H. 2015 *Metropolitan Food Supply in Egypt. Market Analysis Hydroponics Production for Selected Urban Areas. Report 1551*. Wageningen UR Food & Biobased Research, Wageningen.

WUR 2018 *Quantitative Information on Dutch Greenhouse Horticulture 2016–2017*. 'KWIN'. English Version. Wageningen University and Research.

Zhang, H., Asutosh, A. & Hu, W. 2018 Implementing vertical farming at university scale to promote sustainable communities: a feasibility analysis. *Sustainability* **10** (12), 4429.

First received 1 April 2020; accepted in revised form 28 September 2020. Available online 20 October 2020

Brines from industrial water recycling: new ways to resource recovery

Malena Kieselbach, Tobias Hogen, Sven-Uwe Geißen IWA, Thomas Track IWA, Dennis Becker IWA, Hans-Jürgen Rapp, Joachim Koschikowski, Joachim Went, Harald Horn, Florencia Saravia, Annika Bauer, Rebecca Schwantes, Daniel Pfeifle, Nicolas Heyn, Miriam Weissroth and Bernd Fitzke

ABSTRACT

Stricter environmental regulation policies and freshwater as an increasingly valuable resource have led to global growth of zero liquid discharge (ZLD) processes in recent years. During this development, in addition to water, the recovery of recyclable materials, e.g. salts, from industrial wastewater and brines is considered more frequently. Within the framework of the HighCon research project, the subject of this study, a new ZLD process with the goal of pure single-salt recovery from industrial wastewater has been developed and investigated in a demonstrational setup at an industrial site. With regard to pure salts recovery, separating organic components is of great importance during the treatment of the concentrate arising from used water recycling. The removal of COD and of ions responsible for scaling worked very well using nanofiltration. The nanofiltration permeate containing the monovalent ions was pre-concentrated using electrodialysis and membrane distillation before selective crystallization for single-salt recovery was performed. An example economic case study for the newly developed ZLD process – based on demonstration results and considering optimization measures for a full-scale design – indicates that the costs are equal to those of a conventional ZLD process, which, however, does not provide *inter alia* the aforementioned benefit of single-salt recovery.

Key words | brine management, cost analysis, industrial wastewater, resource recovery, water reuse, zero liquid discharge

Malena Kieselbach
Tobias Hogen
Sven-Uwe Geißen IWA (corresponding author)
Technische Universität Berlin,
Department of Environmental Technology, Chair of Environmental Process Engineering, Secr. KF 2, Straße des 17. Juni 135, 10623 Berlin, Germany
E-mail: *sven.geissen@tu-berlin.de*

Thomas Track IWA
Dennis Becker IWA
DECHEMA – Gesellschaft für Chemische Technik und Biotechnologie e.V.,
Forschungsförderung und Tagungen, Theodor-Heuss-Allee 25, 60486 Frankfurt am Main, Germany

Hans-Jürgen Rapp
DEUKUM GmbH,
Maybachstraße 5, 72636 Frickenhausen, Germany

Joachim Koschikowski
Joachim Went
Fraunhofer Institute for Solar Energy Systems ISE,
Heidenhofstr. 2, 79110 Freiburg,
Germany

Harald Horn
Florencia Saravia
Annika Bauer
DVGW-Research Center at the Engler-Bunte-Institut, Water Chemistry and Water Technology, Karlsruhe Institute of Technology,
Engler-Bunte-Ring 9a, 76131 Karlsruhe,
Germany

Rebecca Schwantes
Daniel Pfeifle
SolarSpring GmbH, Christaweg 40, 79114 Freiburg, Germany

Nicolas Heyn
Terrawater GmbH,
Wischhofstraße 1-3, Gebäude 11, 24148 Kiel,
Germany

HIGHLIGHTS

- Stricter environmental regulation policies and freshwater as an increasingly valuable resource are leading to a global growth of zero liquid discharge (ZLD) processes in recent years.
- In the study, the results of a demonstrational setup at an industrial site for a newly developed concentrate treatment process are presented. This new ZLD process, named HighCon, has the goal of pure single-salt recovery from industrial wastewater.
- The demonstration results show that separating organic components and monovalent ions worked out very well with nanofiltration in the HighCon process and recovering an organic-free

doi: 10.2166/wrd.2020.033

salt mixture or – using selective crystallization – the separation of an organic-free single salt with >90% purity has been enabled.

- In a cost analysis, the total specific costs have been calculated for the newly developed HighCon process and a comparable ZLD process. With approximately 4 €/m³, the results show that the specific costs are in the same range for both processes.
- The HighCon process has the advantage of pure (single-)salt recovery over conventional ZLD generating the mixture of solid wastes. Recovery of resources is becoming increasingly important – economically and with regard to the closing of recoverable substance cycles.

Miriam Weissroth
Bernd Fitzke
WEHRLE Umwelt GmbH,
Bismarckstr. 1-11, 79312 Emmendingen,
Germany

GRAPHICAL ABSTRACT

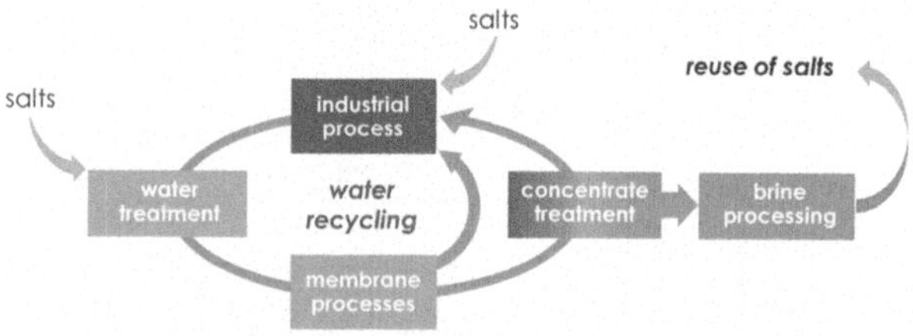

NOMENCLATURE

AAS	Atomic absorption spectroscopy
AEM	Anion exchange membrane
AIC	Annual investment costs
CAPEX	Capital expenditure
CEM	Cation exchange membrane
COD	Chemical oxygen demand
CRF	Capital recovery factor
Cryst.	Crystallization
ED	Electrodialysis
EDR	Electrodialysis reversal
Evap.	Evaporation
FC	Frequency control
FO	Forward osmosis
GOR	Gained output ratio
HP	High-pressure pump
IC	Ion chromatography
ICP OES	Inductively coupled plasma optical emission spectroscopy
IX	Ion exchange
KPI	Key performance indicator
MBR	Membrane bioreactor
MD	Membrane distillation
MEH	Multi-effect humidification
MWCO	Molecular weight cut-off
NF	Nanofiltration
OC	Operating costs
OPEX	Operational expenditure
PF	Plate and frame module for feed gap air gap
FGAGMD	Membrane distillation
PV	Pressure vessel
RO	Reverse osmosis
RR	Recovery ratio
SW AGMD	Spiral wound module for air gap membrane distillation
TC	Total costs
TIC	Total inorganic carbon
TMP	Transmembrane pressure
TNb	Total nitrogen bound
TOC	Total organic carbon
TRL	Technology readiness level
TS	Total solids
UF	Ultrafiltration
XRD	X-ray diffractometer
ZLD	Zero liquid discharge

INTRODUCTION

Freshwater is one of our most important resources. This applies not only to agricultural and potable uses but also to industrial production (Götz *et al.* 2019) in which water is required for cooling and as a solvent. Therefore, in many industrial sectors, water recycling plants have been in operation for decades. Conventional recycling processes are often limited by the discharge options and the costs of disposing of residual streams. For this reason, water recycling rates are often kept artificially low at only 20–50%.

Due to water scarcity and the protection of water bodies, wastewater, and residual streams (e.g. brines from membrane processes) of industrial processes have received more and more attention worldwide (Yaqub & Lee 2019). Increased public environmental awareness and increasingly strict environmental regulation policies are resulting in a change in industrial wastewater treatment; the discussion about wastewater-free industrial production has gained considerable momentum in recent years, and zero liquid discharge (ZLD) has become a serious alternative to conventional systems for wastewater treatment and water recycling (Tong & Elimelech 2016; Götz *et al.* 2019). Furthermore, the increasing value of freshwater – not least due to the globally increasing freshwater scarcity – and the rising cost of wastewater disposal is driving ZLD forward as an attractive strategy for used water management (Tong & Elimelech 2016; Yaqub & Lee 2019). Even where there is no regulatory or financial pressure, companies are striving for ZLD to obtain recognized seals of approval or awards confirming resource-efficient and sustainable production (green labeling) (Götz *et al.* 2019).

During the last decade, there has been a tremendous development of ZLD systems, particularly in North America, Europe, and Asia (Götz *et al.* 2019). According to MarketsandMarkets (2019), the ZLD market was estimated at USD 5.4 billion in 2018. With a compound annual growth rate of 8.3%, the ZLD market is projected to reach USD 8.1 billion by 2023 (MarketsandMarkets 2019).

Although ZLD is becoming increasingly important in industrial wastewater treatment, it is still associated with increased energy consumption and additional costs for investment and operation. Conventional ZLD systems consist of a series of thermal processes alone or, more common recently, combined with a reverse osmosis (RO) system for a pre-concentration. To decrease the energy demand and therefore the operational costs, further enhanced pre-concentration with a lower energy consumption than conventional brine concentrators is targeted (Tong & Elimelech 2016). Applications such as electrodialysis (ED) or electrodialysis reversal (EDR), forward osmosis (FO) and membrane distillation (MD) are emerging as alternative ZLD technologies to further concentrate used water, e.g. after the RO stage (Tong & Elimelech 2016; Pramanik *et al.* 2017; Yaqub & Lee 2019). Since membrane fouling/scaling is a limiting factor in all membrane-based processes, pre-treatment, e.g. for reducing organic components or softening, can enable a further concentration as well. Semblante *et al.* (2018) suggest a wide range of methods/technologies, such as chemical precipitation and coagulation, electrocoagulation, ion exchange (IX), nanofiltration (NF), adsorption, or advanced oxidation, as potential pre-treatment for ZLD. Currently, a change is taking place in the design of ZLD processes; instead of focusing exclusively on volume minimization with well-established technologies – often accepting high energy consumption – innovative technologies or process designs are being considered for an optimized ZLD process.

Within this development, the recovery of recyclable materials from RO concentrates is also being considered more frequently. Considering that the storage or disposal of the solid waste mixed salts generated by conventional ZLD processes can have negative impacts on the environment, other options, such as the recovery of several high-purity salts in ZLD applications, are increasingly becoming an objective (Yaqub & Lee 2019). According to Kim (2011), salt recovery is realized most commonly by evaporation and cooling, but membrane separation processes have become more and more attractive as well. In contrast to salt recovery from brackish water or seawater, the separation of organic and inorganic substances can be of particular importance for salt recovery during industrial wastewater treatment. For this reason, NF has been applied, e.g. in the textile industry to separate dyes and salts (Dasgupta *et al.* 2015). NF has also been investigated for recovering NaCl from a regeneration effluent for ion exchange at a sugar decolorization plant (Wadley *et al.*

1995) and for separating NaCl and lactic acid (Freger *et al.* 2000), as well as retaining organic solutes (glucose) affected by NaCl in a highly concentrated solution (Luo & Wan 2011). In a recent study, Linden *et al.* (2020) investigated the separation of natural organic matter and NaCl by NF, but also by ED and IX in RO brine from a ZLD process in the petrochemical industry.

Since salts and recalcitrant organics are released into wastewater not only through production processes but also through water treatment (coagulants, flocculants, biocides, antiscalants, etc.), an integrative approach and optimization are of great importance. If, for example, water chemicals or substances added in the production process could be substituted or completely omitted, this would have a major impact on concentrate treatment.

The HighCon research project aims to develop a holistic solution for concentrates arising from industrial wastewater reuse (Figure 1) while taking into account the various technical possibilities, production-integrated measures, and the energy efficiency and sustainability of concentrate treatment processes.

In this study, the results of one selected process from one of the demonstration sites within the HighCon project are presented. This includes experimental results of the pilot demonstration as well as an example economic case study for the newly developed concentrate treatment process.

MATERIALS AND METHODS

Demonstration site

The demonstration plant developed within the HighCon project was piloted at two industrial production sites. The results from the food industry site are presented in this study.

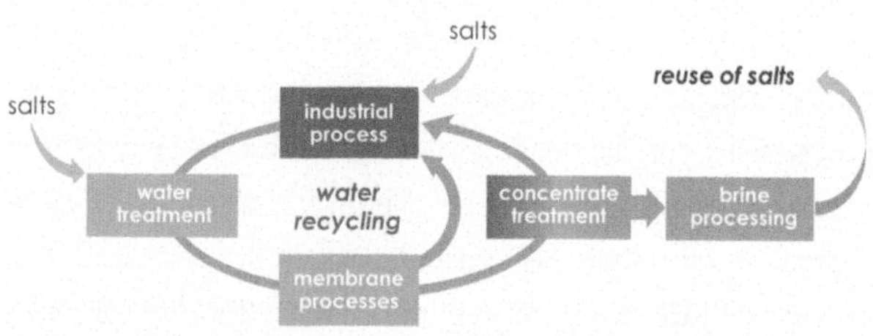

Figure 1 | Scheme of the HighCon concept for concentrate treatment.

The DEK Deutsche Extrakt Kaffee GmbH produces a range of instant coffee products and liquid coffee extracts at their manufacturing facility in Berlin. Here, along with other steps, the roasting, extraction, drying, and agglomeration are carried out. During these process steps and, especially, the system cleaning, process wastewater with high organic contamination accumulates. Apart from good biodegradable substances, this used water contains coffee components that are hardly or non-biodegradable and inorganic salts, e.g. Na^+ and Cl^-, from the cleaning processes. The existing wastewater treatment on-site consists of a mechanical pre-treatment and a membrane bioreactor (MBR) with submerged ultrafiltration (UF) membranes. As process water recycling at the site of DEK Berlin can only be realized when the accumulated concentrate is treated as well, the company participated in the HighCon research project to find new solutions for concentrate handling.

Demonstration plant

Besides the investigation of the overall process combination, one focus was on the separation of organic and inorganic components in the RO concentrate by NF. A second focus was on the further development of single innovative technologies, such as an MD module with a novel channel configuration.

The UF permeate from the DEK's MBR served as influent for the HighCon demonstration plant. The demonstration plant consists of an RO system for investigating water recycling and the following subsequent technologies for further processing the RO concentrate:

- NF for separating organic components and polyvalent ions in NF concentrate, and monovalent ions in the NF permeate;
- ED; and
- MD, both for further concentration of NF permeate near to saturation limits.

The final steps to ZLD, evaporation and crystallization (Evap. & Cryst.) were investigated at lab-scale at TU Berlin. A simplified process flow sheet of the full-scale wastewater treatment plant and the HighCon demonstration plant for water recycling and RO concentrate treatment is shown in Figure 2.

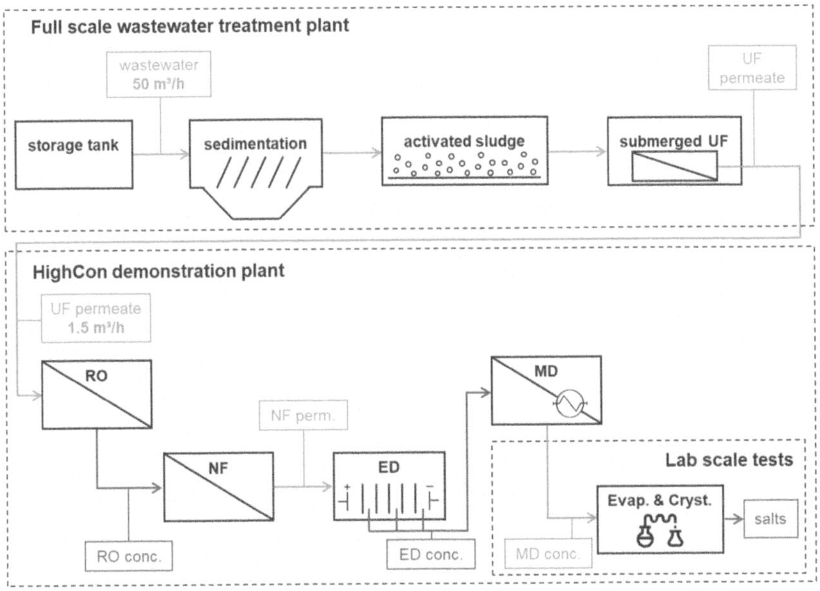

Figure 2 | Simplified process scheme of full scale and demonstration plant at the site of DEK Berlin.

The UF permeate was temporarily stored in a 1 m^3 intermediate bulk container (IBC) before entering the RO system. To protect the membranes in all the following steps, a pre-filter was inserted between the storage tank and the RO plant. RO, NF, and ED were operated for approximately 6–10 h/d, 5 days a week. The MD was operated 24 h/d if required. All plants were shut down over the weekend. System cleaning was carried out at the end of the week, either with clear tap water, distillate, or, if necessary, with cleaning agents. The lab-scale experiments for salt recovery by evaporation and selective crystallization were performed in a batch mode with samples collected from the demonstration plant.

Sampling

During operation days, random samples of all demonstration process flows were taken at least once throughout the day. After sampling, the following parameters were analyzed immediately on-site at the DEK: chemical oxygen demand (COD), total solids (TS), pH, conductivity, and temperature. As these parameters were intended for process monitoring of the single technologies, they were taken at different times. Once a week, samples for analyzing the total organic and inorganic carbon (TOC and TIC), total nitrogen bound (TNb), and ions/elements were taken contemporaneously in the flows of all demonstration plants. The samples were preserved if necessary and stored at 8 °C until analyses had been finalized. These measurements form the data basis for evaluating the HighCon process.

Analytical methods

The following parameters were analyzed to characterize process water flows from the DEK Berlin demonstration site and/or feed, permeate and concentrate samples during demonstrational experiments: pH, conductivity, TS, COD, TOC, TIC, TNb, and a variety of relevant ions respectively.

Conductivity and pH were measured by using a FiveGo Mettler Toledo conductivity meter, Switzerland conductivity F3 with the corresponding measuring sensor Mettler Toledo LE703, and pH meter FiveGo Mettler Toledo pH F2 with the pH electrode LE438-IP67, respectively. COD measurements were achieved using cuvette tests LCK 014, LCK 514, and LCK 1414; a thermostat LT 200; and a UV–VIS spectrophotometer DR 5000 (all HACH LANGE, Germany) according to standard methods (DIN ISO 15705:2003-01). TS was measured using a DBS 60-3 Moisture analyzer (KERN & SOHN, Germany). These parameters were measured directly after sampling at DEK Berlin. Additionally, depending on the level of automatization of the demonstration technology, the pH, conductivity, and temperature were recorded continuously.

The analyses of TOC, TIC, and TNb were performed at TU Berlin using an Analytik Jena, Germany TOC analyzer multi N/C 3100 according to standard methods (DIN EN 1484:1997-08).

Ions and elements were measured by the DVGW Research Center at the Engler-Bunte-Institute (DVGW-EBI), Karlsruhe. Ion chromatography (IC) was used to quantify the amount of selected anions (Metrohm 790 Personal System, Germany). Cations and elements were analyzed with an ICP-OES (inductively coupled plasma optical emission spectroscopy) spectrometer (Agilent Technologies, Germany ICP-OES 5110). Due to high salinity contents of feed and concentrate samples, dilutions up to 1:1000 were applied to perform IC and ICP-OES measurements properly. Ions or elements present in small amounts might be underestimated, caused by the required dilutions.

Solid matter fractions produced in the crystallization tests were analyzed at TU Berlin via Rigaku, Germany SmartLab 3 kW X-ray diffractometer (XRD). The corresponding ion analyses were performed using atomic absorption spectroscopy (AAS, Perkin Elmer, Germany PinAAcleTM) and a Metrohm, Germany 790 Personal IC fed with resolved sample solutions.

Experimental procedure

RO for water recovery and concentrate production at demonstration scale

The RO pilot plant was equipped with four spiral wound membrane modules FilmTec™ BW30-4040, USA, connected in series. The operating pressure was applied using a high-pressure pump, which was designed for a maximum operating pressure of 40 bar at 1.5 m^3/h. In addition, a recirculation pump was installed to ensure a sufficient cross-flow velocity.

For investigations of the concentrate treatment, the RO should be operated continuously and at a maximum possible permeate recovery rate. Therefore, the permeate flux was kept constant (as far as possible due to changes in the wastewater composition) at an average value of 33 L/(m^2 h). As the RO was run in a single-pass operation, the corresponding permeate recovery rate was nearly constant at about 65%, on average. To keep the flux and recovery rates constant, the transmembrane pressure (TMP) was adjusted within the 21–34 bar range. The average TMP was 26 bar.

The temperature during demonstration ranged between 14 and 43 °C. This temperature range also applies to the operation of the other demonstration plants.

Treatment of RO concentrate by nanofiltration

The RO concentrate was subsequently treated by an NF pilot plant. The main task for the NF was rejecting the organic compounds and, thus, preventing fouling in the downstream processes ED/MD. The secondary goal was to achieve a maximum passage of salts to optimize salt recovery.

The NF pilot plant's design is identical to that of a full-sized NF and operated in a 'feed-and-bleed' mode. It is designed to treat between 200 and 800 L/h, with the actual feed volume being regulated via frequency control (FC) of the high-pressure pump (HP). The NF's HP is a Grundfos, Denmark multi-stage centrifugal pump with a nominal pressure head of about 20 bar. There are two pressure vessels (PVs), one 4″ and one 8″, both of which can hold one membrane element each. The different PVs are required to ensure the plant's optimal performance under varying feed flows and/or filtration rates. During the DEK-trial, the NF was operated using DOW-FilmTec™ NF270 elements, USA, first a 4″-element (7.6 m^2), and later, as feed increased, an 8″-element (37 m^2). Feed temperatures ranged between 14 and 43 °C, with an average of 27 °C. The average pH in the feed was 8.3 and the average conductivity was about 5.2 mS/cm. Since hardness and concentrations of multivalent ions, in general, were expected to be very low, no acid dosing was provided during the DEK-trial. This proved to be disadvantageous because the hardness/scaling potential of the effluent was higher than expected, and consequently, the plant suffered from scaling. Yet despite the problems with scaling, the average flux ranged at approximately 18 L/(m^2 h). The yield varied between 30 and 90%, depending on the tested process conditions.

Further concentration of nanofiltration permeate by ED and MD

Electrodialysis. The ED was designed with four fluid cycles, one for the electrode rinsing, which was operated with a 0.1 mol/L Na_2SO_4 solution, the second for the feed, and two cycles for different concentrates. The membrane module itself consists of a membrane stack in the characteristic plate and frame design by the company DEUKUM, Germany. The flow channels were equipped with a special spacer layer with integrated sealing. The membrane module consisted of 40 cells, respective pairs of anion and cation exchange membranes (CEM and AEM). The channel thickness was 270 μm and the average flow velocity in the channels was approximately 2.5 cm/s. The active membrane area was 14.17 m^2 (number of all membranes (CEM and AEM), 51 cm width × 51 cm length). Mono-selective ASV and CSO membranes made by Asahi Glass Co., Japan, and standard ion-exchange membranes FAS and FKS from FUMATECH BWT GmbH, Germany, were used in the module.

The unit was operated in a 'feed-and-bleed' mode on the concentrate side and 'batch' operation on the feed side. The ED module was operated at a constant 15 VDC. A batch started with the conductivity of the NF permeate and 4.2 A and ended at 1.0 mS/cm (2.2 A) in the diluate. Each batch was about 33 L. The conductivity of the concentrate was controlled at a constant value of approximately 20 mS/cm via 'feed-and-bleed' operation by adding deionized water. No classic polarity reversal was applied during operation. Polarity reversal was only used for backflush while refilling the feed tank.

Membrane distillation. The MD pilot system has been designed and constructed by SolarSpring (Germany) and consists of a module pair with a combined membrane area of 16 m^2. In MD, water vapor passes through a hydrophobic membrane due to a vapor pressure difference across the membrane. This vapor pressure difference is induced by applying a temperature difference in a heating and a cooling channel. In the HighCon MD pilot system, a novel channel configuration was implemented that enabled a much higher recovery ratio (RR) of distillate per single pass than previously possible. A depiction of the pilot system with tanks and a control unit is given in Figure 3. Details on the new configuration and other typical MD channel setups can be found in Schwantes *et al.* (2019).

Due to the fully automated control system and onboard thermal energy supply of the plant, flexible system operation could be implemented according to the requirements of the upstream processes. Typically, however, the permeate flow from the system was between 13 and 27 kg/h depending on the salinity of the feed solution. The resulting RR was between 30 and 92%, again depending on the feed solution composition. Flux ranged between ~1 and 2 kg/(m^2 h) depending on the operational setting and feed salinity.

Selective crystallization of MD concentrates at lab-scale

The selective crystallization investigations were conducted at lab-scale using a rotational evaporator (BÜCHI, Switzerland Rotavapor R-300). The feed solution (MD concentrate) was evaporated at a 60 °C heating bath temperature while manually adjusting the operational pressure (in a range of <200 mbar) to maintain smooth boiling. After 90% evaporated, the residue mixture of precipitate and saturated brine was filtered at 60 °C using a glass microfiber filter in a pressurized, heated cartouche. The filtration residue was washed with acetone and subsequently dried at room temperature. The gained solid matter was named 'Fraction I' of the selective crystallization step. The filtrate was fed to a second evaporation step at 60 °C which was stopped after a sample weight change was no longer detected. The sample was

Figure 3 | Perspective view of the MD pilot system and modules: (1) Rig with pumps, valves, and controls; (2) Module 1; (3) Module 2; (4) Feed, concentrate, and distillate tanks (Schwantes *et al.* 2019).

then dried for 24 h at 60 °C; the resulting solid matter was named 'Fraction II'. The testing parameters had previously been assessed using PHREEQC software applying the provided Pitzer-Model database (USGS 2017) to simulate solid–liquid equilibria with regard to the different salt solubilities.

Based on the model assessment, Fraction I was assumed to comprise a big share of sodium hydrogen carbonate with a low percentage of chlorides or nitrates. Fraction II was assumed to contain all dissolved solids of the feed solution that were not present in Fraction I. The solid matter fractions were analyzed according to the section 'Analytical methods'. In this test sequence, dissolved organic solids were neglected due to their low mass share of <0.2% compared to the inorganic solids.

RESULTS AND DISCUSSION

Main influencing factors of the concentrate treatment process

Reverse osmosis

RO permeate and concentrate flow have a huge impact on the entire HighCon process. The RO permeate is decisive for water recycling due to its major part in recovery, and the concentrate determines the design of the entire following concentrate process chain. The characteristic parameters of these two flows during the demonstration are shown in Table 1. Ammonia was not analyzed due to its low concentration.

As expected, the RO membrane rejected organic components >99.5% for COD and TOC, as well as all analyzed cations >99.0%. Anions and TIC passed the membrane somewhat better, at a rejection of 99.0% for Cl^-, 98.5% for SO_4^{2-}, and 97.1% for NO_3^-, which represents the lowest rejection of all measured parameters.

The recovered water can be reused as cooling water, for rinsing purposes, or as boiler feed water (after ion-exchange polishing). Potentially, further processing would be required. In any case, the RO permeate has a higher purity than the used tap water, so the process water pre-treatment can be reduced or even eliminated.

Nanofiltration

The main goal of the NF, the removal of the COD, was successfully achieved; the average COD-removal rate was 99.5% with a standard deviation of 0.33%. With an average COD of 2,400 mg/L in the feed and a COD of usually <10 mg/L in the permeate, the original treatment target value of 100 mg/L COD was reliably reached (Figure 4).

The excellent COD-removal rate can be explained by the composition of the refractory COD, which was mainly complex, color-giving molecules, the so-called melanoidins that arise during coffee roasting. These have high molecular weights of 1,000–100,000 Da, so that their rejection by NF270 (molecular weight cutoff (MWCO) by 200–400 Da) is nearly perfect.

The salt passage was good to excellent for the monovalent ions and poor to very poor for the multivalent ions (Figure 5 and Table 2). In this respect, the pilot system confirmed all expectations with regard to the performance of an NF membrane. Only the rather high passage of silicon was surprising, the concentration in the permeate being about 65% of that in the feed.

The following average concentrations were determined in the NF feed and permeate during the demonstration:

However, although the salt passage was acceptable, the overall salt recovery was unsatisfactory during the trial period because the water yield was only ~50% and much lower than expected. This was a result of the missing acid dosing (see Section 'Experimental procedure') and the constant scaling on the NF, which could not be prevented by antiscalant dosing alone.

To assess the potential total recovery of the most interesting ions – Cl^-, Na^+, and K^+ – the data from a 3-week period

Table 1 | Characteristic composition of RO permeate and concentrate at a 65% water recovery rate

	pH	Cond. (mS/cm)	COD (mg/L)	TOC	TIC	Ca^{2+}	Na^+	K^+	Cl^-	NO_3^-	SO_4^{2-}	Si
Permeate	5.8	0.02	<5	<0.5	3.4	<0.05	2.7	0.7	1.3	3.3	0.5	<0.05
Concentrate	8.0	6.5	2,957	1,035	605	86	888	184	323	298	104	18

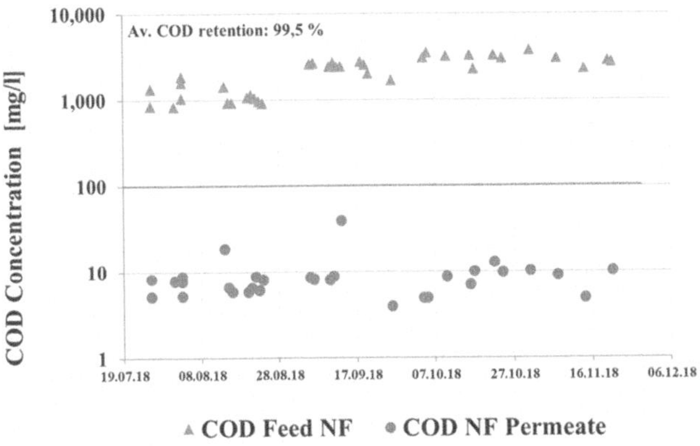

Figure 4 | COD of feed and permeate; permeate target value of 100 mg COD/L.

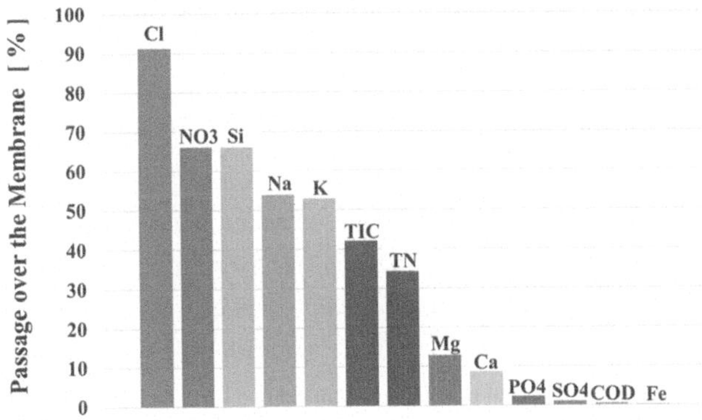

Figure 5 | Average passage of ions over the membrane, expressed as % of feed concentration.

with an ~80% yield were examined more closely. During this period, an average of 40% of K^+, 51% of Na^+, and ~61% of Cl^- were recovered with the NF permeate (Figure 6).

For the cations K^+ and Na^+, the expected and the measured recovery were in good accordance. For Cl^-, however, the measured recovery was remarkably lower than expected (61% instead of ~70%). This is caused by a lower passage of chloride during this 3-week period trial compared to the average passage during the complete trial. This was most likely caused by the relatively high concentration and the retention of the divalent anions (Donnan effect). SO_4^{2-}, in particular, seems to have an influence, since high SO_4^{2-}-concentrate and Cl^--permeate concentration correspond. Thus, the passage or retention of chloride is governed by the anion concentration in the concentrate.

In conclusion, the rejection of COD and polyvalent ions responsible for scaling was very good. This is very favorable as COD as well as these ions can impact the process stability of the subsequent ED and MD. Unfortunately, this advantage for process operation is a disadvantage in salt recovery since the polyvalent ions end up together with the organic components in the NF concentrate and cannot currently be reused. Further investigations are necessary for NF concentrate treatment. Additionally, monovalent ions recovery was not optimal during the demonstration phase because the plant operation had not yet been optimized (pH control, choice of antiscalant). But these conditions can be enhanced effortlessly in future plant operations.

Electrodialysis and membrane distillation

Electrodialysis. Figure 7 shows the current over conductivity for the ED system at the end of the demonstration phase in different states of cleanliness. The relationship in this value range is almost linear. It is measured with pure NaCl solutions of different concentrations as feeds and a constant 15 VDC. The initial value of 4.2 A at 2 mS/cm could not be restored, even after intensive cleaning. Significant degradation of performance already occurred after 3 days of operation with the NF permeate as feed.

At low diluate conductivities, the electrical resistance rises. Batch operation at the feed respectively diluate side was chosen to utilize the high conductivity periods and reduce the periods of low conductivity. Concerning specific energy consumption, the reference parameter has to be selected carefully. Usually, in desalination, the energy demand is linked to the production of desalinated water. However, the diluate flow with ~1 mS/cm is not really

Table 2 | Average NF feed and permeate concentration

	pH	Cond. (mS/cm)	COD (mg/L)	TOC	TIC	Ca^{2+}	Na^+	K^+	Cl^-	NO_3^-	SO_4^{2-}	Si
Feed	8.3	5.2	2,355	815	588	72	663	154	194	233	82	19
Permeate	7.9	3.0	8.1	1.8	238	6.4	333	98	160	161	1.1	13

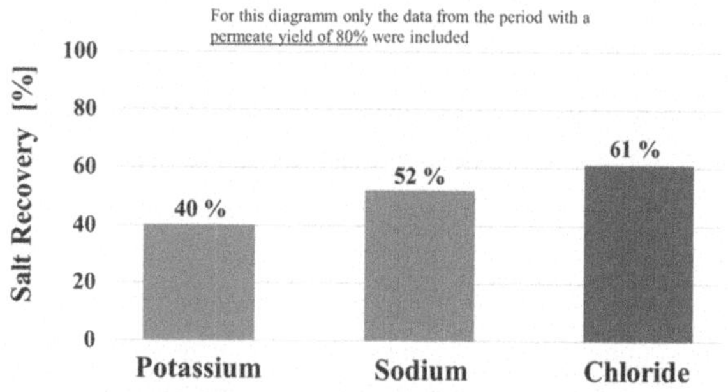

Figure 6 | K^+, Na^+, and Cl^- recovery in the permeate; mass in the permeate compared to the NF feed for 3 weeks.

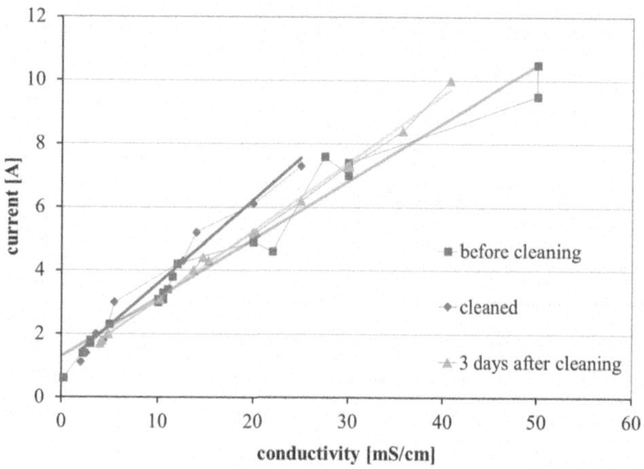

Figure 7 | Characteristic desalination performance of ED system.

desalinated water and must be further treated, e.g. with RO. Diluate and feed flow are almost the same since the trans-membrane water flux is low. The specific energy demand is related to the inlet flow. The operation of the four pumps and the blower had an energy demand of 430 W. Together with all consumers in the control box, the peripheral devices required 460 W of standby power. During a batch process, the direct current energy demand started at 63 W and ended at 33 W. An average total energy demand of 520 W was measured. This corresponds to an average measured feed flow rate of 55 L/h resulting in a specific energy demand of 9.45 Wh/L for this ED system. The peripheral devices make up more than 85% of the energy demand for this small-scale device. This fraction can be significantly reduced in a production scale ED unit, so 2–4 kWh/m^3 can be used as a realistic assumption.

During the demonstration experiments, some potential for optimization was found as necessary with regard to a real operation. First, as in NF operation, the concentration factor can be significantly improved with an optimized pH adjustment and antiscalant dosage. Furthermore, thicker channels in the ED membrane module can avoid the high sensitivity of the system to organic fouling and calcite scaling.

Membrane distillation. For interpreting the operational results from the MD pilot stage within the HighCon system, the following key performance indicators (KPIs) are introduced:

Flux is a well-known value for the evaluation of distillate/permeate production per m^2 of active membrane surface in membrane technologies. It is calculated by dividing the distillate output $\dot{m}_d$ by the membrane surface A, as shown below:

$$j_d = \frac{\dot{m}_d}{A} (\mathrm{kg/m^2h}) \tag{1}$$

Gained output ratio (GOR) provides a dimensionless coefficient for evaluating the heat recovery inside the MD module by giving a relationship between the thermal energy amount needed for the evaporation process of the distillate (numerator) and the amount of heat introduced externally (denominator).

In Equation (2), $\dot{m}_{H/C}$ (kg/h) is the mass flow rate of the heating and cooling solution, $c_{p,H/C}$ (kJ/kg K) is the specific heat capacity of the heating and cooling solution, T_{hi} is the heating inlet temperature, and T_{co} is the cooling outlet temperature. h_v (kJ/kg) is the specific evaporation enthalpy

$$\mathrm{GOR} = \frac{\dot{m}_d \times \Delta h_v}{\dot{m}_{H/C} \times c_p \times (T_{hi} - T_{co})} \tag{2}$$

Thermal efficiency η_{th} is a percentage value for the ratio of latent heat used for the phase change, to the total heat transported through the membrane, and is shown in Equation (3). T_{ho} is the heating outlet temperature.

RR (Equation (4)) is the quotient of the distillate mass flow rate and H/C flow rate:

$$\eta_{th} = \frac{\dot{m}_d \times h_v}{\dot{m}_{H/C} \times c_{p,H/C} \times (T_{hi} - T_{ho})} \times 100\ (\%) \tag{3}$$

$$\mathrm{RR} = \frac{\dot{m}_d}{\dot{m}_f} \times 100\ (\%) \tag{4}$$

Figures 8 and 9 can be analyzed with these KPIs. During the piloting phases, an array of different parameter settings were applied to evaluate and characterize the novel MD system to gain reliable data for cost calculations and upscale the pilot system. A detailed analysis is given in Schwantes *et al.* (2019). Since MD is primarily used for concentrating feed solutions that are of higher salinity than other technologies are capable of treating, the performance of a selected analysis in correlation with salinity increase is presented. The general trend in both graphs and for all values of flux, GOR, RR, and η_{th} is a decreasing one. This effect can be assigned to the fact that as more salt ions are present in the feed solution, more thermal energy will be required to evaporate the same amount of distillate due to the decrease in vapor pressure. Thus, under the same operating conditions, the flux will reduce with increasing salinity in the feed solution, as seen in Figure 8. As an indicator of internal heat recovery, which is coupled to the distillate production, GOR values will decrease correspondingly.

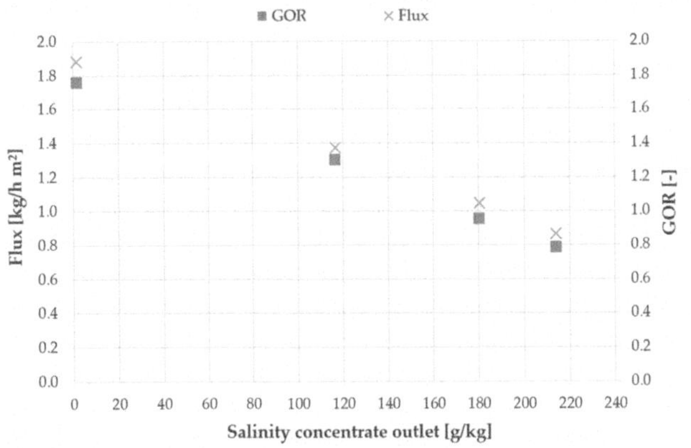

Figure 8 | Influence of concentrate salinity on flux and GOR; MD1 + MD2 combined; feed inlet salinity 0.3/55/95/143 g NaCl/kg; $\dot{V}_f$ = 40 kg/h, $\dot{V}_{H/C}$ = 600 L/h, $T_{hi\ 1/2}$ = 80 °C, $T_{ci\ 1/2}$ = 25 °C (Schwantes *et al.* 2019).

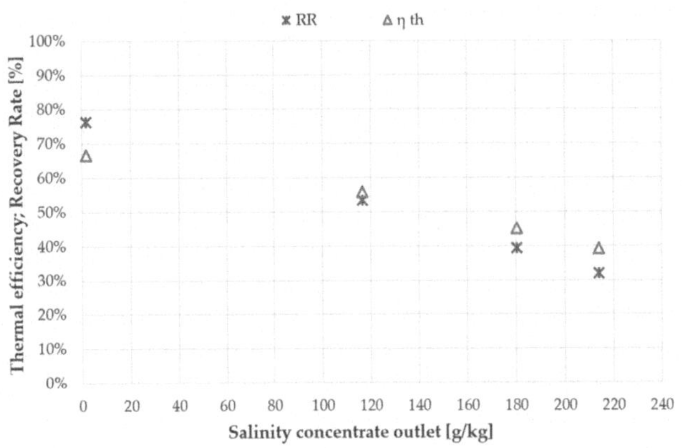

Figure 9 | Influence of concentrate salinity on RR and thermal efficiency η_{th}; MD1 + MD2 combined; feed inlet salinity 0.3/55/95/143 g NaCl/kg ;$\dot{V}_f$ = 40 kg/h, $\dot{V}_{H/C}$ = 600 L/h, $T_{hi\ 1/2}$ = 80 °C, $T_{ci\ 1/2}$ = 25 °C (Schwantes *et al.* 2019).

Thermal efficiency η_{th}, which is shown together with RR in Figure 9, is also affected by the increase in salinity. The effective driving force available for the actual evaporation is reduced with the increase in salt ions in the solution. Alongside this reduction, the fraction of total heat used for evaporation is subsequently also reduced, which can be seen in the η_{th} trend. In comparison to 67% at tap water salinity, only 39% of the incoming thermal energy is used to bring about the phase change when reaching a 214 g/kg salinity. Since one of the main design goals of the novel MD module was a significant increase in RR compared to previous module concepts, the analysis of RR is of specific interest. Although the decrease in RR is high from 76 to 32%, the absolute value is a huge improvement in comparison to earlier MD modules used in high salinity applications. Table 3 provides a comparison of KPIs from the novel plate and frame module (PF FGAGMD) and the previous configuration, an Air Gap MD spiral wound module (SW AGMD; Schwantes *et al.* 2018). At a typical average conductivity for an MD concentration stage of 155 mS/cm, the RR in the novel module type was 15× higher (45%) than in the spiral wound module (3%). The flux was similar, the thermal efficiency was slightly higher, and the GOR was lower by 0.3. The reduced GOR can be explained by the suboptimal flow regime in the first-generation plate and frame prototype in which the main construction focus was more upon general functionality than optimal heat transfer. It is expected that equal or higher GOR values can be achieved after one iteration of the internal flow geometry redesign.

Table 3 | Average KPI values for plate and frame FGAGMD and spiral wound AGMD module, mean concentrate salinity ~155 mS/cm; $\dot{V}_f$ = 40 kg/h, $\dot{V}_{H/C}$ = 300 L/h per module; $\dot{V}_{fsw}$ = 300 L/h; $T_{hi\ 1/2}$ = 80 °C; $T_{ci\ 1/2}$ = 25 °C

	Flux (kg/m² h)	GOR (–)	RR (%)	η_{th} (%)
PF FGAGMD	1.2	1.1	45	50
SW AGMD	1.1	1.4	3	46

Thus, the performance of the MD pilot stage was successful with regard to the desired goals.

Selective crystallization

Figure 10 displays the mineral compositions of Fraction I and Fraction II gathered from the selective crystallization test. The minerals were assessed on ion mass shares, stoichiometric balances, and simulated saturation indices; the respective values can be taken from Table 4.

Table 5 shows the quantitative results of the detected crystalline minerals mass share by XRD.

When comparing the results of ion analyses and XRD, it must be noted that the XRD analysis does not identify the stoichiometry of the amorphous solid matter; thus, this method does not deliver a full ion mass balance. Nevertheless, both results prove the feasibility of using selective crystallization to separate sodium hydrogen carbonate into a new solid, Fraction I, from other dissolved solids with a lab-scale purity of approximately 90%. Other (previously dissolved) solids end up in Fraction II. The occurrence of a remarkable share of NaCl and KNO_3 in Fraction I is not plausible compared to their low saturation in the feed solution. It is assumed that these precipitates are the result of local effects due to an inhomogeneous temperature and concentration distribution through the sample in the reaction flask; near the sample surface, the flask wall was moistened by a thin liquid film that is supposed to deviate from the theoretical perfectly stirred sample in the bulk solution.

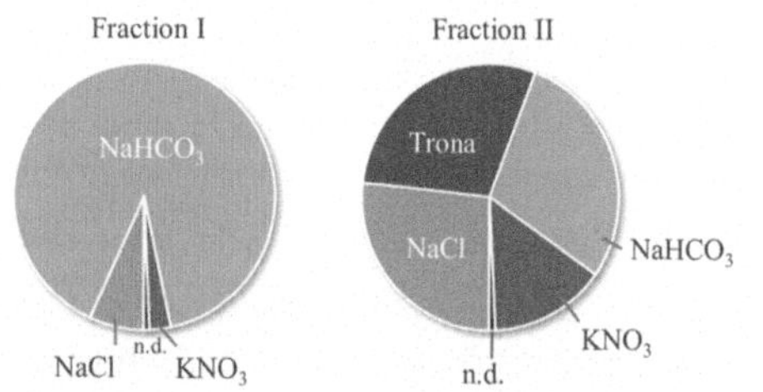

Figure 10 | Solid fractions composition assessment from ion analysis.

Table 4 | Mass share of minerals in Fractions I and II via ion analysis

	Trona	$NaHCO_3$	Na_2CO_3	NaCl	KNO_3	Unknown (n.d.)
Fraction I (%)	0	89.9	0	6.8	2.6	0.7
Fraction II (%)	29.0	29.0	0	26.7	14.5	0.8

Table 5 | Mass share of crystalline minerals in Fractions I and II via XRD

	Trona	$NaHCO_3$	Na_2CO_3	NaCl	KNO_3	Unknown (n.d.)
Fraction I (%)	0	94.0	0	6.0	0	0
Fraction II (%)	27.4	0	13.3	19.8	39.5	0

The prediction of the formed phases in Fraction II is still subject to improvement. The XRD measurement shows a formation of sodium carbonate (Na_2CO_3) instead of sodium hydrogen carbonate, which had not been predicted by the PHREEQC simulation. Two aspects have been identified as the main influencing factors for the deviation. On the one hand, the high ion strength occurring in the final evaporation process of Fraction II leads to inaccuracies in the simulation model. On the other hand, the ternary system Na^+–CO_3^{2-}–H_2O provides numerous stable solid phases, also hydrated, that are quite close with regard to equilibrium activities (Wegscheider & Mehl 1928). In addition, the conducted assessments covering steady-state equilibria had not considered kinetics or possible hydrate conversions during cool-down.

The salt recovery yield that is achievable via selective crystallization is dependent on the difference in salts (sub-) saturation in the feed solution. If the saturation of the target salt (in this case sodium hydrogen carbonate) is reached during evaporation and starts to precipitate, the process can be continued until the next salt (in this case sodium chloride) reaches saturation. If the evaporation is continued beyond this concentration point, the yield of the target salt is raised at a trade-off in purity as the next salt starts to precipitate and adds up to the target fraction. The maximum achievable yield rates (precipitate mass divided by the corresponding dissolved solids mass in the feed solution) according to PHREEQC simulation are up to 92% for $NaHCO_3$ corresponding to 60% of the Na^+.

Full-scale process design based on the demonstration results

Based on the results of the demonstration at the DEK Berlin site, a full-scale process for water recycling via RO and subsequent concentrate treatment using the HighCon process was designed. In addition to the values determined for flux, rejection, and energy demand, the know-how of the industry partners as experienced plant engineers was drawn upon in developing the full-scale process. The optimization potential identified during the demonstration is also being integrated in this way. The scaling potential was taken into account based on the parameters measured during the demonstration for determining the recovery rates. The following optimizations were considered for the full-scale plant design:

Increasing RO permeate recovery rate: The recovery rate is an important influencing factor for the economic feasibility of the process because all the following technologies in concentrate treatment need to handle smaller volumetric flows. For the overall process design, a high recovery rate of 85% was assumed, but with a lower flux and slightly lower rejection of all components due to the higher concentrations reached. With regard to the energy demand, a lower TMP was calculated, based on an optimized multi-stage RO design instead of the single-pass design and including an optimized antiscalant dosage.

Increasing NF permeate recovery rate: With respect to the salt recovery rate of the whole process, the NF permeate recovery needs to be increased as far as possible. A recovery rate of 80% was estimated to be realistic with optimized additive dosage, thus pH control and optimization of antiscalant.

Increasing recovery rate for ED and MD: To optimize the pre-concentration before entering the evaporation process, the water recovery by ED and MD must be increased as well. Their operation will improve with the optimized additive dosage for the NF due to pH control, but an adapted additive dosing is necessary as well.

Taking these steps into account, a full-scale water recycling and concentrate treatment process was designed for an inflow of 1,100 m^3/d UF permeate, as shown in Figure 11.

For the evaporation and crystallization, using multi-effect humidification (MEH) technology was assumed, using the example of technology configuration of the company Terrawater, Germany. In Evap. & Cryst. I (MEH I), a concentrated organic-free salt stream with approximately 10% water content remaining is produced. For the organically loaded residual stream (discharge of MEH II), a remaining water content of approximately 40% was estimated.

Water recovery up to 98% is possible and about 40% of the salts contained in the used water can be recycled without organic contamination. Using a selective crystallization process, about half of this 40% can be recovered as a high-purity single salt (corresponding 20% of the total amount of salts). The other half of the organic-free salts, thus 20% of the total amount of salts, are contained in a salt mixture, e.g. suitable for the use in mine backfilling what is regarded as reuse (according to German waste management enterprises). And 60% of the salts end up in an organic salt mixture arising from evaporating the NF concentrate stream. To utilize the predominantly polyvalent ions, other treatment processes must be considered; promising results have been achieved at the lab-scale, e.g. for electrolysis in an advanced oxidation process for COD reduction. For an application in practice, however, further investigations are required. Nevertheless, in contrast to a conventional ZLD process, 40% of the salts can be recycled. In addition, using energy-optimized technologies reduces the overall energy consumption of the process.

Cost analysis

Based on the process design, costs for water recycling and concentrate treatment were calculated. Investment costs were considered as well as operating costs, such as energy and chemicals used, personnel costs, and costs for residue disposal.

Boundary conditions and database

The cost analysis was carried out for an inflow of 1,100 m^3/d for the water recycling and concentrate treatment process.

In accordance with the industry partners, the boundary conditions such as the annual operating hours, amortization, and personnel deployment, and the most important expendables and energy requirements were defined (Table 6). Furthermore, the specific costs for chemicals, personnel, and energy were also determined based on real costs and practical experience (Table 7). The investment costs for the different technologies were estimated by dimensioning with the help of the flux values and the inlet volume flows (Figure 11 and Table 6).

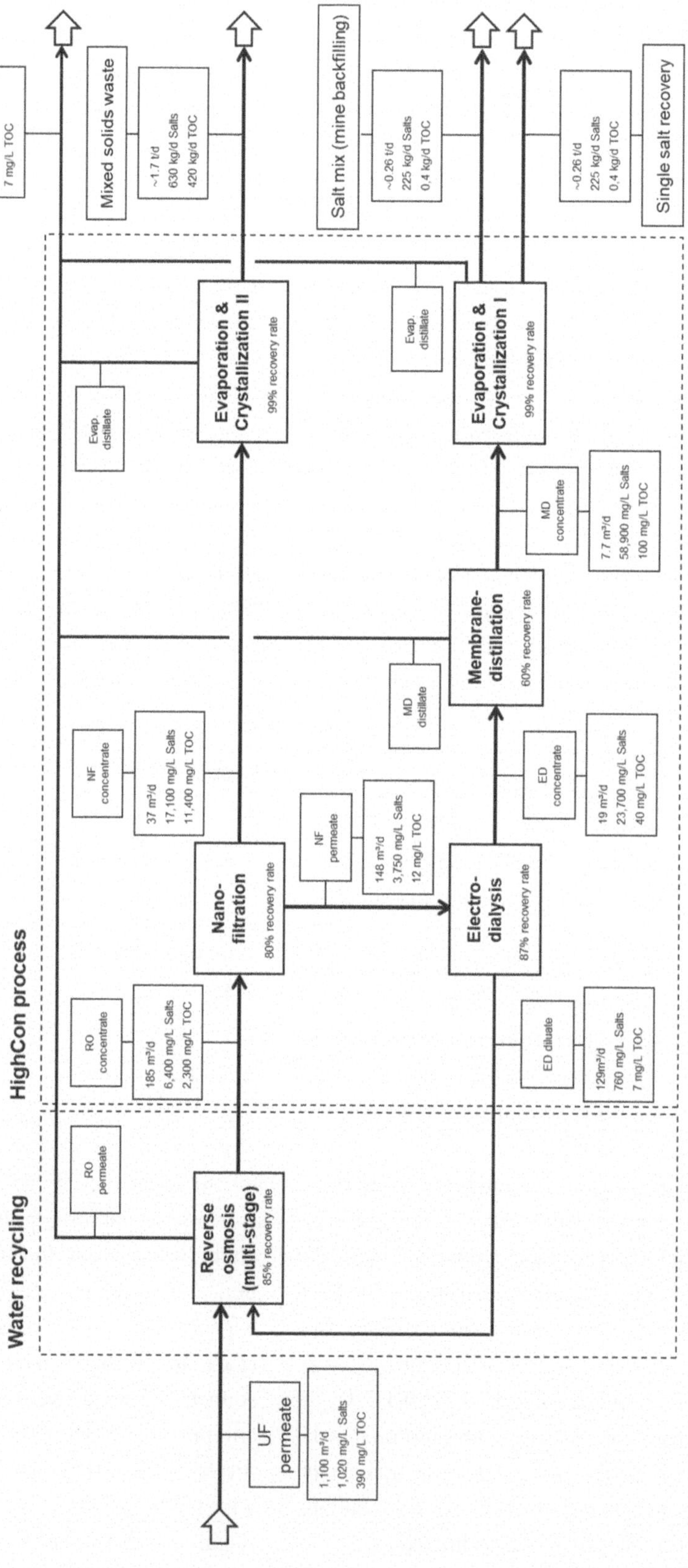

Figure 11 | Simplified process flow sheet for the full-scale water recycling and HighCon concentrate treatment designed for an inflow of 1,100 m³/d MBR UF permeate from coffee production wastewater treatment.

Table 6 | Boundary conditions and cost design parameters for the technologies of water recycling and the HighCon concentrate treatment process

		RO	NF	ED	MD	MEH I	MEH II
Annual operating hours	(h/a)				8,000		
Real interest rate	(–)				0.03		
Amortization period	(a)				7		
Personnel deployment	(h/a)	440	220	200	200	120	600
Membrane lifetime	(a)	2	3	5	2.5	–	–
Specific consumption electric energy	(kWh/m^3)	5.0	2.5	3.0	0.7	24	32
Specific consumption thermal energy	(kWh/m^3)	–	–	–	250	400	540

Table 7 | Real costs and experience values for selected positions of the water recycling and concentrate treatment process

		RO	NF	ED	MD	MEH I	MEH II
Investment costs	(€)	2,670,000	340,000	350,000	300,000	400,000	1,910,000
Specific costs membrane	(€/m^2)	20	22	100	100	–	–
Chemical costs per feed	(€/m^3)	0.35	0.35	1.00	1.00	1.00	1.00
Personnel hour	(€/h)			50			
Electric energy	(€/kWh)			0.1			
Thermal energy	(€/kWh)			0.02			

Table 8 | Costs for freshwater supply, wastewater, and residual waste disposal

		Costs
Freshwater	(€/d)	1.813 (€/m^3) $\cdot Q$ (m^3/d) + 19.26 (€/d)
Wastewater discharge (to municipal sewage treatment plant)	(€/d)	2.21 (€/m^3) $\cdot Q$ (m^3/d) + 18 (€/d)
Landfill disposal	(€/t)	300
Mine backfill disposal	(€/t)	100

The costs for freshwater, wastewater, and especially the landfill or mine backfill disposal of residue streams are given in Table 8.

Calculation of costs

From the investment costs and the various operating costs, the annual costs for water recycling and concentrate treatment were calculated as follows.

The annual costs incurred by investment in the plants or their depreciation were calculated using the method of the German federal states' working group on water issues (LAWA) (DWA 2012).

In accordance with this approach, a capital recovery factor (CRF) was calculated from a real interest rate i and an assumed amortization period n using the following equation:

$$\mathrm{CRF}(i, n) = \frac{i(1+i)^n}{(1+i)^n - 1} \tag{5}$$

All operating costs, such as energy and chemicals used, and personnel costs for operation and maintenance are summarized in the operating costs (OC) and added to the annual share of the investment costs (AIC). This results in the following definition for the total annual costs (TC):

$$\mathrm{TC} = \mathrm{AIC} \cdot \mathrm{CRF} + \mathrm{OC}\ (/\mathrm{a}) \tag{6}$$

The results for the total annual costs of the whole water recycling and concentrate treatment process are shown in Table 9. For comparison, the costs for the indirect discharge of the UF permeate to the municipal sewage treatment plant

Table 9 | Results for the annual costs of the HighCon process and indirect discharge

		Indirect discharge	Conventional ZLD	HighCon process
Total annual costs TC	(€/a)	820,000	1,590,000	1,410,000
Total specific costs	(€/m^3)	2.2	4.3	3.8
Disposal costs	(€/m^3)	2.2	0.7	0.5
OPEX[a]	(€/m^3)	–	1.2	0.6
CAPEX	(€/m^3)	–	2.4	2.7

[a]Excluding disposal costs.

and the costs for a conventional ZLD process are presented alongside.

For the conventional ZLD process, a multi-stage RO, a conventional evaporator, and a second evaporation step with crystallization were assumed. The cost calculation was carried out using the method presented above. As with the organically loaded discharge of the HighCon process, the residual stream of the second evaporation step and crystallization contains about 40% water. Thus, the water recovery is comparable to the HighCon process, with up to 98%. Since only non-selective technologies are used in the conventional ZLD, an organic salt mixture is produced, containing 420 kg/d organic components and about 1,360 kg/d mixed salts. For calculating the disposal costs, landfill disposal was assumed since the waste mixture cannot be utilized. In contrast, in the HighCon process, only 60% of the residuals are not suitable for reuse, and therefore, landfill disposal was assumed. 20% of the salt content can be recovered as a reusable single salt without organic contamination. Therefore, no disposal costs were calculated. The remaining organic-free mixed salts are useable for mine backfilling.

The results of the economic analyses indicate that both ZLD procedures result in comparable total specific costs of about 4 €/m^3. While disposal costs are approximately 30% lower in the HighCon process, the capital expenditure (CAPEX) is slightly higher due to the additional technological effort of the specialized process. The major difference is in the operating costs (disposal costs not included) since the operation of the new technologies like ED and MD is less expensive compared to the conventional thermal processes; pre-concentration before entering the evaporator reduces operational expenditure (OPEX) drastically. Whether in the HighCon process or conventional ZLD, an extensive volume reduction before entering the evaporation process is the key to economic feasibility.

Furthermore, the results show that indirect discharge is significantly low-priced, with total specific costs of about 2 €/m^3 instead of 4 €/m^3. However, it requires sufficient capacity of the municipal sewage treatment system, does not reduce the discharge of recalcitrant substances into water bodies, and does not lead to a reduction of freshwater use. The savings of thermal energy related to recycling warm water are an additional benefit. By using recycled water with a temperature 20 °C above that of the replaced tap water, savings up to 20.000 kWh_{th} per day could be possible.

The costs presented in this study can only give an indication, as the following limitations apply to their calculation:

- The costs were determined by extrapolating pilot plant results in a full-scale process. Operation optimizations were not fully investigated; their impact has been estimated by the industry partners as experienced plant engineers and based on demonstration results.
- Generally, it must be considered that cost analyses are not directly transferable since location-dependent cost factors, such as the water price or disposal costs, are included in the calculation.
- The different technology readiness levels (TRLs) must be considered. For emerging technologies, such as ED or MD, investment and membrane costs can be comparably high at the present moment, but they will decrease in the future with the growing application.

Comparison of conventional ZLD and the HighCon process

The separation of dissolved inorganic salts is a challenge but also an important process step in concentrate treatment. On the one hand, this is the only way to prevent their entry into natural water bodies. Loads of salts and organic components, which are non- or hardly biodegradable, enter the (municipal) wastewater treatment plant and subsequently the receiving waterbody. On the other hand, resource recovery is becoming increasingly important both economically and with regard to the closing of recoverable substance cycles.

The yield of recycled water is also increased by concentrate treatment. Both conventional ZLD and the HighCon ZLD process are estimated to generate residual product streams with 10% water. Hence, the water recycling rate is up to 98% for both processes. Water recycling enables cost savings for freshwater withdrawal and wastewater charges. Above all, the industrial site gains independence with regard to its water supply and wastewater discharge. Overall, ZLD processes cause new costs due to operation and disposal of residuals, so there must be other drivers than just costs savings.

The main advantage of the HighCon process is the separation of organic components in one stream and monovalent ions in another. Combined with selective crystallization, pure salt recovery is possible even for organically highly loaded used water. The HighCon process is applicable for salt recovery where conventional ZLD would result in a waste solid mixture unsuitable for reuse. Thus, disposal quantities are reduced, and disposal safety can be increased by separating the different fractions.

Resource recovery by the HighCon process

The selective crystallization tests demonstrated that it is possible to separate single salts as solid crystals from a complex wastewater matrix resulting from industrial water recycling RO brines. In the first step of the lab-scale experiments, using the concentrate from the DEK demonstration site, sodium hydrogen carbonate salts could be produced with a purity of 90%. As the simple experiments have not yet exploited influence factors like kinetics or volume effects, there is much room for optimization left. It appears probable that technical quality requirements for reusing the recovered salts for defined industrial applications could be met. Another reuse chance could be blending recovered salts with crude material for further processing.

As described in the section 'Full-scale process design based on the demonstration results', the final HighCon process design aims to maximize RO and NF recovery rates, which implies pH adjustment via acidification. This leads to a notable change in the ion concentrations impinging the selective crystallization.

For example, acidic electrolytes will add to the anions, e.g. NO_3^-, when nitric acid is applied. Additionally, along with lower pH, dissolved carbonates will associate, leading to a higher super-saturation of carbonic acid. This leads to CO_2 degassing in the evaporator device, where the specific contact surface of solution and air is maximized for heat and mass transfer efficiency.

The influence of a pH reduction from 8 to 6.5 by nitric acid addition upstream of the RO on the solid–liquid equilibria in the crystallization was assessed for the DEK demonstration site. The NO_3^- would be raised by <1%. Thus, its effect on selective crystallization can be neglected. Nevertheless, the influence on carbonate dissociation cannot be disregarded. Speciation calculations indicate that (compared to the conducted demonstration) approximately 50% of the carbonates could be lost to gaseous CO_2 and, in consequence, will not be available for resource recovery.

Each process step upstream of the selective crystallization contributes directly or indirectly to the water or solids recovery yield of the whole process. The RO provides 85% of the produced water for direct reuse at a comparably low-energy investment. The achievable RO yield rate is one of the main levers to enable an energy-efficient water recycling process. NF was applied due to its selectivity with regard to organics and multivalent ions. As it rejected >99% of the dissolved organic matter from the main process stream, this technology lays the foundation for subsequent separation of inorganic salts nearly cleaned from organic impurities. The ED provides an indirect contribution to the process of water recovery yield. Due to its technologically inherent minimal required diluate conductivity, this desalinated water stream is not supposed to be reused directly. By recycling it to the RO feed, the ED diluate indirectly adds to the water recycling yield of the process. The thermal processes MD and 'Evaporation/Crystallization I' produce distillate flows at a high quality that allows direct reuse of the produced water. Nevertheless, the share of these distillates in the whole process water yield rate appears quite low with approximately 2% compared to the feed. However, this is reasonable because due to energy efficiency reasons, the process stream is reduced as far as possible before entering the thermal process steps. For full-scale application, an 'Evaporation/Crystallization II' was postulated to process the NF concentrate (see also Figure 11). The quality of the distillate produced here has not been investigated in this project. Hence, this should be a subject for further investigation.

One side effect of the applied brine processing technologies is the use of additives, e.g. antiscalants or nitric acid that will add to the dissolved matter in the process stream and end up in the produced solid fractions. In a holistic process optimization, the application of additives is reduced to the required minimum (supported, e.g. via automated dosing).

CONCLUSION

(1) The separation of organic components and monovalent ions worked out very well in the HighCon process and enabled the recovery of an organic-free salt mixture or – using selective crystallization – the separation of an organic-free single salt with >90% purity.

(2) The recovery of resources is becoming increasingly important both economically and with regard to closing recoverable substance cycles. This refers not only to water recovery but also to other valuable components of industrial wastewater, e.g. salts. With regard to pure salt recovery, the HighCon process, therefore, has an advantage over conventional ZLD. Since additives from the production process and water treatment (e.g. antiscalants) usually end up in the produced solids, holistic process optimization is essential.

(3) Both conventional ZLD and the HighCon process result in comparable total specific costs of about 4 €/m^3. In comparison, indirect discharge is significantly cheaper with total specific costs of about 2 €/m^3, but requires sufficient capacity in the municipal sewage treatment system, does not reduce the discharge of recalcitrant substances into water bodies, does not lead to a reduction of freshwater use, and does not recover the thermal energy of the used water. Furthermore, increasingly stricter environmental regulation policies also concern indirect discharges. ZLD processes increase independence with regard to water supply and wastewater discharge for industrial sites.

ACKNOWLEDGEMENTS

The Federal Ministry of Education and Research (BMBF) funded the joint research project 'Concentrates from Wastewater Reuse (HighCon)' – the content of this study – from 2016 to 2019 within the funding measure 'Future-oriented Technologies and Concepts to Increase Water Availability by Water Reuse and Desalination (WavE)'; HighCon funding reference number: 02WAV1406. The company Endress + Hauser supported the HighCon project by supplying various measurement technologies.

Dr Stefan Berendts from the Institute of Chemistry, TU Berlin, provided X-ray diffraction crystal analyses for the selective crystallization tests.

DATA AVAILABILITY STATEMENT

Data cannot be made publicly available; readers should contact the corresponding author for details.

REFERENCES

Dasgupta, J., Sikder, J., Chakraborty, S., Curcio, S. & Drioli, E. 2015 Remediation of textile effluents by membrane based treatment techniques: a state of the art review. *Journal of Environmental Management* **147**, 55–72. doi:10.1016/j.jenvman.2014.08.008.

DIN EN 1484:1997-08 1997 *Water Analysis – Guidelines for the Determination of Total Organic Carbon (TOC) and Dissolved Organic Carbon (DOC); German Version EN 1484:1997*. Deutsches Institut für Normung e.V. (DIN), Berlin, Germany.

DIN ISO 15705:2003-01 2003 *German Standard Methods for Water Quality – Determination of the Chemical Oxygen Demand Index (ST-COD) – Small-Scale Sealed Tube Method (ISO 15705:2002)*. Deutsches Institut für Normung e.V. (DIN), Berlin, Germany.

DWA 2012 *Leitlinien zur Durchführung dynamischer Kostenvergleichsrechnungen (KVR-Leitlinien) (Guidelines for the Performance of Dynamic Cost Comparison Calculations (CCC Guidelines))*, 8th edn. DWA Deutsche Vereinigung für Wasserwirtschaft, Abwasser und Abfall e. V., Hennef, Germany.

Freger, V., Arnot, T. C. & Howell, J. A. 2000 Separation of concentrated organic/inorganic salt mixtures by nanofiltration. *Journal of Membrane Science* **178**, 185–193.

Götz, G., Teschke, P., Netzel, L. & Geißen, S.-U. 2019 Zero Liquid Discharge für eine abwasserfreie Industrieproduktion (Zero liquid discharge for a wastewater free industrial production). *Chemie Ingenieur Technik* **91** (10), 1480–1485. doi:10.1002/cite.201900047.

Kim, D. H. 2011 A review of desalting process techniques and economic analysis of the recovery of salts from retentates. *Desalination* **270**, 1–8. doi:10.1016/j.desal.2010.12.041.

Linden, N. v., Shang, R., Stockinger, G., Heijman, B. & Spanjers, H. 2020 Separation of natural organic matter and sodium chloride for salt recovery purposes in zero liquid discharge. *Water Resources and Industry* **23**, 1–11. doi:10.1016/j.wri.2019.100117.

Luo, J. & Wan, Y. 2011 Effect of highly concentrated salt on retention of organic solutes by nanofiltration polymeric membranes. *Journal of Membrane Science* **372**, 145–153. doi:10.1016/j.memsci.2011.01.066.

MarketsandMarkets 2019 *Zero Liquid Discharge Systems Market (Preview to Market Report, Code: CH 5288) Published Date Feb 2019*. Available from: https://www.marketsandmarkets.com/Market-Reports/zero-liquid-discharge-system-market (accessed 11 March 2020).

Pramanik, B. K., Shu, L. & Jegatheesan, V. 2017 A review of the management and treatment of brine solutions. *Environmental Science: Water Research & Technology* **3**, 625–658. doi:10.1039/c6ew00339 g.

Schwantes, R., Bauer, L., Chavan, K., Dücker, D., Felsmann, C. & Pfafferott, J. 2018 Air gap membrane distillation for hypersaline brine concentration: operational analysis of a full-scale module – new strategies for wetting mitigation. *Desalination* **444**, 13–25. doi:10.1016/j.desal.2018.06.012.

Schwantes, R., Seger, J., Bauer, L., Winter, D., Hogen, T., Koschikowski, J. & Geißen, S.-U. 2019 Characterization and assessment of a novel plate and frame MD module for single pass wastewater concentration–FEED gap air gap membrane distillation. *Membranes* **9**, 118. doi:10.3390/membranes9090118.

Semblante, G. U., Lee, J. Z., Lee, L. Y., Ong, S. L. & Ng, H. Y. 2018 Brine pre-treatment technologies for zero liquid discharge systems. *Desalination* **44**, 96–111. doi:10.1016/j.desal.2018.04.006.

Tong, T. & Elimelech, M. 2016 The global rise of zero liquid discharge for wastewater management: drivers, technologies, and future directions. *Environmental Science & Technology* **50**, 6846–6855. doi:10.1021/acs.est.6b01000.

USGS 2017 *PHREEQC Interactive 3.3.10.12220 [Software]*. Available from: https://www.usgs.gov/software/phreeqc-version-3.

Wadley, S., Brouckaert, C. J., Baddock, L. A. D. & Buckley, C. A. 1995 Modelling of nanofiltration applied to the recovery of salt from waste brine at a sugar decolourisation plant. *Journal of Membrane Science* **102**, 163–175.

Wegscheider, R. & Mehl, J. 1928 Über Systeme Na_2CO_3-$NaHCO_3$-H_2O und das Existenzgebiet der Trona (About systems Na_2CO_3-$NaHCO_3$-H_2O and the existence region of Trona). *Monatsheft für Chemie* **49**, 283–315. doi:10.1007/BF01518402.

Yaqub, M. & Lee, W. 2019 Zero-liquid discharge (ZLD) technology for resource recovery from wastewater: a review. *Science of the Total Environment* **681**, 551–563. doi:10.1016/j.scitotenv.2019.05.062.

First received 3 April 2020; accepted in revised form 20 June 2020. Available online 31 August 2020

Sustainable industrial wastewater reuse using ceramic nanofiltration: results from two pilot projects in the oil and gas and the ceramics industries

Matan Beery, Christian Pflieger and Marcus Weyd

ABSTRACT

The federal research project, PAkmem, deals with the recycling of industrial wastewater. The aim of the project is to develop and pilot an innovative integrative process for produced water treatment in the oil and gas industry utilizing flotation and ceramic micro-nano-membrane filtrations (MF-NF membranes) as well as the wastewater treatment of the ceramic industry with ceramic NF-membranes and electrodialysis (ED). The process utilized should remove fine particles, organic matter and divalent ions in order to make the water dischargeable or reusable (direct disposal or reuse as process water in the ceramic industry and the enhanced oil recovery reinjection in the oil and gas industry in which the water is conditioned in order to increase the oil production yield). Three pilot plants were designed and built according to strict safety standards and were operated on industrial manufacturing sites in Germany in 2019. Two innovative optical fine particle measuring techniques (inline and online) have been specially adapted for the project and integrated into the pilot plants. The results show promising technical potential for the use of ceramic membranes in the above-mentioned applications.

Key words | ceramic membranes, flotation, nanofiltration, produced water

Matan Beery (corresponding author)
akvola Technologies GmbH,
Am Borsigturm 100, 13507 Berlin
Germany
E-mail: *beery@akvola.com*

Christian Pflieger
Fraunhofer-Institut für Keramische Technologien und Systeme IKTS,
Nougat-Allee 3, 98574 Schmalkalden,
Germany

Marcus Weyd
Fraunhofer-Institut für Keramische Technologien und Systeme IKTS,
Michael-Faraday-Straße 1, 07629 Hermsdorf,
Germany

HIGHLIGHTS

- Produced water treatment using ceramic membranes shows promising results in removing both oils and solids.
- Economics of flotation in combination with ceramic membrane filtration prove more advantageous than external disposal.
- MF, ED and RO remove COD, suspended solids and conductivity from industrial ceramic wastewaters.

INTRODUCTION

With the UN's sustainable development goals in mind, increasing water availability worldwide, combined with the search for constant industrial improvement, increasing efficiency and conserving resources, are all on the top of the agenda of the global water industry. Industrial wastewater especially poses major challenges for the relevant stakeholder groups. Due to their great complexity and diversity, the treatment and/or recycling of such waters is often associated with high costs, high energy consumptions and non-trivial safety issues. One of the major obstacles are organic and inorganic contaminants of loads and kinds that are rarely seen in municipal or agricultural waters. Industries with a particularly high water consumption, such as the oil

doi: 10.2166/wrd.2020.029

and gas industry, are more of a subject to scrutiny than others.

The federal grant-financed R&D project PAkmem was launched in late 2016 with the goal of developing and piloting multi-stage separation processes for the treatment of industrial wastewaters with the goal of recycle or reuse in mind. The processes are based around ceramic nanofiltration with state-of-the-art membranes developed and produced in Germany. Additional pre-treatment by means of induced gas flotation and ceramic microfiltration, as well as post-treatment by means of reverse osmosis or electrodialysis, were tested in conjunction with novel optical measurement methods for water quality (Figure 1). A total of nine project partners comprising representatives of different stakeholders in the value chain, including research institutes, small and large companies, have joined the national consortium. The project was financed by the German Ministry of Education and Research (BMBF) as part of the WavE-initiative and concluded in March 2020.

Produced water in the oil and gas industry

When oil is extracted from the ground, water and gases are also brought to the surface in addition to the hydrocarbons. The so-called water cut of this three-phase mixture can be as high as 95% by weight. This necessitates the further processing of the extracted oil using different fluid separation technologies. After the oil and gas have been separated, the remaining water that is usually reinjected into the reservoir in several spots around the production well maintains the pressure underground and continues to push out more hydrocarbons to the surface, in a process known as water flooding. The injected water often still has considerable amounts of hydrocarbons and dissolved salts. Fine suspended solids are typically also present, which entails a risk of the fine formation pores becoming blocked. The remaining hydrocarbons inside the produced water also reduce the overall oil yield as some of the product (oil) is simply being put back in the ground. Since the produced water does not belong to the natural water cycle reservoirs (the earth layers are typically much deeper than the ones where ground water is found) there are often no or very lax water quality values for reinjection into the reservoir water. Nonetheless, when the reinjection is done for disposal purposes, the regulation becomes stricter and less favorable for the producers and requires more advanced treatment techniques (Fink 2012). In the old onshore oil fields of Germany, the typical treatment chain of the produced water–hydrocarbon mixture will only include three phase gravity separators followed by large settling tanks with sufficient residence time before most of the water is re-injected into the ground. However, due to old infrastructure and pipelines, as well as varying oil-water composition, the actual water quality at the reinjection spot could be of poor quality, containing too many hydrocarbons or fine solids. A better solution, such as one employing ceramic membranes (Ebrahimi *et al.* 2008, 2010; Pedenaud *et al.* 2011) for this wastewater reuse scheme is therefore needed to increase productivity and reduce the water footprint.

In the context of enhanced oil recovery (EOR), partial demineralization of the reinjection water by means of nanofiltration for divalent ion removal was also set out to be investigated in the project. Oil molecules bind to rock particles in the deposit by bridging divalent ions. Removing at least some of these ions (especially Mg^{2+} and Ca^{2+}) by means of nanofiltration can lead to relaxation of the

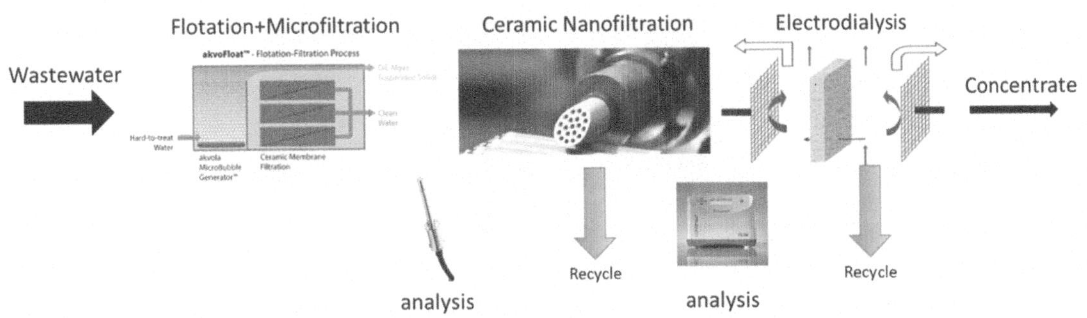

Figure 1 | The integrated process steps being inspected in the PAkmem project (*sources*: Fraunhofer IKTS, akvola, LUM, SOPAT).

ion-bridges, expanding the electrical double layer and enabling the bridges to be replaced by non-bridging monovalent ions, which in turn can mobilize more oil molecules and increase the overall production yield (Robbana *et al.* 2012; Mainwaring 2015). Additionally, the use of nanofiltration for produced water treatment can avoid blocking of the porous oil formation upon injection and potentially inhibit corrosion and biofouling on the internal wall of the injection piping (Gamal Khedr 2015). These additional benefits in yield and productivity make the process at hand economically interesting for the oil industry and create an additional incentive to the environmental one.

Ceramics industry

Similar separation challenges come about in the treatment of wastewater from the ceramics industry. Ceramic raw materials such as Al_2O_3, SiO_2 or TiO_2, as well as the organic ingredients used in the processing of the ceramic bodies and fluorescent organics used for quality assurance, are typically the main components in these wastewaters.

Classical wastewater treatment processes for the removal of fine inorganic suspended solids such as ceramic particles, glazes and various salts by physical-chemical processes (flocculation/precipitation, sedimentation, flotation, dewatering in filter press) usually require the addition of conditioning chemicals to ensure adequate separation of the inorganic (fine) fraction. Besides the additional components, all the substances contained in the wastewater are mixed unselectively with the minerals in the mass produced. The recycling of such materials has so far not seen general acceptance in the industry. Furthermore, the processes used to date do not usually provide adequate water quality that can be reused in some way in the plant (e.g. process or washing waters). In particular, the remaining COD and salt amounts are most problematic. This conventional cleaning must be upgraded through some demineralization technique if the water cycle is to be closed or wastewaters reused. The investigation of polymeric nanofiltration membranes for this purpose is documented in the literature (Moliner Salvador *et al.* 2012). Besides the use of ceramic nanofiltration for the upgrade of such wastewater treatment processes, it was also decided to evaluate the use of electrodialysis with bipolar membranes in the process as introduced by Fraunhofer IKTS in Dresden (Friedrich 2015). The bipolar membranes used consist of cation and anion exchange membranes and a catalytic intermediate layer to accelerate water dissociation in between. By applying an electric field, chloride and sulfate ions migrate through the anion exchange membrane into the anolyte stream, are retained on the cation-selective side of the bipolar membrane and form hydrochloric and sulphuric acid together with the protons produced in the bipolar membrane. In the same way, the anion-selective side of the bipolar membrane produces a sodium hydroxide solution and a desalinated diluate is formed in the inlet chamber.

METHODS

Oil and gas industry

The treatment of the produced water from the E&P site Barnstorf in Lower Saxony (Figure 2) by means of flotation and microfiltration followed by nanofiltration was first

Figure 2 | Main treatment facility of the oil field Barnstorf in Lower saxony, Germany (*source*: Google, Wintershall Dea).

Table 1 | Typical water composition of the produced water on the feed to the pilot

Sampling location	Reinjection water Barnstorf	Unit
Date	05.11.2018	
Density at 20 °C	1.1213	g/cm^3
Max temperature	35	°C
pH	6.1	
Calculated total hardness	2,451	°dH
Iron	78	mg/L
Manganese	5.6	mg/L
Aluminium	<1	mg/L
Calcium	13,500	mg/L
Strontium	943	mg/L
Barium	91	mg/L
Magnesium	2,160	mg/L
Sodium	49,800	mg/L
Potassium	337	mg/L
Lithium	5	mg/L
Ammonium	113	mg/L
Zinc	<0.5	mg/L
Lead	<0.5	mg/L
Chloride	107,500	mg/L
Bromide	732	mg/L
Iodide	13.2	mg/L
Nitrate	1.8	mg/L
Bicarbonate	146	mg/L
Borate	166	mg/L
Phosphate	<2	mg/L

tested on a laboratory scale. The flotation-microfiltration process, akvoFloat®, can remove suspended solids and hydrocarbons from the produced water and is described in greater detail elsewhere (Ludwig *et al.* 2015). The treated produced water came from the primary separation treatment in-site following a three-phase-separator and sedimentation tanks to remove most of the gases, solids, and crude oil.

Based on water analysis tests, the expected feed water quality and composition is shown in Tables 1 and 2. The produced water is highly saline, having especially high mass concentrations of iron, calcium, magnesium, sodium and chloride. As seen in Table 2, an oil content of approximately 95 mg/L and a solids content of approximately 73 mg/L (mostly sand particles) were to be expected in the raw wastewater.

The concentration of particles and organic load, especially hydrocarbons, was to be reduced in the project using induced gas flotation and ceramic microfiltration. The generated permeate should have a maximum oil content of 25 mg/L according to DIN ISO 9377-2 (H53 method) and a particle size proportion of less than 1% for particles larger than 3 μm (according to the internal specifications of Wintershall DEA for reinjection). An analysis of the organic load based on HS gas chromatography has shown it to be composed of 13% BTX, 0.23% PAH and 86.75% other hydrocarbons.

The samples collected at the beginning of the project were put to a lab flotation-filtration test using ceramic microfiltration membranes. The oil content could be consistently reduced by more than 95% but the solids contents were reduced by only 25%, the reason being that the produced water, when coming into contact with oxygen, immediately became turbid orange, indicating iron oxidation and precipitation. This phenomenon was well documented by the oil field services team on-site, deeming it critical for the actual treatment in the pilot plant (and later in the commercial application) to be oxygen-free for more than just the common explosive safety reasons.

The samples were then filtered by tubular ceramic nanofiltration membranes (Figure 3) having a molecular weight cutoff (MWCO) of 200 Da. These multicoated membranes are based on previous developments (Puhlfürß *et al.* 2000; Voigt *et al.* 2013; Pflieger *et al.* 2018) and have shown the

Table 2 | Expected feed water quality in terms of oil and solids content

			Particle size (μm)					
	Oil content (mg/L)	Total solids (mg/L)	2	5	15	25	50	100
Mean	94.4	72.6	957.436	627.958	44.607	8.101	430	40
Standard deviation	40.8	149.1	53.520	66.499	32.133	7.467	313	28

Figure 3 | Ceramic tubular NF membranes used in the pilots (*source*: IKTS, Rauschert).

removal of up to 90% of divalent ions from saline solutions in lab tests.

According to the promising laboratory results, a pilot project was carried out at Wintershall DEA in Barnstorf (Bockstedt) over a period of 23 weeks. The pilot plant was built according to strict industrial safety regulations and was operated in an oxygen-free environment with the flotation operated with nitrogen in a stainless steel sealed tank with a positive over-pressure including an off-gas activated carbon treatment and all equipment being inertized before coming into contact with the liquid medium. The produced water was taken via a bypass from the reinjection line and was fed into the pilot plant. The cleaned water (permeate) was returned to the reinjection line with the possibility of bypassing the NF while sludge and retentate were collected and disposed of in separate IBCs (see the pilot plant's process flowchart including the corresponding volumetric flow rate ranges and the detailed process and instrumentation diagram in Figure 4).

The raw wastewater (produced water post primary treatment) had a pre-pressure of approximately 11–12 bar. Both for safety and operational reasons, the pressure was reduced to 1 bar using the pressure reducer PC-10/V-107. In the event of a technical malfunction of the pressure reducer, a safety shut-off valve was installed to go off above 4.5 bar. The downstream pneumatic valve V-102 was used to regulate the level of the tank T-1 of the flotation-MF system, but also to shut the feed line in case the maximum level in the tank was reached (measured by the level sensor LS 01). The chemical tanks T-12 and T-13 with the dosing pumps P-12 and P-13 made it possible to add pre-treatment chemicals such as acids, coagulants and emulsion breakers into the process, if necessary. A static mixer was installed for better mixing of these chemicals. The first treatment step using flotation occurred in the T-1 tank using MicroGas, an induced gas fine bubbler generation device. Nitrogen was used as the flotating gas to avoid the previously mentioned problem of Fe^{2+} ions oxidizing to Fe^{3+} ions resulting in the precipitation of iron (III) hydroxide complexes. In the next treatment step, the wastewater was filtered under negative pressure through the submerged flat sheet ceramic microfiltration membranes (SiC, 0.1 μm) that were integrated into the flotation tank, T-1. For this purpose, the water was sucked through the membranes by the pump P-21 and pumped through the permeate line, feeding the backwash tank, T-21, and the NF feed tank, B-51. The pump P-22, in addition to the pumps P-23 and P-24, were periodically used to backwash the MF membranes with or without chemicals stored in T-23 and T-24. As previously mentioned, a bypass was placed in the permeate line for direct integration into the reinjection line. An on-line particle characterization was carried out both before and after the NF system. The NF was composed of three stainless steel pressure vessels holding the membrane modules with a surface area of 1.2 m^2 each. Pump P-01 was the main feed pump building-up the pressure (up to 20 bar) with P-02 serving as a circulation pump assuring the most efficient crossflow regime in every module. The permeate produced by the NF was fed into the tank B-52 and could then be pumped through the pump P-04 into the reinjection line or used with the pump P-03 to backwash the nanofiltration membranes. The sludge and drainage (emptying) of the flotation-MF system, as well as the retentate of the nanofiltration, were collected as waste in the two IBCs T-25 and T-26 and transported to a nearby disposal facility. A comprehensive HAZOP analysis was executed before starting up the pilot plant.

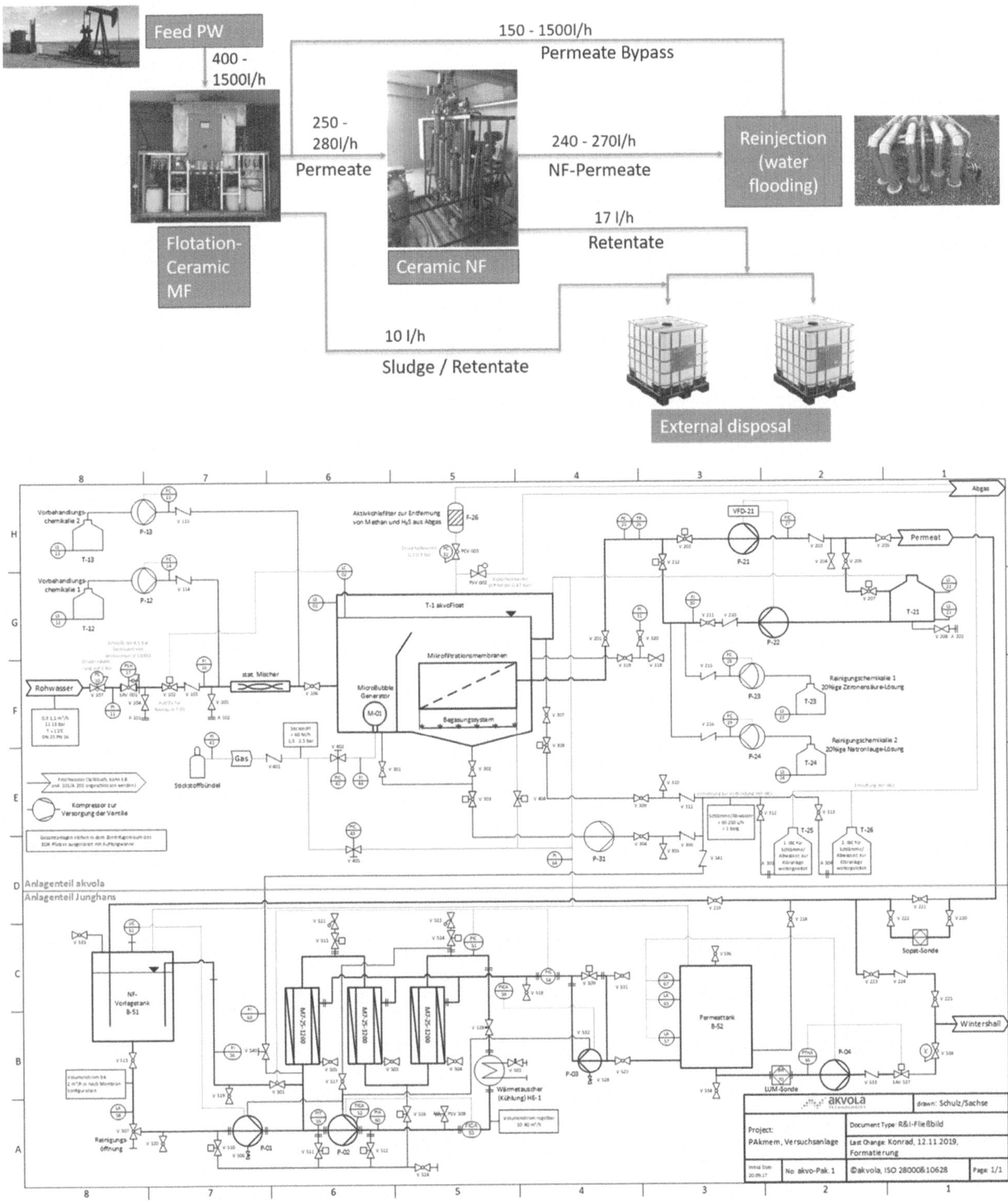

Figure 4 | Flowchart (top) and P&ID (bottom) of the produced water pilot plant (*source*: akvola, Junghans).

Ceramics industry

The central concern of the industrial partner in the ceramic industry is the treatment of industrial water to replace the use of municipal drinking water in production. Solids must be completely removed (also important for the electro-membrane process technology) and the conductivity of the water thus obtained should be in the order of 250 µS/cm. Higher and lower conductivities are unfavorable as they will inevitably lead to changes in the design and operation parameters of currently installed unit operations in the process line.

Preliminary tests on single-channel tubular geometries and selected 19-channel tubular geometries were promising, but the target value for conductivity is challengingly low. At Fraunhofer IKTS, a plant was set up for the on-site tests. Figure 5 shows the pilot plant consisting of: (1) ceramic MF/UF/NF pressurized tubular membrane system (partially automated, automatic filling device), (2) backwash system (operating mode: pulse or volume method), (3) industrial cooler (approximately. 5 KW), (4) buffer tank (180 L) and (5) RO system (brackish water spiral wound module).

For the first on-site tests, 19-channel tube geometries with the lowest cut-off, i.e. the smallest pore size and highest possible selectivity, were used. These initially did not yield a sufficient reduction in conductivity. Permeates produced in this way were therefore further processed by a brackish water reverse osmosis system (BWRO). The final waters produced in this way had, however, too low conductivities. In the pilot phase, process water was therefore produced for preparation tests by blending NF and RO permeates.

The on-site tests were also used to produce test water for use with electrodialysis (ED). The permeate water of the NF stage has already been successfully treated with ED in advance. For the extended test series on ED, two new ED stacks, each with 10 membrane pairs and with a membrane area of approximately 2 m^2 each, were designed for the pilot tests during the period under investigation (Figure 6).

RESULTS AND DISCUSSION

Oil and gas industry

The akvoFloat® flotation-MF system was initially operated without the following NF and the permeate produced was pumped directly into the reinjection system. This gave the opportunity to test the plant at higher fluxes and to evaluate the operating behavior. Extensive jar tests on-site as well as during the first live pilot runs with coagulants and emulsions breakers did not show any improvements and so the further use of pre-treatment chemicals was deemed as unnecessary for the rest of the pilot. In total, after the start-up phase the pilot plant was running for a period of 13 weeks while constantly going through operational optimization to maintain stable operation at the highest possible fluxes. During that time, the feed water quality of the produced water changed frequently in composition. Figure 7 shows the visual

Figure 5 | NF pilot plant used in the ceramics industry.

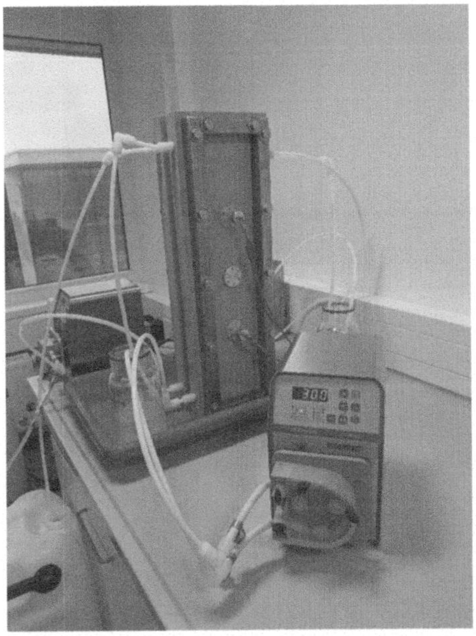

Figure 6 | The ED system used for treatment of the membrane permeates out of the pilot in the ceramic industry (*source*: Friedrich, IKTS).

Figure 7 | Samples taken from the feed to the plant (left bottles, produced water after primary treatment) and post flotation-MF (right bottles, always clear water). The two images were taken on two consecutive days.

difference in feed water quality on two consecutive days in September 2019 and the constantly high quality permeate produced by the flotation-MF system despite the changing feed water quality. Figure 8 shows a typical development of the trans-membrane pressure-loss (TMP, bar) over time (hours) with two constant filtration fluxes used at 125 and 66 L/m^2/h ('lmh'). The peaks in the diagram depict backwash events with the red lines showing chemically enhanced backwashes (CEB) or cleaning in place (CIP) events performed in the 42-hour period. A recovery of 98% could be reached. The crossflow nanofiltration system finally reached stable operation at constant pressures of 20 bar with hourly backwashes (Figure 9).

Samples of the feed and the permeate were occasionally taken for lab analysis. The hydrocarbons and suspended solids were efficiently reduced by the integrated process to below detection levels utilized by the analytical methods [DIN EN ISO 9377-2 : 2001-07] and [DIN EN 872 : 2005-04] accordingly. The divalent ions detected by atomic emission spectroscopy and ion-exchange chromatography could only be reduced to some extent with the best lab removals listed in Table 3. On most occasions, however, only lower or even zero ion removal by the NF membrane was detected, suggesting membrane defects.

Cost analysis

Based on the data collected during the pilot phase, a business case calculation was performed for a site in which the current reinjection system is not able to treat the full capacity of the produced water being generated and about two-thirds of the water is currently being externally disposed of at 80 €/m^3. The over-capacity should be treated by a new integrated flotation-ceramic membrane filtration plant (MF + NF/RO) with a plant capacity of 46,667 m^3/year, operating on 50 weeks/year and 7 days/week resulting in 133.33 m^3/d or (at 24 h/d) a nominal capacity of 5.56 m^3/h. For the sake of simplification, a new plant capacity of 6 m^3/h was considered. Assuming the membrane concentrate would still have to be externally disposed of (at 600 €/m^3) and the permeate could be indirectly discharged into the local sewage system (at 3 €/m^3), the overall yearly OPEX savings of €568 k could pay back the estimated CAPEX investment of €1.2M in just over two years (see Table 4 for details). It is worth pointing out that such a business case is only economically attractive if the alternatives are too expensive (as in the case of external disposal) or if the environmental regulator forces such a complex treatment scheme.

Ceramics industry

On average, approximately 14% conductivity reduction could be achieved for the ceramic NF stage (specific permeate flow in the range of 8 L/(m^2 h bar)). Further

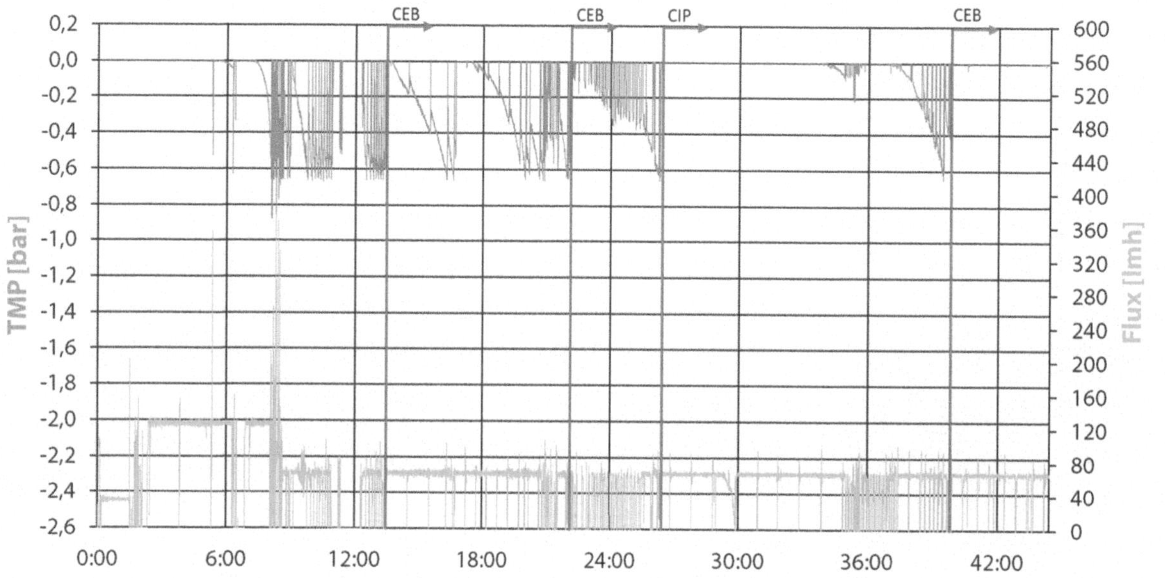

Figure 8 | TMP-flux diagram for the microfiltration system. Please refer to the online version of this paper to see this figure in color: http://dx.doi.org/10.2166/wrd.2020.029.

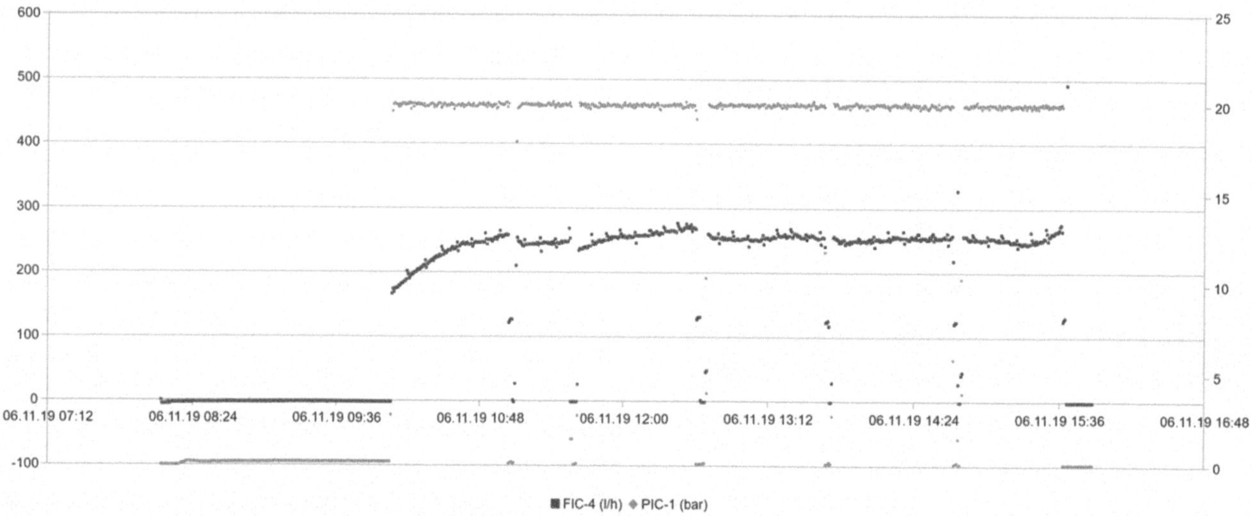

Figure 9 | Flow (L/h, left axis) – Pressure (bar, right axis) diagram for the nanofiltration system.

Table 3 | Outtake of water analysis results from the produced water membrane filtration system

Parameter	Value in feed (mg/L)	Value in permeate (mg/L)	Removal efficiency
Hydrocarbons (C10-C40)	<30	<0.1	>99%
Total solids	>20	n.d.	>99%
Ca^{2+}	13,700	11,700	14.6%
Mg^{2+}	2,290	1,340	42.5%
SO_4^{2-}	117	35	70.1%

processing by means of RO showed retention in the range of 98% (related to the initial conductivity of the NF permeate). By mixing permeates, the industrial partner, Duravit, was provided with treated wastewater with a conductivity of 243 μS/cm that could be re-used in the industrial processes. As filtration tests were carried out, a new batch of prototype membranes was manufactured and tested at Duravit. In comparison, the new batch delivered higher specific permeate flows in the range of

Table 4 | Business case calculation for a new flotation-ceramic membrane filtration (CMF) plant as an alternative to external disposal of produced water

Current situation				
External disposal				
	Description	**m³/y**	**€/m³**	**€/y**
1	Produced water total	70,000	0€	0€
2	Produced water being externally disposed	46,667	80€	3,733,333€
	Tot (€/y)			3,733,333€
OPEX				
General				
	Produced water to be treated	46,667	m^3/y	
	Recovery	90,0%		
Flotation + CMF plant				
		kWh/m^3	€/m^3	€/y
1	Energy + compressed air (@0.2€/kWh)	5.0	1.0€	46,667€
		m^3/y	€/m^3	€/y
2	Cleaning	46,667	0.3€	14,000€
3	Chemicals	46,667	3.0€	140,000€
4	Concentrate disposal	4,667	600€	2,800,000€
5	Maintenance			
6	Indirect discharge fees	42,000	3.0€	126,000€
		h/week	€/h	€/y
7	Labor	40	20€	38,400€
Tot. OPEX Flotation + CMF (€/y)				3,165,067€
Total costs calculation				
Running costs				
	Description	Current situation: external disposal		New situation: flotation+ CMF
1	External disposal	3,733,333 €		2,800,000 €
2	Operation (energy, cleaning, labor, etc.)			365,067 €
	OPEX (€/y)	3,733,333 €		3,165,067 €
Investment costs				
				New situation: flotation+ CMF
1	Flotation + CMF plant			1,200,000€
	CAPEX (€)			1,200,000€
Summary				
OPEX savings (€/y)				568,267€
Payback time (y): CAPEX/(OPEX savings)				2.1

12 L/(m^2 h bar) and achieved a conductivity removal of 10%. The permeates obtained in this way were also made available to Duravit for testing as batch water. As far as the COD removal goes, the only way to achieve stable operation and high removal efficiency was by combining the NF filtration with MF pre-treatment. The

results showed a reduction from COD levels of 950 to <50 mg/L. Both conductivity and COD removal are depicted in Figure 10.

During on-site tests, various NF and MF permeates were collected for use in the testing of the of ED. Around 800 L of the filtrate produced in this way (0.85 mS/cm) was treated by electrodialysis by IKTS and treated waters with three different conductivities (0.32, 0.38 and 0.42 mS/cm) were produced (see Figure 11). Duravit was supplied with 50 L of each of these for preparation trials. For the extended series of tests in the field of electro-membrane process engineering, no further optimization was necessary of the ED module. However, the membrane area was doubled to 10 membrane pairs and the throughput increased as a result.

CONCLUSIONS

The pilot tests showed that an integrated membrane system containing ceramic nanofiltration membranes could

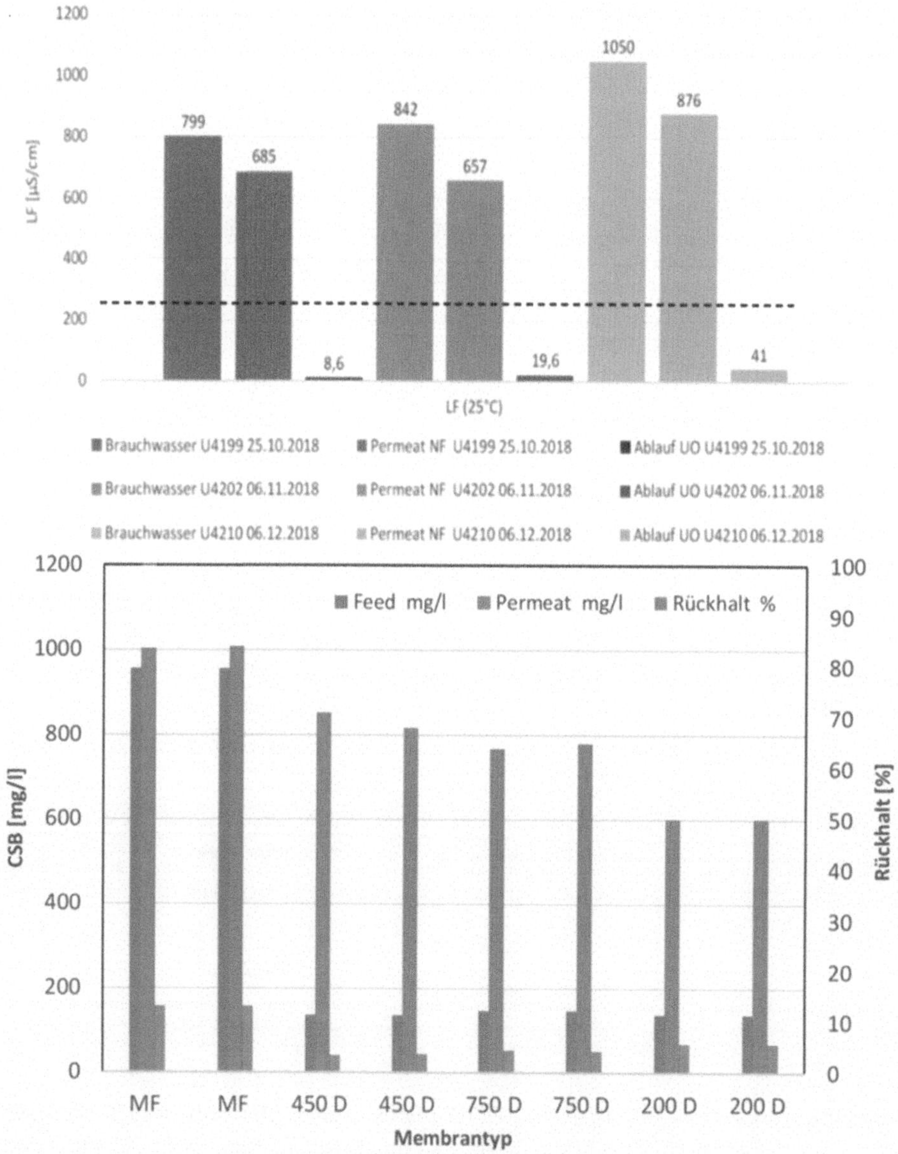

Figure 10 | Conductivity removal using NF + RO (top) and COD removal ('CSB'£') using MF + NF (bottom).

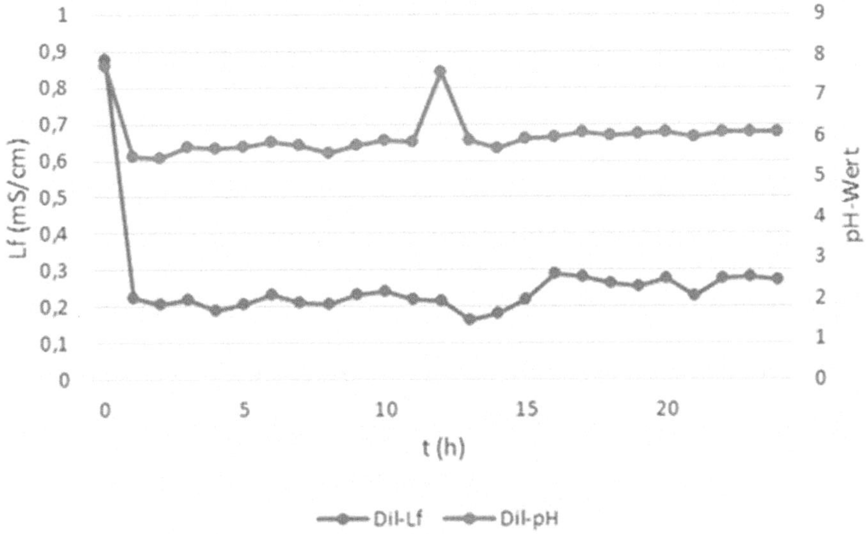

Figure 11 | Treatment of NF filtrate by ED: Reduction of conductivity (blue) and pH (orange).

effectively treat industrial process waters to such an extent that:

1. Oil, particles, and divalent ions (to some extent) could be removed in continuous operation with real on-shore produced waters in an oxygen-free environment using flotation and MF as pre-treatment.
2. COD, suspended particles and conductivity could be removed from industrial ceramic wastewaters using MF as pre-treatment and ED or RO as post-treatment.

As a result, the water quality was such that industrial wastewater reuse schemes could be technically feasible on a large scale. Their economic merits can, however, only be established if the local drinking water and/or the effluent discharge are heavily restricted and costly. One should also add the caveat that the capacity of the membranes to sustain long stable operations with manageable fouling and efficient CIP cleanings while the operators successfully manage the handling of the waste streams (membrane concentrates) is recommended to be proven in a longer term (12 months) test run.

ACKNOWLEDGEMENTS

The authors would like to thank the partners and affiliates of the project consortium as well as their devoted employees: akvola Technologies GmbH, Fraunhofer IKTS, DBI Gas und Umwelttechnik GmbH, SOPAT GmbH, Andreas Junghans GmbH, Rauschert Kloster Veilsdorf GmbH, LUM GmbH, DURAVIT Sanitärporzellan Meißen GmbH, Wintershall Dea GmbH. This R&D project was financed and supported by the German Ministry of Education and Research (BMBF) as part of the WavE initiative (Grant allocation 02WAV1407).

DATA AVAILABILITY STATEMENT

All relevant data are included in the paper or its Supplementary Information.

REFERENCES

Ebrahimi, M., Shams Ashaghi, K., Engel, L. & Czermak, P. 2008 Characterization and application of different ceramic membranes for the oil-field produced water treatment. In: *Proceedings Engineering with Membranes*, 25–28.5.2008, Vale do Lobo, Portugal, pp. 231–232.

Ebrahimi, M., Willershausen, D., Ashaghi, K. S., Engel, L., Placido, L., Mund, P. & Czermak, P. 2010 Investigations on the use of different ceramic membranes for efficient oil-field produced water treatment. *Desalination* **250** (3), 991–996.

Fink, J. 2012 *Petroleum Engineers Guide to Oil Field Chemicals and Fluids*. Gulf Professional Publishing, Amsterdam.

Friedrich, H. J. 2015 *Elektrochemische Verfahren für Wasserbehandlung und Rohstoffrecycling*. Fraunhofer IKTS

Jahresbericht 2014/15. Fraunhofer-Institut für Keramische Technologien und Systeme IKTS, Dresden.

Gamal Khedr, M. 2015 Nanofiltration of oil field-produced water for reinjection and optimum protection of oil formation. *Desalination and Water Treatment* **55** (12), 3460–3468. doi: 10.1080/19443994.2014.939497.

Ludwig, J., Beery, M., León, L. & Ripperger, S. 2015 *Flotation-filtration-process with Ceramic Membranes for Water Treatment*. Filtrieren und Separieren. F&S International edition. VDL-Verlag GmbH, Rödermark.

Mainwaring, J. 2015 *LoSal: BP's Low-Salinity Enhanced Oil Recovery Technology*. Rigzone Magazine. Available from: www.rigzone.com/news/oil_gas/a/140671/losal_bps_lowsalinity_enhanced_oil_recovery_technology/

Moliner Salvador, R., Deratani, A., Palmeri, J. & Sánchez-Vilches, E. 2012 Use of nanofiltration membrane technology for ceramic industry wastewater treatment. *Boletín de la Sociedad Española de Cerámica y Vidrio* **51** (2), 103–110.

Pedenaud, P., Wayne, E., Samuel, H. & Didier, B. 2011 Ceramic membrane and core pilot results for produced water management. In *OTC Brasil*. Offshore Technology Conference. 10.4043/22371-MS.

Pflieger, C., Weyd, M., Richter, H., Puhlfürß, P., Voigt, I., Beery, M., Prehn, V. & Junghans, A. 2018 Development ceramic nanofiltration membranes with high volume-specific membrane area and their application in the effective treatment of problematic process waste waters from the oil and gas industry as well as mining. In *Euromembrane Conference Proceedings, Valencia Spain*. Universitat Politècnica de València, Valencia, Spain.

Puhlfürß, P., Voigt, A., Weber, R. & Morbé, M. 2000 Microporous TiO_2 membranes with a cut off <500 Da. *Journal of Membrane Science* **174** (1), 123–133.

Robbana, E., Buikema, T., Mair, C., Williams, D., Mercer, D., Webb, K., Hewson, A. & Reddick, C. 2012 *Low Salinity Enhanced Oil Recovery – Laboratory to Day One Field Implementation – LoSal EOR Into the Clair Ridge Project*. Society of Petroleum Engineers, Richardson, TX, USA. 10.2118/161750-MS.

Voigt, I., Adler, J., Weyd, M. & Kriegel, R. 2013 Ceramic filters and membranes. In *Ceramics Science and Technology*. WILEY-VCH, Weinheim, Germany, pp. 117–167.

First received 3 April 2020; accepted in revised form 29 September 2020. Available online 25 November 2020

Economic and environmental assessment of water reuse in industrial parks: case study based on a Model Industrial Park

Birte Boysen, Jorge Cristóbal, Jens Hilbig IWA, Almut Güldemund, Liselotte Schebek and Karl-Ulrich Rudolph IWA

ABSTRACT

Industrial wastewater reuse is a major measure to mitigate the depletion of available freshwater resources in the catchments around industrial areas and to prevent possible future water shortages and the resulting problems for industry, economy and society. Combining a set of environmental aspects and economic aspects of different wastewater treatment technologies, the authors developed a model-based approach for planning and evaluating water reuse concepts in industrial parks. This paper is based on an exemplary Model Industrial Park. The results based on data primarily calculated for Germany show that, for the majority of the indicators, the installation of the Water Reuse Plant seems to be beneficial for all examined reuse options. Considering the economic dimension, due to economies of scale, reuse options with larger volumes of treated water are preferable since the costs per m^3 of reuse water are reduced by up to 33%. On the other hand, the environmentally preferable option depends on the respective indicator, e.g. for freshwater eutrophication, the higher the reuse factor, the lower the impact, leading to reductions between 8 and 12%. For climate change, the best option is dependent on the reuse purpose leading to reductions between 8 and 52%.

Key words | cost function, economic evaluation, environmental assessment, industrial park, LCA, water reuse

Birte Boysen (corresponding author)
Jens Hilbig IWA
Karl-Ulrich Rudolph IWA
IEEM gGmbH – Institute of Environmental Engineering and Management at the Witten/ Herdecke University,
Alfred-Herrhausen-Straße 44, 58455 Witten, Germany
E-mail: *birte.boysen@uni-wh-ieem.de*

Jorge Cristóbal
Almut Güldemund
Liselotte Schebek
Chair of Material Flow Management and Resource Economy,
Technical University of Darmstadt, Institute IWAR,
Franziska-Braun-Straße 7, 64287 Darmstadt, Germany

Jorge Cristóbal
Chemical and Biomolecular Engineering Department,
University of Cantabria,
Avd. De los Castros s/n, 39005 Santander, Spain

HIGHLIGHTS

- Industrial water reuse can reduce future water shortages.
- Development of a model-based approach for planning and evaluating reuse concepts in industrial parks (IPs).
- Economic and environmental evaluation of a Model IP located in Germany.
- Economic profitability of reuse depends on wastewater effluent discharge and possible alternatives.
- In the evaluated case, reuse was shown to be environmentally beneficial.

doi: 10.2166/wrd.2020.034

INTRODUCTION

Freshwater supply is one of the most urgent challenges in many regions worldwide, presumably posing significant risks to societies in the 21st century if left unaddressed. On average worldwide, the majority of freshwater is consumed in agriculture; however, industrial processes also require large amounts of water. In 2015, the industrial share corresponded to 19% of total global water withdrawal (FAO 2016). In industrialised nations, this share can be considerably higher, accounting for between 50% and more than 90% of the national water withdrawal (FAO 2016). Thus, industrial wastewater reuse is a major measure to mitigate the depletion of available freshwater resources in the catchments around industrial areas and to prevent possible future water shortages and the resulting problems for the industry.

Industrial parks (IPs) are equipped with a wastewater treatment plant (WWTP) whose objective is to remove contaminants from the wastewater (mainly suspended solids, organic matter and, in certain areas, nutrients), so that it complies with legislation and can be discharged back into the environment. When treated wastewater is to be reused, additional treatment is needed in order to ensure its quality for the intended purpose and to minimise health and environmental risks. The additional treatment is called reclamation water treatment (as well as reuse or recycling) and is carried out in a Water Reuse Plant (WRP) (Alcalde Sanz & Gawlik 2014).

It is clear that the need for more treatment beyond the WWTP is associated with environmental and economic impacts that might outweigh the benefits from avoided water production and wastewater discharge. For that reason, water reuse in IPs must be evaluated from a life cycle perspective. Yet very few research papers concerning this topic appear in the literature, and they draw different conclusions. For the environmental dimension, Pasqualino *et al.* (2011) assessed the stages of a Spanish municipal WWTP and compared four urban wastewater reuse alternatives with a no-reuse scenario. While the WWTP of the no-reuse scenario is characterised by a primary and secondary treatment and the subsequent disposal of wastewater to the sea, the reuse scenario incorporates an additional tertiary treatment that includes coagulation, flocculation, sand filtration, chlorination and UV disinfection with subsequent replacement of potable water for non-potable uses. The authors concluded that the addition of a tertiary treatment to the traditional two-stage WWTP and the subsequent utilisation of the treated water for potable water replacement increases the environmental impact of the plant. Tong *et al.* (2013) also compared different reuse scenarios in an IP: no reuse, reuse as industrial process water and reuse for horticulture, and concluded that water reuse in IPs is beneficial from the environmental perspective.

Considering the economic dimension, Giurco *et al.* (2011) evaluated the cost effectiveness of water treatment and industrial reuse concluding that the studied options are not self-sufficient from a cost perspective and, therefore, would not be implemented by companies in the absence of financial assistance or other incentives. On the other hand, in Eco-Industrial Parks such as Kalundborg (Denmark), water reuse has been successfully applied for decades not only due to economically driven actions but also because of water scarcity (Ehrenfeld & Gertler 1997). The economic benefit of water reuse in IPs depends on the costs of the wastewater reuse system compared with the existing water supply. Besides economic and ecological benefits, the implementation of water reuse schemes depends on the regulatory framework, the general availability of water and additional factors like company strategies or demand-related behaviour.

Urkiaga *et al.* (2008) developed a methodology for assessing the performance of water reuse projects including both the environmental and the economic perspective (as well as social factors). That methodology is not particularly focused on industrial reuse but also provides possible industrial reuse purposes. Results have shown possible environmental water quality benefits, whereas an economic analysis of water reuse indicated a potential increase in costs when the WRP is not located near the industrial user.

Based on this literature review, it can be concluded that the implementation of water reuse schemes is dependent on several site-specific framework conditions which determine the economic and environmental impacts. In this context, the joint research project Water Reuse in Industrial Parks (WaReIp) deals with the systematic investigation of wastewater generated within IPs, its treatment and its reuse for various purposes. This paper is written within the framework of WaRelp and its main objective is to assess water management concepts in IPs focusing on water reuse and resource depletion. The main novelty being the updated cost functions used for industrial wastewater in combination with a modular calculation model complemented by an environmental assessment and applied to a theoretical Model Industrial Park (MIP).

This paper is organised as follows: at first, the developed MIP is introduced along with the reuse framework, as well as the reuse options with the specific design of the WRP. Thereupon, the methods for both the economic and environmental assessment are introduced. Results are shown and discussed, and finally, the conclusions of the study are drawn.

MODEL INDUSTRIAL PARK AND REUSE FRAMEWORK

In the context of the WaRelp project, surveys and interviews are conducted in existing IPs in Germany, China and Vietnam. The experiences and results are used for the development of both the composition of a general MIP and the reuse framework shown in Figure 1 in which the wastewater is treated in a central WWTP to an effluent standard adhering to legal requirements. Subsequently, a defined quantity of wastewater is treated further in a WRP in order to be utilised for different infrastructural water use purposes (e.g. street cleaning and toilet flushing) (for detailed information on the developed MIP, see Bauer *et al.* 2019).

Water reuse in production processes itself is not considered under the WaRelp project (Bauer *et al.* 2019) since this is part of internal process optimisation of the independent companies located in the IP. In addition, feedback from both IP operators and production companies shows that the production plant operators have no experience with the use and possible effects of reduced water qualities in their production processes and are reluctant to modify their processes without intensive preparations.

In the development of the MIP, wastewater from 19 different production plants across various industries with different polluting loads are considered, as well as wastewater of a canteen and sanitary wastewater. The calculated overall wastewater flow of the MIP is 94,658 m^3/d (Bauer *et al.* 2019). For the economic and environmental assessment, the MIP is assumed to be located in Germany.

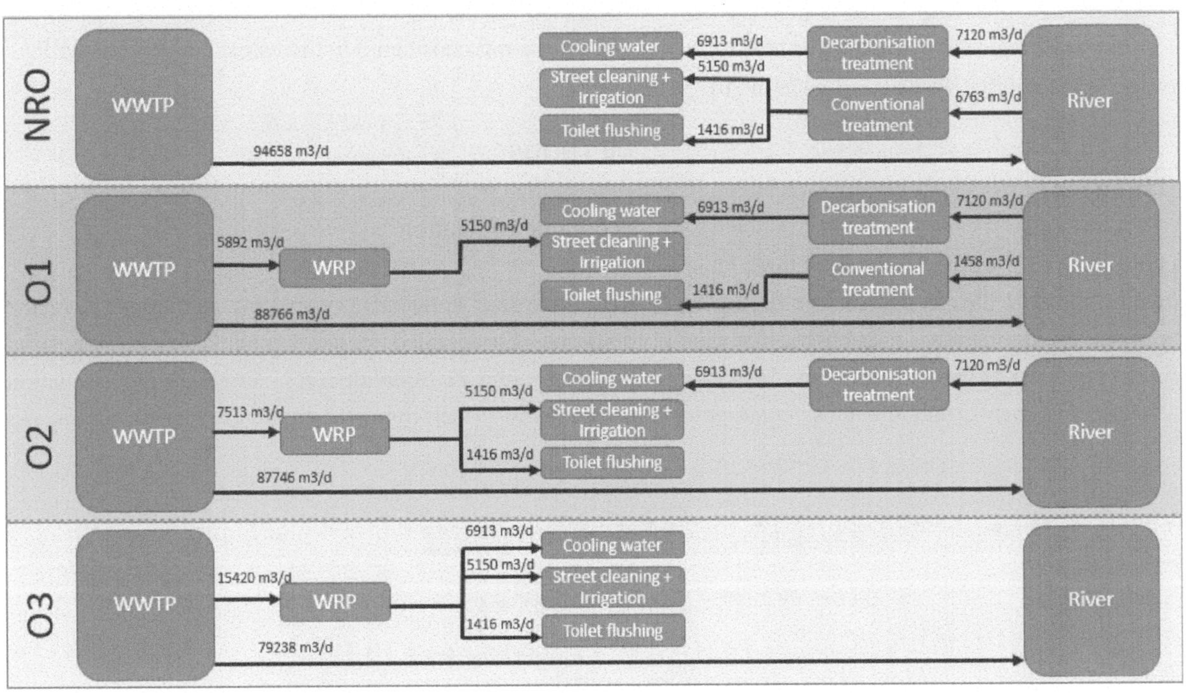

Figure 1 | Model Industrial Park water reuse options.

REUSE OPTIONS

In this paper, three different reuse options within the MIP are evaluated and compared with a non-reuse option (NRO). The reuse options are characterised by the reuse factor (i.e. the percentage of wastewater from the WWTP that is reused) and by the considered reuse purposes, specifically street cleaning and irrigation, toilet flushing as well as cooling water. Whenever the mentioned reuse purposes cannot be supplied by reuse water from the WRP, this demand is assumed to be supplied by groundwater, surface water or external sources treated in a central water treatment plant. Wastewater from the WWTP that is not treated in the WRP for the specific reuse purposes, is considered to be discharged directly into a river. Figure 1 shows the wastewater and reuse water flows of the different reuse options.

The NRO refers to the situation in which the total wastewater flow from the WWTP is discharged into a river. In Option 1 (O1), the WRP is installed to treat approximately 5% of the wastewater from the MIP to cover the demand of street cleaning and irrigation water (i.e. 5,150 m^3/d). In Option 2 (O2), the reuse factor is increased to 7% in order to cover the demand of street cleaning and irrigation water, as well as toilet flushing water (i.e. 6,566 m^3/d). Finally, in Option 3 (O3), the reuse factor is increased to 14% in order to cover all the water demands for infrastructural purposes including cooling water (i.e. 13,479 m^3/d).

WATER REUSE PLANT

The WRP includes the same process steps for all three reuse purposes: sandfiltration, UV disinfection and chlorination. These steps are selected, among other possible technologies for WRP such as ozonation or peroxide addition, based on the project WaReIp experience and the most commonly used technologies for tertiary treatment. Figure 2 depicts the technical process modules dimensioned for O1.

These data are based on the MIP and literature research for the process modules. The sandfiltration is dimensioned for a filtration velocity of 8 $m^3/(m^2\ h)$ (Mann *et al.* 2012; Tchobanoglous *et al.* 2014) and a solids loading of 3 kg TSS/(m^3 d), according to DWA-A 203 (2019). The energy demand for sandfiltration is assumed as 0.05 kWh/m^3 (Kraus *et al.* 2016). UV dimensioning is based on a hydraulic loading rate of 100 l/(min lamp) with a redundancy factor of 2 (Tchobanoglous *et al.* 2014). Based on Bischoff & Wagner (2012) and Müller *et al.* (2009), an energy demand of 0.035 kWh/m^3 is used. For the last reuse treatment step, chlorination, a contact time of 90 min and a chlorine dosage of 15 mg/l are chosen (Tchobanoglous *et al.* 2014). As it is assumed that the chlorine gas is not produced on-site, the energy demand for chlorination was assumed to be negligible.

Further assumptions of the MIP include an 8% loss of reuse water in the supply network; in O1, this leads to the treatment of 5,598 m^3/d wastewater from WWTP to satisfy the demand of 5,150 m^3/d. The WRP was oversized in such a way that the treatment of the wastewater is carried out at a mean plant utilisation rate of 75% as shown in Supplementary Material, Table A1. Energy and chemical demand were calculated for the mean flow rate. Furthermore, the sewage sludge was not considered in the calculations of the MIP.

METHODS

Economic evaluation

Economic and financial aspects play a major role in the planning of integrated wastewater concepts for industrial zones (Kreuter & Rudolph 2016). An economic evaluation is necessary to determine whether water reuse within IPs

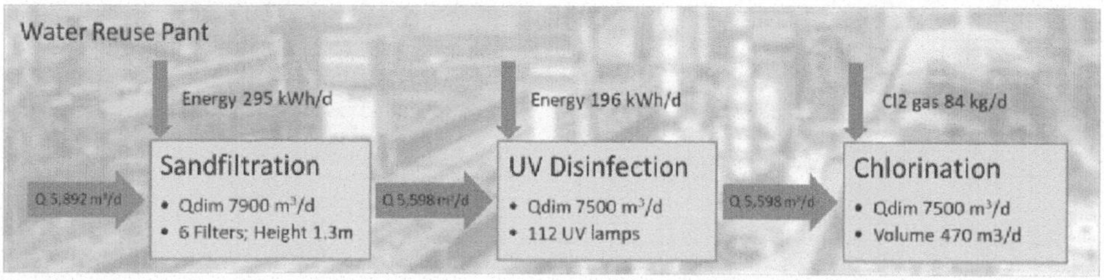

Figure 2 | WaReIp Reuse Plant dimensioned for Option 1.

is viable or not. The economic evaluation here is based on cost functions and a calculation model which can be extended to a financial model. The cost functions are based on literature research of mostly German municipal WWTP including studies like Günthert & Reicherter (2001), Türk *et al.* (2013), and Horstmeyer *et al.* (2014) (For more details, see Hilbig *et al.* (2020), 'Economic Evaluation of Water Reuse in Industrial Parks' in this same special issue). By calculating the arithmetic function of the individual cost functions and introducing a trend line, an average cost function of the municipal WWTP can be derived for individual process steps as shown in Figure 4. This also covers treatment sizes that are not considered in all publications. In order to be able to compare the cost functions of different references, the functions must be converted to one price level and one unit. Cost data from other countries are raised (via price index of the respective country) to a 2017 level and then converted into Euros. The data are mostly based on German WWTPs with exceptions such as data from Hunziker-Betatech (2008) which are from Switzerland and the data for chlorination which are not available for German WWTPs. In order to be comparable, the results are converted to €/m^3/d and €/m^3. For the sake of simplicity, the cost functions of the municipal WWTPs are used for the industrial WRP. As mentioned before, the industrial wastewater is treated in a central WWTP to a legally required effluent standard before it reaches the WRP and, therefore, no longer has higher industry-specific pollution.

The cost functions are split into functions for operational expenses and for investment. As far as possible and useful, investment functions are broken down to civil works, mechanical equipment and, if available electronic instrumentation and control (E/I&C) to later calculate depreciation and to adapt the sections separately to country-specific conditions. If a specific breakdown is not available, a percentage allocation is made.

The investment cost functions of the sandfiltration, UV disinfection and chlorination are shown in Table 1. In order to adapt them to the conditions of the German MIP, they will be validated in expert interviews. The operational expenditures are generally based on the described process modules including energy and consumables; furthermore, they include labour and maintenance costs.

Table 1 | WaReIp WRP relevant investment cost functions

Process	X	y	Cost function investment
Sandfiltration	m^3/d	€/m^3/d	$y = 651.24X^{-0.15}$
UV disinfection	m^3/d	€/m^3/d	$y = 1{,}048.3X^{-0.321}$
Chlorination	m^3/d	€/m^3/d	$y = 1{,}004.4X^{-0.29}$

Examples are the cost functions for UV disinfection, which consists of cost functions provided by Horstmeyer *et al.* (2014), Müller *et al.* (2009) and further data from an internal project database, leading to the average cost function shown in Figure 3. The investment breakdown for sandfiltration is based on Hunziker-Betatech (2008), whereas the breakdown of the UV investment is divided with the help of percentages. The chlorination cost function is based on Das (2002) and U.S. EPA (1999). The potencies in the cost functions (Table 1) show economies of scale for investments. The determined investment cost functions are transferred into the calculation model.

Calculation model

The WaReIp economic calculation model is an assessment tool to model the economic effects of water reuse concepts. The model is built in a modular way to allow for a comparison of different technologies and process chains. The modules integrated in the WaReIp calculation model for example include sandfiltration, UV disinfection, chlorination and a pipeline network. Each of these modules can be modelled independently, so that both the individual effect of a module and its weight in combination with other modules can be evaluated to form a complete water reuse concept. Furthermore, the calculation model is able to calculate three different variants with different technical specifications and volume flows in parallel and compare the results.

Within the model, first of all, global variables which apply to all three scenarios and all modules need to be defined. The time dimension is defined by the number of periods to be considered, the start of the fiscal year and the first period under consideration. For the WaReIp MIP model, a timeframe of 30 periods from 2018 onwards was selected for the German fiscal year structure. This default setting automatically models the reference timeline for all inputs on the following

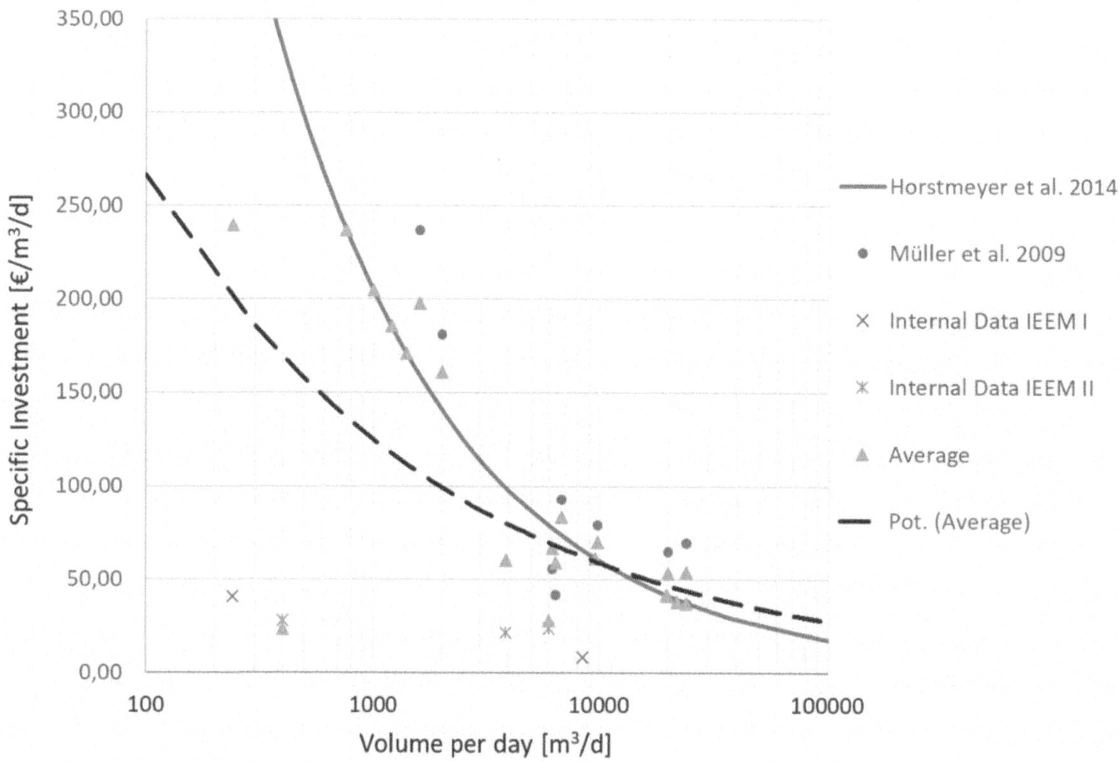

Figure 3 | Specific investment cost function UV disinfection.

calculation worksheets. Furthermore, the underlying inflation rate was set to 1.80% and the applicable value-added tax rate for net inputs to 19%. Monetary inputs are shown in EUR, and the modelled WRP operates 365 days/year.

The next step is to define the parameters of the three reuse options used. These options may differ only in volume sizes or flow rates or may use completely different module combinations at the same time. In the MIP model, the process chains are identical but the volumes change.

The investments are calculated with the previously determined net cost functions, which are (where possible) divided into different subcategories, namely, general costs and fees, groundworks, civil works, mechanical equipment, E/I&C and heating/ventilation/air conditioning. For each subcategory, the depreciation period can be defined individually. General costs and fees, surrounding works and construction are depreciated over 30 years, E/I&C technology over 7 years and heating/ventilation/air conditioning over 10 years. In the case of mechanical equipment, an additional distinction was made between general mechanical equipment (10 years) and UV lamps (5 years). The selected depreciation periods are identical for all three scenarios. The reinvestments depending on the depreciation periods are automatically taken into account in the model over the selected 30-year period.

The operating costs are also divided into subcategories that have been tailored to the underlying calculation formulas: labour costs, maintenance, energy costs and consumables. The inputs needed to calculate the annual operating costs per module and in total are derived on the basis of cost functions and the input of the process modules in combination with results from expert assessments, prices from retail inquiries and hourly wages. For sandfiltration, the labour and maintenance costs are widely based on Hunziker-Betatech (2008). The energy demand of 0.05 kWh/m^3 is defined in the technical process module and multiplied by an assumed company electricity price of 0.12 €/kWh. The UV disinfection energy demand of 0.035 kWh/m^3 is used and automatically multiplied by the defined electricity price. An expert assessment estimated a labour demand of 3 h/day which is multiplied by an hourly wage of 47.40 € (Jekel *et al.* 2016). The price of chlorine gas has been estimated in a retail inquiry to be 2.3 €/kg, and the demand from the process modules is given as 0.015 kg/m^3. For UV disinfection and chlorination, the maintenance costs are calculated via percentages of the investment, for civil works, it is 0.5%, for mechanical equipment 2.5% and for E/I&C 1% of the investment. The extrapolation of the operating costs determined in this way is carried out by simple inflation through the annual inflation rate determined in the global variables of the calculation model.

After entering all necessary information or using the default values, the investment and operational costs are calculated automatically in the model with underlying formulas.

Environmental assessment

As mentioned before, in order to save water resources by reusing water in IPs, a WRP must be installed, which nonetheless entails environmental burdens of its own. A method for the environmental assessment of water reuse concepts must, therefore, be able to highlight the shifting of burdens from one life cycle or process stage to another (e.g. environmental impacts of WRP versus wastewater discharge to river), from one protective good to another (e.g. air, soil and water) as well as from one impact category to another (e.g. climate change, eutrophication and toxicity). With its life cycle approach and the consideration of different impact categories, life cycle assessment (LCA) is an appropriate tool to meet these requirements.

According to the ISO (2006a), the LCA must be performed in the following four well-defined steps: goal and scope definition, inventory analysis, impact assessment and interpretation.

Goal and scope definition

The aim of the environmental assessment is to quantify and compare, for the specific case of the exemplary MIP in Germany, the environmental impacts of the different specified reuse options and the no-reuse reference case in order to identify the environmentally most advantageous alternative.

Through the examination of an exemplary MIP using LCA, generic findings are gained that can be transferred to different water reuse concepts in real IPs. The results of this assessment can, therefore, be used by IPs' operators and planners as well as public institutions to improve the environmental impacts associated with the supply and disposal of water within IPs and to enable scientifically sound decisions between different water reuse concepts.

From the framework of the MIP, the following two functions of the product system can be derived:

- the disposal of the wastewater from the WWTP and
- the supply of water for infrastructural purposes within the MIP.

Therefore, the functional unit selected for this LCA is the disposal of the total wastewater flow from the WWTP and the provision of the required water flows for the different reuse purposes with the required quality.

System expansion is used where one of the above-mentioned functions is not included in the system boundaries. This is the case for those reuse scenarios that do not supply enough reuse water to satisfy all reuse purposes and for the no-reuse reference case. Here, the water for the reuse purposes has to be provided by other means to make a fair comparison. In the environmental assessment, surface water is assumed to be treated by a standard conventional treatment plant for the purposes of street cleaning, irrigation and toilet flushing or by a decarbonisation treatment for cooling water.

The decision, that shall be supported by this assessment, concerns the installation of a WRP. Processes upstream to the WRP are considered not to be modified and are, therefore, outside the system boundaries. Therefore, the operation of the WWTP and the sludge line are not considered. This means that the wastewater from the WWTP enters the system 'burden free'. The provision of energy and chemicals for the WRP are included in the system boundaries as well as the infrastructure of the WRP technologies as shown in Figure 4.

LIFE CYCLE INVENTORY

Detailed data on the Inventory are shown in the Supplementary Material. Inventory of foreground processes is usually obtained by data collection at the plant. In this case, since the study intends to see the suitability of a prospective

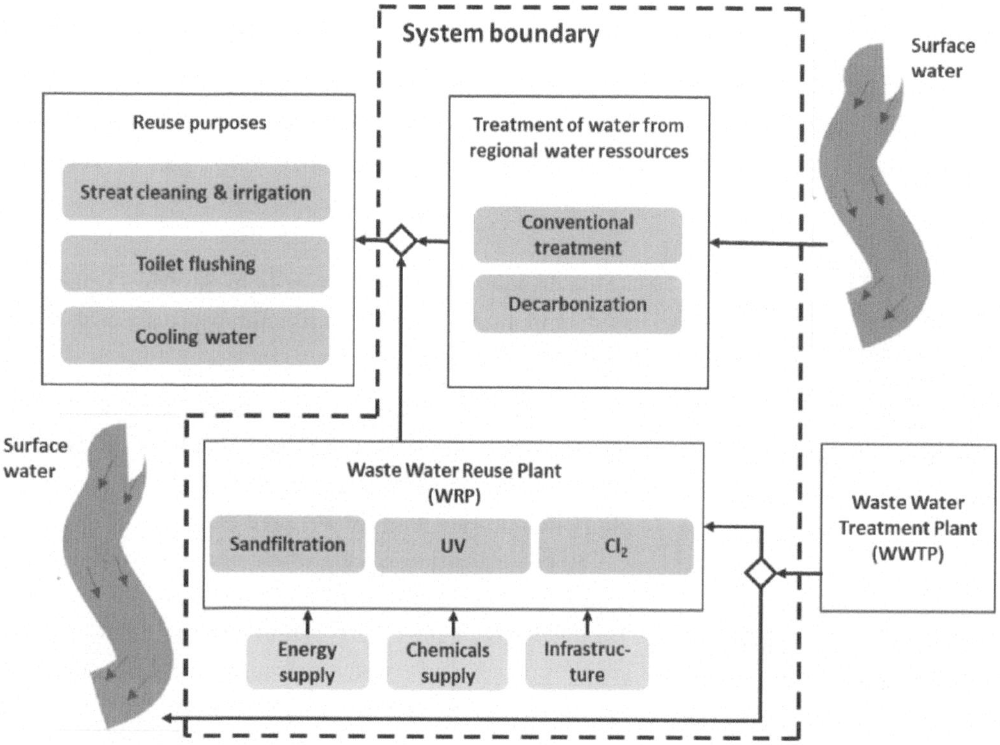

Figure 4 | System boundaries.

WRP (i.e. in the planning stage), data will be obtained through modelling and simulation as proposed by Bisinella *et al.* (2015). Foreground data were mainly provided by the technical process modules and the calculation based on literature and expert estimates from the interviewed stakeholders within the project.

For background data, ecoinvent v3.4 was used. The system model APOS (Allocation at the point of Substitution) was chosen. APOS follows the attributional approach in which burdens are attributed proportionally to specific processes. Further information on this topic can be found in the documentation of the ecoinvent v3 database (Wernet *et al.* 2016). When possible, the data of the background system were chosen to be representative of Europe. Products traded on the global market were modelled with globally representative data sets. The technical infrastructure (e.g. buildings and facilities) is also considered in the system boundaries, but the uncertainty associated with those processes is very high.

Due to the high data uncertainty, several assumptions had to be made, some of which can influence the results considerably. The following assumptions were made within the framework of this LCA:

- It is assumed that the wastewater from the WWTP complies with a minimum quality for emission. Different standards have been used to impose those limits depending on the parameter (MEP 1996; WQA 2013). The values used in this LCA are as follows:
  - COD – 100 mg/l
  - NH_4-N – 15 mg/l
  - P – 0.5 mg/l
  - TSS – 70 mg/l
  - Sulfid – 1 mg/l
  - Nitrite – 1 mg/l
  - Nitrate – 10 mg/l
  - Chloride – 750 mg/l
- In the same line, it is also assumed that the wastewater from the WRP complies with the minimum quality required for the infrastructural purposes (except hygiene parameters).
- The water from the backwash of the sandfilter is sent to the WWTP. Since the WWTP is neither considered

within the system boundaries nor modelled for this case study, an average wastewater treatment process is selected from the ecoinvent database.

- The water losses in the piping system have been included (i.e. 8%) after the WRP. However, since no information is given neither on the fate of those waters nor on the quality, the impact has not been taken into account.
- For the provision of electricity, an electricity mix from Germany was assumed. It can vary from country to country being a determinant for the final results.
- The disposal of sand from the sandfilter is not considered.
- The sewer system is not included in the system boundaries, and the dismantling of the WRP is out of the scope.
- The system expansion accounts for the water provision from conventional processes when reuse is not in place. Those conventional treatments may be more detailed than the modelled technologies in the WRP influencing the results.

LIFE CYCLE IMPACT ASSESSMENT

Due to the lack of primary data on the wastewater quality and the wastewater treatment technologies, only the most important impact categories influenced by the available data have been included within the framework of this LCA. Those impact categories are Climate change, Freshwater eutrophication, Marine eutrophication, Resource depletion – mineral, fossils and renewables, and Resource depletion – water. Other impact categories were neither calculated nor evaluated. The midpoint characterisation methods used are the ones recommended by the ILCD Handbook (Hauschild *et al.* 2011) for life cycle assessments (Table 2).

In particular, the distribution of toxic compounds could not be investigated, as this would require extensive modelling. However, this factor should not be ignored in the decision-making process.

The optional components of the impact assessment according to ISO 2006b, normalisation and weighting were omitted.

RESULTS AND DISCUSSION

Economic assessment

The focus of the calculation model is to determine both the total costs and the costs per m^3 of reuse water per process step and per scenario over the determined period. The modular structure also allows each individual module to be compared with the others and examined for its influence on the overall result.

Reuse O1, with 5,150 m^3/d of reuse water, represents the smallest volume and is only designed for irrigation and street cleaning purposes. The pipeline network has been neglected for this scenario, as a standpipe point at the WRP has been assumed. O2 extends the scope of the water reuse concept by toilet flushing purposes and provides 6,566 m^3/d of reuse water. For O2, investments and operating costs for the pipeline network are added. O3 represents the largest volume size with 13,479 m^3/d of available reuse water. The additional water is intended for cooling purposes in the industrial estate. In Figure 5, a comparison of the undiscounted costs per m^3 of reuse water for the different options is shown.

Table 2 | Impact assessment methods (Hauschild *et al.* 2011)

Impact category	Impact assessment model	Impact category indicators	Source
Climate change	Bern model – global warming potentials (GWP) over a 100 year time horizon	kg CO_2 equivalent	Intergovernmental Panel on Climate Change (2007)
Eutrophication – aquatic	EUTREND model	fresh water: kg P equivalent marine: kg N equivalent	Struijs *et al.* (2009) as implemented in ReCiPe
Resource depletion – water	Swiss Ecoscarcity model	m^3 water use related to local scarcity of water	Frischknecht *et al.* (2008)
Resource depletion – mineral, fossil	CML2002 model	kg antimony (Sb) equivalent	van Oers *et al.* (2002)

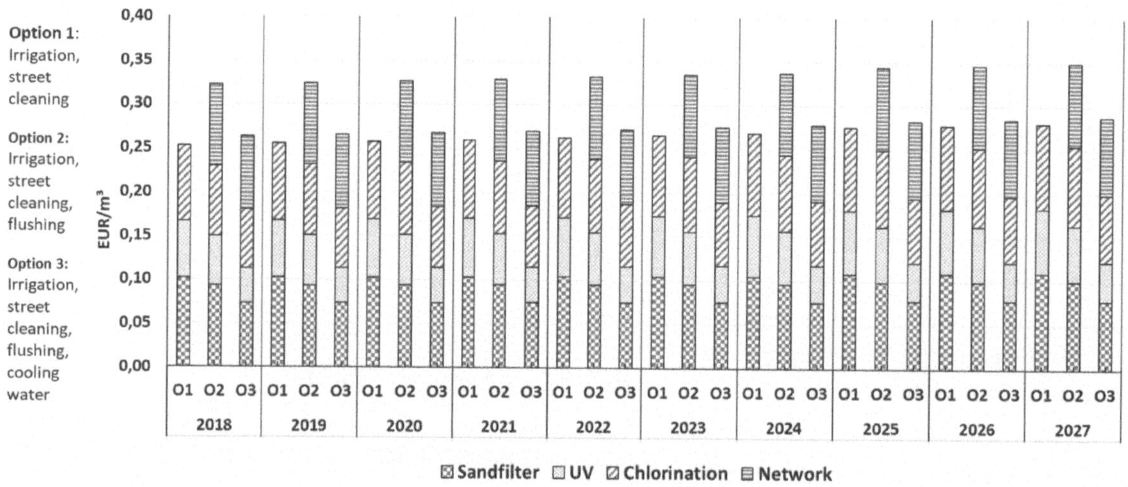

Figure 5 | Cost per m^3 reuse water in each reuse option over 10 years.

Comparing the interim results, O1 has the lowest costs per m^3 of produced reuse water, but mainly because it does not integrate costs for the pipeline network. If the pipeline network costs for O2 and O3 are neglected, there is a trend of slightly decreasing costs per m^3 with increasing size of the water reuse volumes, and a cost degression is identifiable. The economies of scale are an important factor when considering different reuse purposes and options; it might lead to a lower cost per m^3 when an additional application is included as noticeable comparing O1 and O3. UV costs for example amount to 0.06 €/m^3 in O1 and decrease by 33% to 0.04 €/m^3 in O3. The overall costs per m^3 decrease by 28% from 0.25 €/m^3 in O1 to 0.18 €/m^3 in O3. The overall economies of scale almost balance the costs of the pipeline and leave a difference of 0.01 €/m^3 between O1 (0.25 €/m^3) and O3 (0.26 €/m^3). Option 2 with the pipeline network and not profiting as much from economies of scale is the most expensive option.

The second trend is slightly increasing costs per m^3 of reuse water produced over time. This trend has two basic causes. First, the operating costs increase from year to year due to the inflationary projection, and second, renewal investments lead to increased depreciation over time, as the model also inflates the renewal investments, these results need to be included in a potential financial model. In the above time axis over 10 periods, only few renewal investments come into effect, so that it can be expected that this trend will intensify over time as renewal investments increase.

The calculated results are slightly higher than those stated by Bischoff (2013) for example, where UV costs are given with approximately 0.02–0.03 €/m^3. Costs for sandfiltration are given between 0.03 €/m^3 and 0.15 €/m depending on site-specific conditions (Hunziker-Betatech 2008) while chlorination with gas can be as low as 0.03 €/m^3 (Elefritz 2000) and chlorine dioxide as an alternative is given as 0.05–0.13 €/m^3 in Bischoff (2013). In contrast to the cost functions, the results presented here are not inflated and can, therefore, be lower. Furthermore, discounting, a heterogeneous study design and different plant sizes can lead to deviating results.

IPs have to treat their effluent to a specified discharge quality according to legal requirements and their operating licences. For that reason, only the costs of the additional WRP are considered without the costs of the mandatory WWTP. Furthermore, alternatives in the no-reuse option are the usage of tap water or the extraction from the river as cooling water. To extract water from a river in Germany, a permit is required (Wasserhaushaltsgesetz WHG § 8; Federal Water Act article 8). Additionally, a fee is charged for the permit and the water extraction. This fee may vary according to local conditions and state regulations. Given the assumed volumes in the MIP, such a fee would be below 6 cent/m^3 with no further treatment, which is significantly lower than the further treatment at the reuse plant. However, water needs to be available at any time and the extraction has to be permitted. Depending on the duration of the permit, renewals are necessary and planning security may be limited, as the regulating authority may alter the conditions or volumes of abstraction (e.g. due to water stress). Under certain conditions, river water needs to be decarbonised at a cost of 0.46 €/m^3 (ecoinvent v3.4) which is higher than the calculated price for the reuse plant water. Besides the extraction from a river, tap water or lower quality factory water can be used. In an inquiry in Germany, a price of 0.70 €/m^3 was determined for factory water, which in comparison would be more expensive than the cost for the reuse.

Environmental assessment

This sub-section shows the results of the impact assessment for the impact categories previously mentioned. Figure 6 shows the comparison of the three proposed options reusing water with the NRO. Thus, the vertical axis would be the NRO, and the results of the different impact categories show the decrease or increase in the percentage of each option compared with the NRO.

The results for climate change, freshwater eutrophication and marine eutrophication are analysed in detail below. The other impact categories' results are further commented upon within the Supplementary Material.

As shown in Figure 7, the results for the climate change impact category show a general decrease in the impact when reuse is implemented in comparison to the NRO. Thus, the NRO presents an impact of around 2,002 kg CO_2 eq./day, whereas 97% of this impact is due to the production of water from the conventional treatment for the purpose of irrigation, street cleaning and toilet flushing. The main contributor to this impact is electricity production. When the WRP is installed and reuse water is used to cover the demand of irrigation and street cleaning (O1) and additionally toilet flushing (O2), the impact is reduced to 1,181 and 955 kg CO_2 eq./day, respectively. Thus, avoiding the production of water using conventional treatment, the climate change impact is reduced by 41 and 52%, respectively. Finally, for O3, the results for the climate change impact increase (i.e. 1,838 kg CO_2 eq./day) in comparison to O1 and O2. Nonetheless, the impact on climate change is still lower compared with the NRO (i.e. a reduction of 8%).

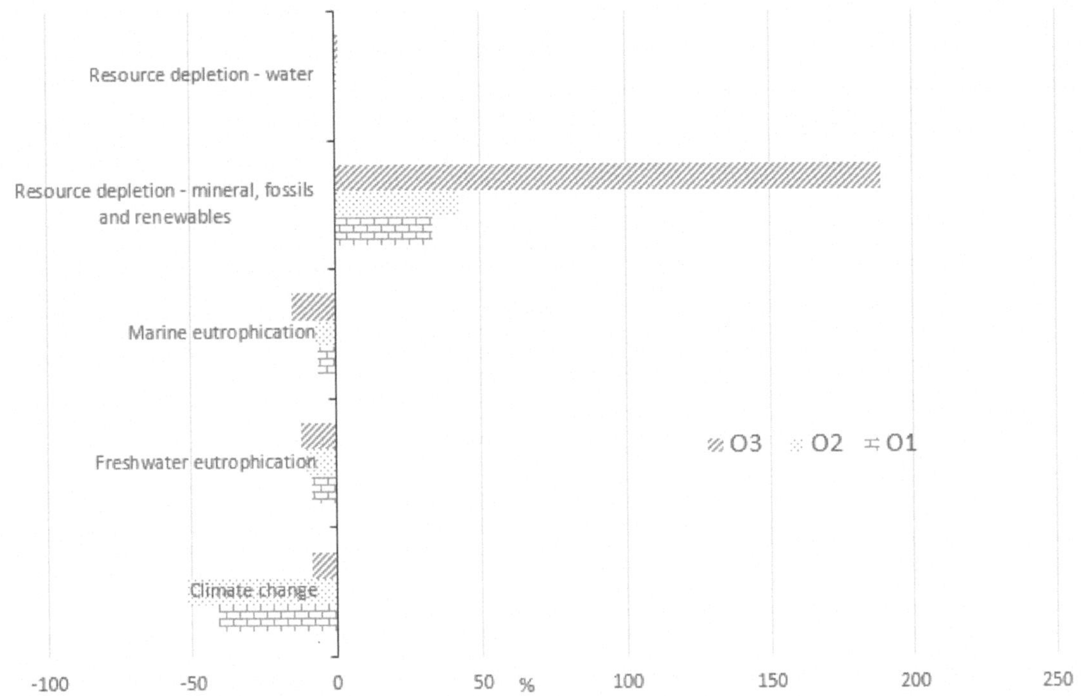

Figure 6 | Impact results for the comparison of the three proposed options reusing water with the NRO.

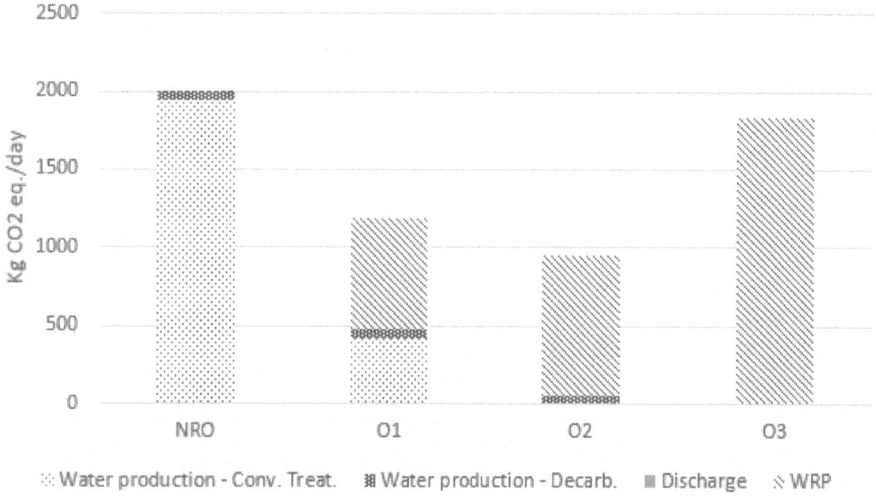

Figure 7 | Results for the impact category climate change.

Replacing the supply of cooling water by the WRP using the decarbonisation technology leads to an increase of GHG emissions mainly due to the electricity production, infrastructure and the treatment of the backwash water in the sandfilter. Those results are in line with Tong *et al.* (2013) who report larger environmental credits for scenarios in which treated water is used for higher value applications.

Freshwater eutrophication and marine eutrophication impact categories follow the same trend for the different options. Thus, for freshwater eutrophication (see Figure 8), the NRO presents an impact of around 17.2 kg P eq./day, with 92% of this impact due to discharge of water into the river and the other 8% mainly due to the production of water from the conventional treatment for purposes of irrigation, street cleaning and toilet flushing. The main contributor to this impact is the phosphate emitted to the river. When the WRP is installed and reuse water is used to cover the demand of irrigation, street cleaning (O1) and additionally toilet flushing (O2), the impact is reduced to 15 and 15.4 kg P eq./day, respectively. By avoiding the discharge of the wastewater into the river, the impact on freshwater eutrophication is reduced by 8 and 10%, respectively. O3 presents an impact of 15.2 kg P eq./day with a small contribution of the WRP of around 8% mainly due

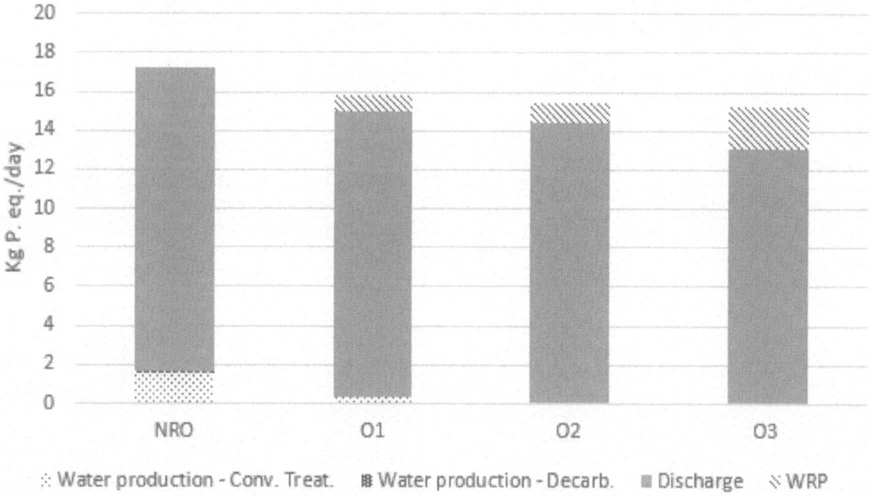

Figure 8 | Results for the impact category freshwater eutrophication.

to the treatment of the backwash water from the sandfilter. The total reduction of O3 for this impact category in comparison to the NRO is around 12%.

Those results are in line with Pasqualino *et al.* (2011), who conclude that the global warming potential and the freshwater eutrophication potential of the reuse scenario are reduced by 14 and 5%, respectively, compared with the no-reuse scenario. Nevertheless, it has to be mentioned that the system boundaries of Pasqualino *et al.* (2011) are slightly different including the sludge treatment and not considering the plant construction.

In the same line, for marine eutrophication (see Supplementary Material, Figure A1), the NRO presents an impact of around 1,349 kg N eq./day, with 100% of this impact due to the discharge of water into the river. The main contributor to this impact is the ammonium emitted to the river. When the WRP is installed and reuse water is used to cover the demand of irrigation and street cleaning (O1) and additionally toilet flushing (O2), the impact is reduced to 1,270 and 1,248 kg N eq./day, respectively. Avoiding the discharge of the wastewater to the river, impact is reduced by 6 and 8% for O1 and O2, respectively. O3 presents an impact of 1,144 kg N eq./day. The total reduction of O3 for this impact category in comparison to the NRO is around 15%.

Finally, LCA results show that impact categories behave in two different ways. On the one hand, for eutrophication impact categories (both marine and freshwater), the higher the reuse factor, the lower the impact in the respective categories. Thus, avoiding wastewater emissions to the river decreases these impacts. On the other hand, climate change experiences a higher decrease for O1 and O2, when the water is used for street cleaning and irrigation, and toilet flushing, compared to O3 in which the water is used additionally for cooling. In that case, results show that using the reclaimed water for low impact value applications is detrimental for the environment.

CONCLUSION

For the WaReIp MIP located in Germany, the economic and environmental assessment leads to promising interim results. The economic evaluation of the reuse options shows higher costs compared with the extraction from a water body with no further treatment but lower costs per m^3 compared with factory water when only the costs of the WRP are calculated. Furthermore, the costs show economies of scale, which is beneficial for O3 and especially relevant when reuse purposes and, therefore, volumes of required reuse water are chosen. Due to the economies of scale, an increase in volumes like in O3 reduces the costs per m^3 of reuse water by 28% from 0.25 €/m^3 in O1 down to 0.18 €/m^3 and would, therefore, be the preferred option. On the other hand, in O1, the pipeline is not needed, while in O3, the pipeline network is included and amounts to 0.08 €/m^3, summing up the cost of O3 to 0.26 €/m^3 in the first year and of O1 to 0.25 €/m^3. Hence, the total costs in O1 are lower even though the actual reuse water costs per m^3 are higher.

Despite the great uncertainty, it can be concluded from the LCA results that the installation of the WRP seems to be environmentally beneficial for all examined reuse options in comparison to the NRO considering all impact categories, except resource depletion. Considering the global warming potential, the reuse options lead to reductions between 8 and 52% for O3 and O2, respectively, compared with the NRO. Freshwater and marine eutrophication are reduced between 8–12% and 6–15%, for the different reuse options, respectively. In contrast to the above-mentioned impact categories, the impact on resource depletion increases for all three reuse options, being for O3 188% higher compared with the NRO.

Depending on the focus of the comparison, different reuse options were identified as favourable. This applies to the economic versus ecological dimension as well as to different environmental impacts. Therefore, decision makers need to weight the results depending on individual situations, circumstances and local requirements of the specific IP. Therefore, the application of further decision-making tools is necessary to evaluate the relative relevance of the two dimensions as well as of each impact category, making it possible to determine a definite preference for decision-making. In addition to the environmental and economic perspective, further factors like legal requirements, planning security and the company's strategy play an important role and must be considered in decision-making concerning water reuse strategies in IPs.

In future research, the data will be further validated and confirmed by industry experts and adapted to China and Vietnam. In water-scarce regions like North-Western China, the economic and environmental assessments of reuse concepts are not the only criteria for decision-making, political decisions to promote economic development and job creation may be an additional driver for water reuse practices.

ACKNOWLEDGEMENTS

The authors express their gratitude to the BMBF for the financial funding of the WaReIp project (Grant No. 02WAV1409), as well as the experts and partners in Germany, Vietnam and China for their cooperation and contribution. J. C. acknowledges financial support from the Spanish Ministry of Science, Innovation and Universities through the 'Beatriz Galindo' grant BEAGAL18/00035. The authors thank Dagmar Pohl and Dr Maike Beier for their contribution and fruitful discussions as well as the reviewers for their valuable comments.

DATA AVAILABILITY STATEMENT

All relevant data are included in the paper or its Supplementary Information.

REFERENCES

Alcalde Sanz, L. & Gawlik, B. 2014 *Water Reuse in Europe: Relevant Guidelines, Needs for and Barriers to Innovation. A Synoptic Overview*. JRC Science and Policy Reports, Publications Office of the European Union.

Arbeitsblatt DWA-A 203 Abwasserfiltration durch Raumfilter nach biologischer Reinigung (Worksheet DWA-A 203 Waste water filtration through spatial filter after biological treatment) 2019 DWA-Arbeitsgruppe KA-8 3, Deutsche Vereinigung für Wasserwirtschaft, Abwasser und Abfall e.V., Hennef, Germany.

Bauer, S., Dell, A., Behnisch, J., Linke, H. J. & Wagner, M. 2019 Sustainability requirements of wastewater management concepts for new industrial park developments in water-stressed regions. In: *International Water Association (IWA) Water Reuse 2019, 12th IWA International Conference on Water Reclamation and Reuse Book of Abstracts: Overcoming Water Stress*, June 16–20, 2019, Berlin, Germany, pp. 274–280.

Bischoff, A. 2013 *Desinfektion von behandeltem Abwasser – Vergleich verschiedener Desinfektionsverfahren (Disinfection of Treated Wastewater – Comparison of Different Disinfection Processes)*. PhD Thesis, Technical University of Darmstadt, Germany.

Bischoff, A. & Wagner, M. 2012 *Vier Wege, ein Ziel – Desinfektion von Abwasser in China (Four Ways, One Goal – Disinfection of Wastewater in China)*. Ressource Wasser- Wasserforschung für eine nachhaltige Zukunft, Federal Ministry of Education and Research (ed.), Bonn, Berlin, Germany.

Bisinella de Faria, A. B., Sperandio, M., Ahmadi, A. & Tiruta-Barna, L. 2015 Evaluation of new alternatives in wastewater treatment plants based on dynamic modelling and life cycle assessment (DM-LCA). *Water Research* **84**, 99–111.

Das, T. K. 2002 Evaluating the life cycle environmental performance of chlorine disinfection and ultraviolet technologies. *Clean Technologies and Environmental Policy* **4**, 32–43.

Ehrenfeld, J. & Gertler, N. 1997 Industrial ecology in practice: the evolution of interdependence at Kalundborg. *Journal of Industrial Ecology* **1** (1), 67–80.

Elefritz, R. A. 2000 Comparison of alternate methods of disinfection for reclaim effluents. *Florida Water Resources Journal* 19–29.

FAO – AQUASTAT 2016. Available from: http://www.fao.org/nr/water/aquastat/tables/WorldData-Withdrawal_eng.pdf (accessed 27 March 2020).

Frischknecht, R., Steiner, R. & Braunschweig, A. 2008 Swiss Ecological Scarcity Method: the new version 2006.

Giurco, D., Bossilkov, A., Patterson, J. & Kazaglis, A. 2011 Developing industrial water reuse synergies in Port Melbourne: cost effectiveness, barriers and opportunities. *Journal of Cleaner Production* **19** (8), 867–876. doi:10.1016/j.jclepro.2010.07.001.

Günthert, F. W. & Reicherter, E. 2001 *Investitionskosten der Abwasserentsorgung: Mit 31 Tabellen (Investment Costs of the Sewage Disposal: With 31 Tables)*. Oldenbourg-Industrieverlag, München, Germany.

Hauschild, M., Goedkoop, M., Guineee, J., Heijungs, R., Huijbregts, M., Jolliet, O., Margni, M. & De Schryver, A. 2011 *Recommendations for Life Cycle Impact Assessment in the European Context – Based on Existing Environmental Impact Assessment Models and Factors (International Reference Life Cycle Data System – ILCD Handbook)*. Publications Office of the European Union, Luxemburg.

Hilbig, J., Boysen, B., Wolfsdorf, P. & Rudolph, K.-U. 2020 Economic evaluation of different treatment options for water reuse in industrial parks using modular cost functions. *Water Reuse and Desalination* (this issue) doi:10.2166/wrd.2020.032.

Horstmeyer, N., Rapp-Fiegle, S., Helmreich, B. & Drewes, J. E. 2014 *Kosten der Abwasserbehandlung: Finanzierung, Kostenstrukturen und Kostenkenndaten der Bereiche Kanal,*

Sonderbauwerke und Kläranlagen (Costs of Wastewater Treatment: Financing, Cost Structures and Cost Data of the Areas Sewer, Special Structures and Wastewater Treatment Plants). DIV Deutscher Industrieverlag (Edition GWF Wasser, Abwasser), München, Germany.

Intergovernmental Panel on Climate Change – IPPC 2007 Climate Change 2007: Working Group I: The Physical Science Basis.

ISO 2006a *ISO 14040 – Environmental Management – Life Cycle Assessment – Principles and Framework (ISO 14040:2006).*

ISO 2006b *ISO 14044 – Environmental Management – Life Cycle Assessment – Requirements and Guidelines (ISO 14044:2006).*

Jekel, M., Altmann, J., Ruhl, A. S., Sperlich, A., Schaller, J., Gnirß, R., Miehe, U., Stapf, M., Remy, C. & Mutz, D. 2016 *Integration der Spurenstoffentfernung in Technologieansätze der 4.* Reinigungsstufe bei Klärwerken (Integrating organic micro-pollutant removal into advanced phosphorus removal). Universitätsverlag TU Berlin, Berlin, Germany.

Kraus, F., Remy, C., Seis, W. & Miehe, U. 2016 *D 3.3: Generic Assessment of Treatment Trains Concerning Their Environmental Impact and Risk Reduction Potential.* Innovation Demonstration for a Competitive and Innovative European Water Reuse Sector (DEMOWARE), Kompetenzzentrum Wasser Berlin gGmbH, Berlin, Germany.

Kreuter, S. & Rudolph, K.-U. 2016 *Best Practice Guide for the Development of Integrated Waste-Water Concepts for Industrial Zones in Tropical Areas*, 2nd edn. Prepared in the Framework of the AKIZ Project, Witten, Germany.

Mann, T., Koglin, B., von Kienle, H., Wegmann, U. & Weisbrodt, W. 2012 Wastewater, 5. Physical treatment. *Ullmann's Encyclopedia of Industrial Chemistry* (38), 685–710.

Massnahmen in ARA zur weitergehenden Elimination von Mikroverunreinigungen – Kostenstudie (Measures in ARA for further elimination of micropollutants – cost study) 2008 *Hunziker-Betatech, Report for the Federal Office for the Environment (FOEN).* Winterthur, Switzerland.

MEP, Ministry of Environmental Protection of the People's Republic of China 1996 Integrated Wastewater Discharge Standard (GB 8978-996).

Müller, K., Bleisteinr, S., Pirchner, A., Gnirß, R. & Hübner, M. 2009 Desinfektion von biologisch gereinigtem abwasser (Disinfection of biologically treated wastewater), KA – Korrespondenz Abwasser. *Abfall* **56** (6), 593–599.

Pasqualino, J. C., Meneses, M. & Castells, F. 2011 Life cycle assessment of urban wastewater reclamation and reuse alternatives. *Journal of Industrial Ecology* **15** (1), 49–63. doi:10.1111/j.1530-9290.2010.00293.x.

Struijs, J., Beusen, A., van Jaarsveld, H. & Huijbregts, M. A. J. 2009 Chapter 6 Aquatic eutrophication. In: M. Goedkoop, R. Heijungs, M. A. J. Huijbregts, A. De Schryver, J. Struijs & R. Van Zelm, R. 2009 ReCiPe 2008 - A life cycle impact assessment method which comprises harmonised category indicators at the midpoint and the endpoint level. Report I: Characterisation factors, first edn. VROM, The Hague.

Tchobanoglous, G., Burton, F. L., Stensel, H. D. & Tsuchihashi, R. 2014 *Wastewater Engineering: Treatment and Resource Recovery*. Metcalf & Eddy Inc., 5th edn, revised. McGraw-Hill Higher Education, New York, NY.

Tong, L., Liu, X., Liu, X., Yuan, Z. & Zhan, Q. 2013 Life cycle assessment of water reuse systems in an industrial park. *Journal of Environmental Management* **129**, 471–478.

Türk, J., Dazio, M., Dinkel, F., Ebben, T., Hassani, V., Herbst, H., Hochstrat, R., Matheja, A., Montag, D., Remmler, F., Schaefer, S., Schramm, E., Vogt, M., Werbeck, N., Wermter, P. & Wintgens, T. 2013 *Abschlussbericht zum Forschungsvorhaben :Volkswirtschaftlicher Nutzen der Ertüchtigung kommunaler Kläranlagen zur Elimination von organischen Spurenstoffen, Arzneimitteln, Industriechemikalien, bakteriologisch relevanten Keimen und Viren (TP 9)(Final Report on the Research Project: Economic Benefit of Upgrading Municipal Wastewater Treatment Plants to Eliminate Organic Trace Substances, Pharmaceuticals, Industrial Chemicals, Bacteriologically Relevant Germs and Viruses (TP 9)).* Report for the Ministry of Environment, Agriculture, Conservation and Consumer Protection of North Rhine-Westphalia (MKULNV), AZ IV-7–042 600 001I, Duisburg, Germany.

United States Environmental Protection Agency (U.S. EPA) 1999 *Chlorine Disinfection Wastewater Technology Fact Sheet.* Available from: https://www3.epa.gov/npdes/pubs/chlo.pdf (accessed 7 July 2020).

Urkiaga, A., las Fuentes, L. d., Bis, B., Chiru, E., Balasz, B. & Hernández, F. 2008 Development of analysis tools for social, economic and ecological effects of water reuse. *Desalination* **218** (1–3), 81–91. doi:10.1016/j.desal.2006.08.023.

van Oers, L., de Koning, A., Guinée, J. B. & Huppes, G. 2002 Abiotic resourcedepletion in LCA – improving characterisation factors from abioticdepletion as recommended in the new Dutch LCA Handbook.

Wernet, G., Bauer, C., Steubing, B., Reinhard, J., Moreno-Ruiz, E. & Weidema, B. 2016 The ecoinvent database version 3 (part I): overview and methodology. *The International Journal of Life Cycle Assessment* 1218–1230. doi:10.1007/s11367-016-1087-8.

WQA 2013 *Technical Fact Sheet: Nitrate/Nitrite.* Water Quality Association. Available from: https://www.wqa.org/Portals/0/Technical/Technical%20Fact%20Sheets/2014_NitrateNitrite.pdf (accessed 27 March 2020).

First received 3 April 2020; accepted in revised form 29 September 2020. Available online 23 October 2020

Sustainability requirements of implementing water-reuse concepts for new industrial park developments in water-stressed regions

S. Bauer IWA, A. Dell, J. Behnisch, H. J. Linke and M. Wagner IWA

ABSTRACT

Requirements for wastewater management and water-reuse concepts concerning sustainability are gaining greater importance, especially in times of climate change. Industrial park developments are often hindered due to water scarcity. Thus, nowadays, the reuse of wastewater is becoming more and more important to increase the availability of water and to enable new developments. The sustainable Industrial WasteWater Management Concept with the focus on Reuse ($IW^2MC \rightarrow R$) provides a solution strategy to produce reuse water from industrial wastewater within production plants. To achieve sustainability, certain requirements are essential since the reuse water can be provided directly via an optimized wastewater treatment process for subsequent use. Hence, it is then 'fit for purpose'. This enables a more efficient application of reuse water. Furthermore, due to environmental issues, it is important to construct space-saving water resource recovery facilities for reducing land consumption within industrial parks.

Key words | 'fit for purpose', industrial parks, industrial wastewater, water reuse, water scarcity

S. Bauer IWA (corresponding author)
Hochschule für Technik Stuttgart,
Stuttgart,
Germany
E-mail: *sonja.bauer@hft-stuttgart.de*

A. Dell
H. J. Linke
Department of Land Management,
Technische Universität Darmstadt,
Darmstadt,
Germany

J. Behnisch
M. Wagner IWA
Department of Wastewater Technology,
Technische Universität Darmstadt,
Darmstadt,
Germany

HIGHLIGHTS

- The reuse of wastewater is becoming more and more important to increase the availability of water and to enable new developments.
- An optimized treatment process reduces the required area inside the industrial park.
- The IRF can be increased by optimized industrial park management by controlling new production plant settlements.
- The $IW^2MC \rightarrow R$ has to fulfill diverse sustainability requirements.

doi: 10.2166/wrd.2020.028

GRAPHICAL ABSTRACT

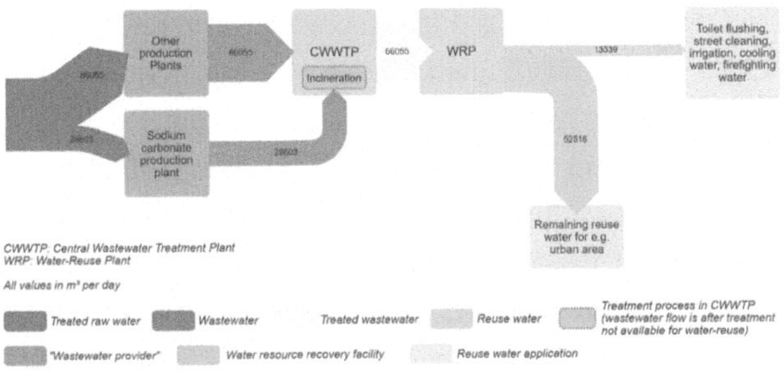

INTRODUCTION

In water-scarce regions, for instance, the northwestern parts of China, huge areas have high levels of water stress due to an uneven distribution of natural water resources and high levels of pollution of water bodies (WRI – World Resource Institute 2018). The development of industrial parks is generally related to the availability of water. Due to increasing water shortages, industrial park developments are often hindered; production plants may have high water requirements depending on their production processes. For instance, food production processes require less water than chemical production processes (Bauer *et al.* 2019). However, water is also needed for infrastructural purposes, such as the irrigation of green spaces or street cleaning in the park. Especially in water-stressed regions, the water demand for production plants increasingly competes with other water requirements such as municipal and agricultural water demand. This is resulting in the limitation of industrial expansions (Bauer *et al.* 2020a).

To enable new industrial park locations or their expansion in water-scarce regions, a sustainable water supply has to be ensured and water consumption from natural resources has to be reduced. In this context, and due to worsening climate change, new sustainable water-reuse concepts are achieving greater importance. Another important element is that industrial park developments drive urban developments (Zhao *et al.* 2017), which usually occur in densely populated urban areas. Thus, in industrial parks, the space for wastewater treatment plants is often limited due to the restricted availability of land as it is mostly provided for production plants. Hence, the reduction of land consumption and a high water-reuse potential is important to implementing water-reuse concepts.

Previous studies in the context of water reuse focus mostly on municipal wastewater treatment and reuse in an urban context. A well-known concept is integrated urban water management (IUWM), which includes all aspects of the water cycle: wastewater, drinking water, stormwater, and evaporation (Sharma *et al.* 2008). IUWM is an approach for urban water utilities to minimize their impact on the environment, to maximize their contribution to social and economic issues, and to generate well-being (Maheepala *et al.* 2010). Another important approach is 'urban metabolism,' which can be holistically understood according to Musango *et al.* (2017) as a 'collection of complex socio-technical and socio-ecological processes by which flows of materials, energy, people, and information shape the city, service the needs of its populace, and impact the surrounding hinterland.' A further approach is the innovative SEMIZENTRAL approach (Tolksdorf *et al.* 2016; Tolksdorf & Cornel 2017). It was developed for a semi-centralized supply and disposal system for fast-growing urban areas. The municipal wastewater is treated in a semi-centralized Resource and Recovery Center (RRC). Instead of only one municipal central wastewater treatment plant in a large urban area, the urban area is divided into smaller units, each of which has an RRC. The RRC provides water 'fit for purpose' via an in-house solution. It integrates the sectors of wastewater treatment, reuse water supply, and waste treatment. The concentration of pollution in industrial wastewater varies greatly and usually depends on the production process. Therefore, studies dealing with the treatment of industrial wastewater are very different and refer to specific case studies. As an example, a project by Whynman *et al.* (2019) dealing with industrial wastewater reuse at San Francisco airport can be mentioned here.

The particular aspect of the $IW^2MC \rightarrow R$ concept presented here is that wastewater flows are clustered according to certain concentrations and can therefore be specifically treated. The wastewater treated according to the 'fit for purpose' principle is reused for infrastructural purposes. This approach is therefore not purely project-specific or production plant-specific and can therefore be transferred to other projects.

MATERIAL AND METHODS

To enable sustainable industrial park developments, an innovative Industrial WasteWater Management Concept with the focus on Reuse (brand name: $IW^2MC \rightarrow R$) is developed. The concept includes the efficient treatment of industrial wastewater in a central wastewater treatment plant (CWWTP) and an additional water-reuse plant (WRP) which provides reuse water 'fit for purpose' (Bauer *et al.* 2020a). The treated wastewater can directly be reused for respective applications, for instance, for the irrigation of green spaces. To achieve further aspects of sustainability, the $IW^2MC \rightarrow R$ aims to reduce land consumption via an area-saving construction type of the CWWTP and the WRP. This aspect is an especially important issue in densely populated areas. To reduce land consumption, the wastewater treatment processes have to be optimized. Furthermore, the concept enables an optimization of water-reuse opportunities by reaching a high industrial park reuse factor (IPRF). The IPRF describes the relation between the wastewater inflow to the CWWTP and all possible reuse water flows. The IPRF includes the infrastructure reuse factor (IRF), which relates to infrastructural reuse applications, e.g., irrigation, street cleaning, and toilet flushing. Additionally, the IPRF includes the production plant reuse factor (PPRF), whereby it is possible to calculate the reuse water requirements for the production plants, such as process water (Bauer *et al.* 2020a). The $IW^2MC \rightarrow R$ is not focusing on treatment processes for internal water reuse for production plants within the industrial park. This is due to the fact that water reuse for infrastructural purposes is a further water-reuse opportunity as the internal reused water is discharged into the water resource recovery facilities at the end (Bauer *et al.* 2020b). Furthermore, production processes require a higher quality water than, for example, irrigation does. Consequently, the cost of treatment processes is higher as well. The concept focuses on the outflow of the WRP which is treated 'fit for purpose' for infrastructural purposes. Remaining reuse water can be discharged to the adjacent urban area since the water demand for irrigation and street cleaning cannot be covered by the municipal water resource recovery facility (Bauer *et al.* 2020b). Hence, it is not necessary to implement further cost-intensive treatment tracks for the treatment of the industrial wastewater for an internal production plant water reuse.

To calculate and evaluate area requirements for the CWWTP and WRP and to calculate the IRF, the creation of a model industrial park (MIP) is essential. Therefore, the case study of China was chosen due to its dynamic in developing greenfield areas as new industrial park locations (Bauer *et al.* 2020a). Furthermore, due to the uneven distribution of natural water resources, and the relatively high levels of pollution of water bodies, water shortage is often a challenge (WRI – World Resource Institute 2018). Another reason for choosing this case study area is that besides the United States, China is the country with the highest worldwide water requirements for industrial purposes (FAO 2017).

For the first IRF calculation, six exemplary production plants were taken into account. Accordingly, the IRF results in ~25% when including three infrastructural reuse applications: irrigation of green spaces, street cleaning, and toilet flushing (Bauer *et al.* 2020a). To get more viable results, an extended calculation that considered 19 production plants instead of six production plants was conducted. Using a MIP as methodology allows the adjustment parameters concerning wastewater quantity and quality. Even if the IRF varies due to the variety of production plants with different water demands, the MIP shows a methodology which calculates the reuse potential of a specific industrial site. If this is applied on real industrial parks, all parameters can be adjusted. The IRF shows a specific reuse potential for further applications as well as the volume of remaining treated wastewater flows that can be discharged into the water body or can be applied for further purposes such as for the irrigation of city parks in an adjacent urban area.

Besides the MIP of six production plants, the reuse applications for cooling water and firefighting water were further considered. Concerning the land consumption of the water resource recovery facilities, the needed treatment technology for the MIP is identified and average sizes of CWWTPs and WRPs are analyzed. However, it should be noted in this context that a MBR plant is not necessarily needed for processing reuse water for infrastructural purposes, even if it takes up little space. If the use of reuse water has higher requirements, it would be possible to include the MBR, even if the costs are higher.

RESULTS AND DISCUSSION

Characteristics of the MIP with 19 production plants

For the MIP, the 19 production plants included several types of industries, such as chemical, paper, food, and beverage production facilities, as they are statistically the most water-intensive industries and various production types are often localized in industrial parks. Additionally, the wastewater of a canteen and generally sanitary wastewater are considered.

Wastewater flows were divided into categories (A to C) according to their carbon, nitrogen, and phosphorus concentration (see Table 1). These parameters were chosen because they are common in the characterization of wastewater flows and are often used to determine the qualities to be maintained for the treatment of wastewater. It was assumed that the wastewater would be of constant quality throughout the entire year. This is due to the fact that in industrial CWWTPs, the equalization tank homogenizes the loads, concentrations, and wastewater flow.

The criterion for characterization was a minimum nutrient supply for aerobic biological wastewater treatment with COD:N:P ratio of 100:5:1 (Sahm *et al.* 2013), as it is the central treatment process of each CWWTP (Bauer *et al.* 2020a). Wastewater flows with special characteristics, such as sodium carbonate (see No. 21, Table 1), which have a very high salt concentration, have to be considered separately and should not be mixed with others. Incineration could be one solution for those flows. Based on this consideration of the basic parameters for wastewater characterization, a treatment track was defined, which must ensure a cleaning performance to comply with all required limit values. Those limit values are based on two Chinese water reuse standards. The 'Water quality standard for urban miscellaneous water consumption' (GB/T 18920-2002; MoHURD 2003) contains quality requirements for toilet flushing, street cleaning, firefighting, irrigation, vehicle washing, and construction. The standard addressing industrial usage, in particular, is the 'Water quality standard for industrial water consumption' (GB/T 19923-2005; MoHURD 2006). In this standard, requirements relating to reuse water qualities for cooling water, washing water, boiler feedwater, and process/product water are prescribed (for extracts see Table 2).

Due to the many food industries in the MIP, which produce highly loaded organic wastewater (Ranade & Bhandari 2014), in categories A to C the amount of COD (2,488,010 kg/d) is relatively high compared to the amounts of nitrogen (124,987 kg/d) or phosphorus (10,528 kg/d). It is, therefore, appropriate that the CWWTP provides at least two treatment tracks for wastewater treatment: anaerobic treatment to reduce carbon concentration and aerobic treatment to eliminate nutrients and further reduce COD (see Figure 1).

The anaerobic treatment makes it possible to obtain biogas and thus achieve the most energy-efficient treatment of wastewater. A detailed consideration of the wastewater flows shows that ethanol (see No. 10, Table 1) must be pre-treated to reduce high concentrations of solids. In this case, a simple sedimentation stage would be the best solution. The anaerobically pre-treated wastewater has to be discharged further into the aerobic process line to achieve further elimination of COD and nutrients. After the aerobic biological treatment, low concentrations of COD and nutrients are reached. To meet the requirements for the reuse standards, the treated wastewater is discharged to the WRP, filtered, disinfected by means of UV, and mixed with chlorine. It can then be reused for different applications. As the Chinese quality requirements of reuse water for irrigation, toilet flushing, firefighting, street

Table 1 | Production plants in the MIP with the characteristics of the different wastewater flows (Bauer *et al.* 2019)

No.	Wastewater origin (production process/ other)	Annual output [t/a]	Wastewater characteristics Flow [m³/d]	COD [kg/d]	N [kg/d]	P [kg/d]	TSS [kg/d]	TDS [kg/d]	TS [kg/d]	Reference
CATEGORY A: Wastewater with carbon as a main component										
1	H_2O_2	230,000	945	1,207	0	0	58	nda	nda	BVT Large Volume Organic Chemical Industry (2017)
2	Polystyrene	300,000	904	33	0	0	8	nda	nda	BVT Production of Polymers (2007)
3	Chlorine	215,000	365	26	0	0	nda	212	nda	BVT Chlor-Alkali Manufacturing Industry (2014)
4	Fruit juice	40,000	147	250	3	1	nda	nda	nda	DWA-M766 and Rosenwinkel *et al.* (2015)
5	Maize starch	1,100,000	1,658	3,315	10	0	nda	nda	nda	DWA-M 776
6	Wheat starch	182,500	700	19,250	700	140	nda	nda	nda	DWA-M 776
7	Yeast	55,000	1,243	26,541	230	34	nda	7,087	nda	DWA M 778 and Rosenwinkel *et al.* (2015)
8	Sugar (beets)	1,000,000	2,740	23,973	370	0	nda	nda	nda	DWA-M 713 and Rosenwinkel *et al.* (2015) andd Südzucker Werk-Offenau
9	Silicone	50,000	3,288	1,973	0	0	nda	29,451	nda	BVT Production of Speciality Inorganic Chemicals (2007)
10	Ethanol	800,000	25,001	2,387,624	118,756	8,876	nda	nda	2,250,117	Rosenwinkel *et al.* (2015)
11	Paper	300,000	10,274	8,527	0	0	nda	nda	nda	DWA-M 731
12	Canteen	–	375	709	4	2	nda	nda	nda	DWA M 775
CATEGORY B: Wastewater with carbon and nitrogen as the main component										
13	Butchery (cattle)	100,000	260	1,822	137	14	nda	nda	nda	DWA-M 767 and Rosenwinkel *et al.* (2015)
14	Potato	175,200	1,920	6,985	568	48	nda	nda	nda	DWA M 753
15	Animal by-products	15,000	41	339	53	0	52	nda	nda	DWA-M 710
16	Filament glass fibre	500,000	8,904	601	89	0	nda	11,037	nda	BVT Manufacture of Glass (2013)
CATEGORY C: Wastewater with carbon, nitrogen and phosphorus as the main component										
17	Superphosphate	850,000	2,911	1,397	3,959	1,374	nda	nda	nda	BVT Large Volume Inorganic Chemicals – Ammonia, Acids and Fertilizer Industries (2007)
18	Soft drinks	600,000	2,564	1,749	34	18	nda	nda	nda	Rosenwinkel *et al.* (2015) and DWA M 766 and Nachhaltigkeits-bericht 2015 Coca Cola Deutschland
19	Potato starch	10,000	45	633	57	7	nda	nda	nda	DWA-M 776 and Rosenwinkel *et al.* (2015)
20	Sanitary wastewater	–	1,769	1,056	18	14	nda	nda	nda	
CATEGORY D: Other										
21	Sodium carbonate	1,200,000	28,603	1,216	1,106	205	nda	4,138,852	nda	BVT Large Volume Inorganic Chemicals – Solids and Others industry (2007)
	Sum		**94,658**							

nda: no data available; COD: chemical oxygen demand; N: total nitrogen; P: total phosphorus; TSS: total suspended solids; TDS: total dissolved solids; TS: total solids.

Table 2 | Quality parameters for water reuse according to GB/T 18920-2002 and GB/T 19923-2005

	GB/T 18920-2002			GB/T 19923-2005	
Parameter	**Toilet flushing**	**Street cleaning/Firefighting**	**Irrigation**	**Parameter**	**Cooling water (recirculating)**
pH	6.0–9.0			pH	6.5–8.5
Color	30			Color	30
Olfactory	No odor			SS [mg/l]	–
NTU	5	10	10	NTU	5
TDS [mg/l]	1,500	1,500	1,000	COD [mg/l]	60
BOD_5 [mg/l]	10	15	20	BOD_5 [mg/l]	10
NH_4-N [mg N/l]	10	10	20	Chlorid [mg/l]	250
LAS [mg/l]	1.0	1.0	1.0	SiO_2	50
Fe [mg/l]	0,3	–	–	Fe [mg/l]	0.3
Mn [mg/l]	0.1	–	–	Mn [mg/l]	0.1
DO [mg O_2/l]	1.0			$CaCO_3$ [mg/l]	450
Total residual chlorine [mg Cl/l]	Contact 30 min later $\geq$1.0; endpoint of pipe network $\geq$0.2				
Total coliforms [1/l]	3				

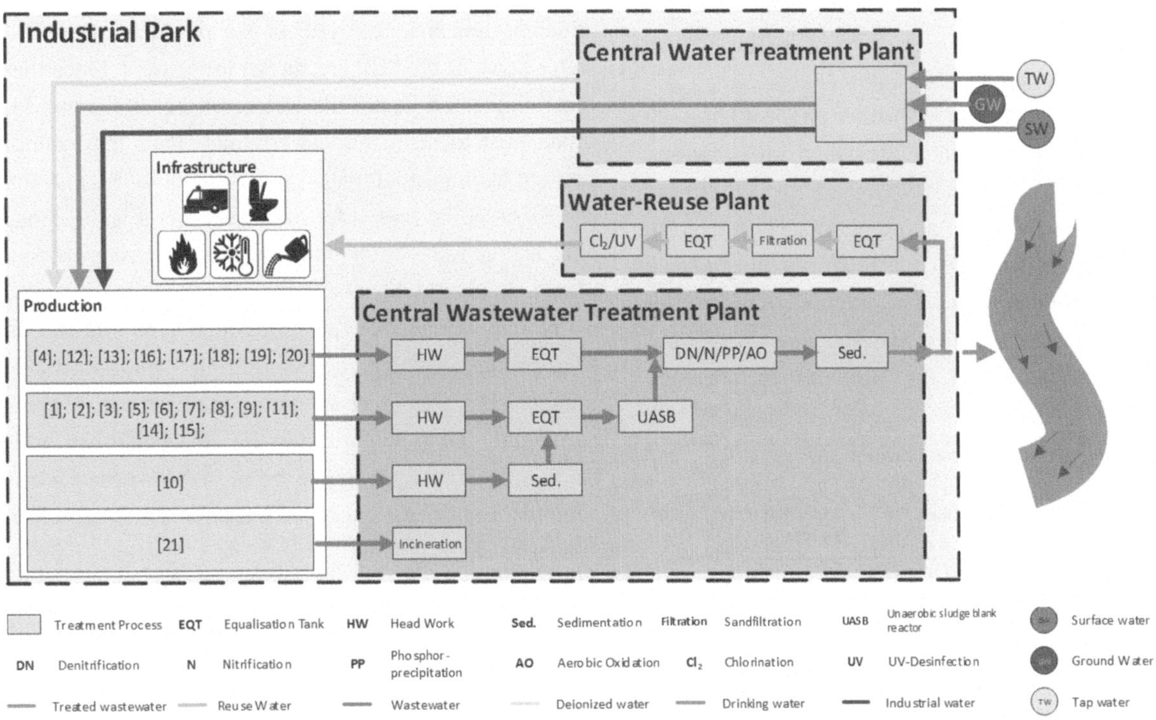

Figure 1 | The IW[2]MC→ R with respect to 19 production plants (Bauer *et al.* 2019).

cleaning, and cooling are very similar, the effluent of the CWWTP/WRP is expected to remain within the limits. Hence, to reach the required water quality, only one WRP treatment technology is necessary.

Calculation of the (enlarged) infrastructure reuse factor (IRF)

Concerning the dimension of the MIP with 19 production plants, different average values of 12 existing Chinese industrial parks and governmental design regulations served as a basis. To determine the park size, the average site size per company in units/ha was calculated by dividing the park size of the investigated parks through the number of companies, assuming that the park areas are fully occupied by production plants (no free construction areas) and the area is evenly distributed. The achieved data should, therefore, be regarded as a first approximation to be verified via real production plant sizes with their respective capacities. The number of employees in the different parks divided by the respective park size was the second essential average value (employees/ha) to be gathered from the investigated parks. Consequently, the total MIP has a size of 823 ha, the green spaces take up 165 ha (20%), and the area of roads amounts to 74 ha (9%) (see Table 3).

For this enlarged calculation, further applications such as cooling and firefighting water are taken into account. For the implementation of the $IW^2MC \rightarrow R$, it is premised that with respect to the water shortage in several regions, the industrial park enables only recirculating cooling instead of once-through cooling systems. The latter needs high water flows which cannot be provided in water-stressed regions. Therefore, the calculation of the IRF only includes the make-up water for cooling towers. Due to the German directives for cooling water (VGB R455 2000; VGB R135 2001), the make-up water is estimated as 2% of the total cooling water demand. It is assumed that this value can also be transferred to other countries, such as China, due to similar technical conditions. Minor deviations according to Chinese conditions would not have a noticeable effect on the result. For four of the 19 production plants, specific cooling water demand is given (see Table 4). For the first approximation of the others, it is calculated with the average cooling water demand of two existing industrial parks. As in water-stressed regions where the provision of firefighting water could be a challenge, (underground) fire water storage tanks are considered. To maintain the water quality, a partial water exchange in these tanks is expected, which results in permanent water demand of 60 m^3/d (see Table 4) even if there is no fire. This shows the example of the industrial park InfraServ Gendorf (2018). For toilet flushing, street cleaning, and irrigation of green spaces, the calculation of the water demand is based on Bauer *et al.* (2020a) (see Table 4). For the MIP with 19 production plants in China, the result is an IRF of ~14.3% (wastewater flows: 94,685 m^3/d; reuse water demand: 13,539 m^3/d), which is less than the IRF of six production plants. The water flows of the MIP are shown in Figure 2. Since there is remaining reuse water, the water can be discharged to the urban area to meet the water requirements that cannot be covered by municipal treated wastewater (Bauer *et al.* 2020b).

Table 3 | Dimensioning principles of the MIP (own table)

	Average values for China	Calculated dimension of the MIP
Number of production plants		19
Surface area	43.3 ha/company	823 ha
Employees	43 employees/ha	35,389 employees
Area of green spaces	min. 20%	165 ha
Area of roads	9%	74 ha

An essential reason for this lower factor is the considerably higher wastewater effluents of the chosen production plants. It should be noted that the IRF depends on the water requirements of the production type and the quantities produced which can be controlled via industrial park management (Bauer *et al.* 2019). Furthermore, a low IRF can equally be a potential for the adjacent urban area, as it allows remaining water to be reused for urban purposes such as the irrigation of city parks or street cleaning (Bauer *et al.* 2020b).

Reaching sustainability by an area saving construction type of the water reuse plant

Due to the restricted availability of land, as it is mostly provided for production plants, the size of the CWWTP and

Table 4 | Calculation of the reuse water flows in the MIP with 19 production plants (own table)

	Specific water demand	Data referring to the MIP	Calculated reuse water demand (m^3/d)
Toilet flushing	40 L/employee × day	35,389 employees	1,416
Street cleaning	2.5 L/m^2 × day	74 ha (=0.09 × 823 ha)	1,850
Irrigation of green spaces	2.0 L/m^2 × day	165 ha (=0.2 × 823 ha)	3,300
Firefighting			60
Cooling tower make-up water (2%) for:			
H_2O_2 production	9 $m^3/t_{Product}$	230,000 t/a	113
Polystyrene production	50 $m^3/t_{Product}$	300,000 t/a	822
Sodium carbonate	75 $m^3/t_{Product}$	1,200,000 t/a	4,932
Filament glass fibre	1.5 $m^3/t_{Product}$	500,000 t/a	41
Cooling tower make-up water average for other production plants	67 m^3/production plant × day*	15 production plants	1,005
Sum reuse water demand			**13,539**

**Source*: Average cooling tower make-up water demand of two industrial parks assuming evaporation of 2%.

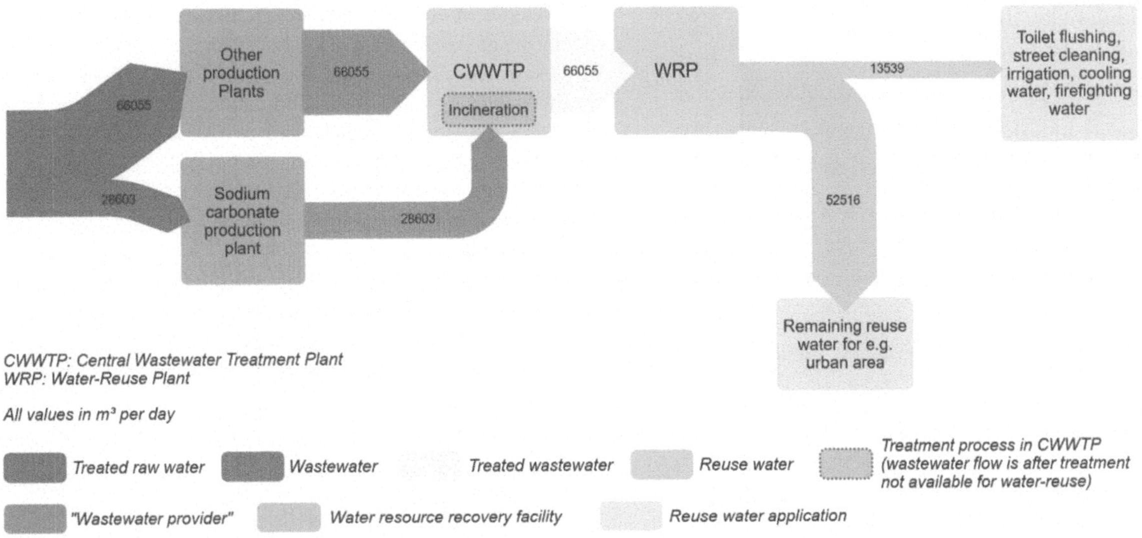

Figure 2 | Water balance for the MIP.

WRP have to be minimized via optimized treatment processes. Analyses of ten Chinese CWWTPs in different industrial parks show that the average size in relation to their treatment capacity is about 2.0 m^2/m^3. Calculated with the sum of the wastewater flow of the MIP (94,658 m^3/d, see Table 1) the CWWTP would have a size of approximately 19 ha (~2% of the total industrial park area). Due to low inflow loads from the CWWTP to the WRP, and thus due to an appropriate treatment process to provide reuse water 'fit for purpose', the area of the latter is expected to be substantially smaller than the area of the CWWTP. For a first appraisal, aerial photographs of three existing water-reuse projects in different parts of the world are analyzed. These projects also use additional filters and disinfection after the conventional treatment processes to provide reuse water with a similar quantity, e.g., for irrigation or toilet flushing. The average capacity is about 0.5 m^2/m^3. With a reuse water flow of 13,539 m^3/d (see

Table 4), the WRP would have a size of less than 5% of the calculated CWWTP size.

CONCLUSIONS

Regarding the required surface of the CWWTP and WRP, an optimized treatment process reduces the required area inside the industrial park. A technical upgrade of a CWWTP with a WRP could be conceivable due to its low land consumption. The results, regarding the IRF, show that depending on the production plants, the reuse factor can be lower despite further reuse applications, which is due to the different wastewater flows of the respective production plants. However, a previous study showed that reuse opportunities can increase by additional reuse applications in the adjacent urban areas which is required for enabling such developments in water-scarce regions (Bauer *et al.* 2020b). Additionally, the IRF can be increased by optimized industrial park management by controlling new production plant settlements concerning their wastewater effluents. Optimized industrial park management can thus compensate the production plants higher or lower wastewater effluents (Bauer *et al.* 2019). Providing a solution to enable new developments in water-stressed regions, the $IW^2MC \rightarrow R$ has to fulfill diverse sustainability requirements. Hence, via an optimized treatment technology providing reuse water 'fit for purpose,' the concept tackles these challenges.

DATA AVAILABILITY STATEMENT

All relevant data are included in the paper or its Supplementary Information.

REFERENCES

Bauer, S., Dell, A., Behnisch, J., Linke, H. J. & Wagner, M. 2019 Optimizing water-reuse opportunities for industrial parks. In *Water Environment Federation (WEF) (Hg.): 92nd Water Environment Federation Technical Exhibition and Conference (WEFTEC 2019)*, Chicago, IL, USA, pp. 470–484.

Bauer, S., Dell, A., Behnisch, J., Chen, H., Bi, X., Nguyen, V. A., Linke, H. J. & Wagner, M. 2020a Water-reuse concepts for industrial parks in water-stressed regions in South East Asia. *Water Supply* **20**, 296–306. https://doi.org/10.2166/ws.2019.162.

Bauer, S., Linke, H. J. & Wagner, M. 2020b Combining industrial and urban water-reuse concepts for increasing the water resources in water-scarce regions. *Water Environment Research* **92**, 1027–1041. https://doi.org/10.1002/wer.1298.

BVT Production of Polymers 2007.

BVT Large Volume Inorganic Chemicals – Ammonia, Acids and Fertilizer Industries 2007.

BVT Large Volume Inorganic Chemicals – Solids and Others industry 2007.

BVT Production of Speciality Inorganic Chemicals 2007.

BVT Manufacture of Glass 2013.

BVT Chlor-Alkali Manufacturing Industry 2014.

BVT Large Volume Organic Chemical Industry 2017.

FAO – Food and Agriculture Organization of the United Nations 2017 *AQUASTAT Main Database – Water Use.* Water withdrawal by sector, Industrial water with-drawal, Latest values. Available from: http://www.fao.org/nr/water/aquastat/data/query/index.html (accessed 2 August 2018).

GB/T 18920-2002 Ministry of Housing and Urban-Rural Development of the P. R. China (MoHURD) (formerly Ministry of Construction of the P. R. China), China. The reuse of urban recycling water – Water quality standard for urban miscellaneous water consumption 2003.

GB/T 19923-2005 Ministry of Housing and Urban-Rural Development of the P. R. China (MoHURD) (formerly Ministry of Construction of the P. R. China), China. The reuse of urban recycling water – Water quality standard for industrial water consumption 2006.

InfraServ Gendorf 2018 *Umwelterklärung (Environmental Declaration). Available* from: https://www.emas.de/fileadmin/user_upload/umwelterklaerungen/reg/DE-155-00047_155-00052_Industriepark-InfraServ-Gendorf.pdf (accessed 14 September 2018).

Maheepala, S., Blackmore, J., Diaper, C., Moglia, M., Sharma, A. & Kenway, S. 2010 Towards the adoption of integrated urban water management approach for planning. *Proceedings of the Water Environment Federation* **9**, 6734–6753. https://doi.org/10.2175/193864710798207017.

Musango, K., Currie, P. & Robinson, B. 2017 Urban metabolism for resource-efficient cities: from theory to implementation. In: *24th International Sustainable Development Research Society Conference*, 13–15 June, Paris, France.

Ranade, V. V. & Bhandari, V. M. 2014 *Industrial Wastewater Treatment, Recycling, and Reuse.* Elsevier, Philadelphia, PA, USA.

Rosenwinkel, K.-H., Kroiss, H., Dichtl, N., Seyfried, C.-F. & Weiland, P. 2015 *Anaerobtechnik. Abwasser-, Schlamm- und Reststoffbehandlung, Biogasgewinnung (Anaerobic Technique. Waste Water, Sludge and Residual Material Treatment, Biogas Production).* 3., Auflage Springer, Berlin, Germany.

Sahm, H., Stahmann, K.-P., Antranikian, G. & Takors, R. 2013 *Industrielle Mikrobiologie (Industrial Microbiology)*. Springer Spektrum, Berlin, Germany.

Sharma, A. K., Gray, S., Diaper, C., Liston, P. & Howe, C. 2008 Assessing integrated water management options for urban developments – Canberra case study. *Urban Water Journal* **5** (2), 147–159. https://doi.org/10.1080/15730 62070 1736829.

Tolksdorf, J. & Cornel, P. 2017 Semicentralized greywater and blackwater treatment for fast growing cities: how uncertain influent characteristics might affect the treatment processes. *Water Science and Technology* **75** (7–8), 1722–1731.

Tolksdorf, J., Lu, D. & Cornel, P. 2016 First implementation of a SEMIZENTRAL resource recovery center. *Journal of Water Reuse and Desalination* **6** (4), 466–475. https://doi.org/10. 2166/wrd.2016.129.

VGB-R 135 2001 *Richtlinie für die Planung von Kühltürmen (Guidelines for the Planning of Cooling Towers)*. VGB Power Tech Service, Essen, Germany.

VGB-R 455 2000 *Kühlwasser-Richtlinie – Wasserbehandlung und Werkstoffeinsatz in Kühlsystemen (Cooling Water Directive – Water Treatment and Material use in Cooling Systems)*. VGB Power Tech Service, Essen, Germany.

Whynman, D., Costanzo, M., Dresang, J., Salveson, A., Tikalsky, D. & Engel, G. 2019 Necessity is the mother of invention – Treating a diverse industrial waste stream for water reuse at the San Francisco Airport. In *Water Environment Federation (WEF) (Hg.): 92nd Water Environment Federation Technical Exhibition and Conference (WEFTEC 2019)*, Chicago, IL, USA, pp. 5078–5084.

WRI – World Resource Institute (Hg.) 2018 *Drop by Drop, Better Management Makes Dents in China's Water Stress*. Available from: https://www.wri.org/blog/2018/04/drop-drop-better-management-makes-dents-chinas-water-stress (accessed 23 June 2020).

Zhao, S., Bi, X., Zhong, Y. & Li, L. 2017 Chinese industrial park planning strategies informed by American edge cities' development path – case study of China (Chongzuo)-Thailand Industrial Park. *Procedia Engineering* **180**, 832–840.

First received 3 April 2020; accepted in revised form 29 September 2020. Available online 3 November 2020

Enhancement of overloaded waste stabilization ponds using different pretreatment technologies: a comparative study from Namibia

Jochen Sinn and Susanne Lackner IWA

ABSTRACT

Waste stabilization ponds (WSP) are a well-established wastewater treatment technology in Namibia. However, they are often overloaded and we still lack concepts and technologies for improvement. Therefore, this study presents the full-scale implementation of two pretreatment technologies to reduce the inflow of organic and solid loads into a facultative pond. We specifically compared the effects of anaerobic biological and mechanical pretreatment by an upstream anaerobic sludge blanket (UASB) reactor and a 250 μm micro sieve (MS). Not only in Namibia but also in most sub-Saharan countries, there is little experience with these technologies for the treatment of municipal wastewater in small and fast-growing local communities. Both technologies were tested in parallel for a period of 17 months and proved operational. While the UASB achieved better removal results with respect to chemical oxygen demand (COD) and suspended solids (TSS), the MS was more flexible in handling changing inflow patterns and had a much smaller footprint. The average total COD reductions of the MS and the UASB were 22 and 50%, respectively. TSS were removed by 45% with the MS and by 57% with the UASB reactor. Therefore, UASB and MS are viable options for the enhancement of existing WSP to reach better effluent values of the facultative pond.

Key words | agricultural reuse, developing country, load reduction, micro sieve, UASB, waste stabilization ponds

Jochen Sinn
Susanne Lackner IWA (corresponding author)
Department of Civil and Environmental Engineering Sciences, Institute IWAR, Chair of Wastewater Engineering,
Technical University of Darmstadt,
Franziska-Braun-Strasse 7, 64287 Darmstadt, Germany
E-mail: *s.lackner@iwar.tu-darmstadt.de*

HIGHLIGHTS

- Extensive data acquisition at waste stabilization ponds (WSP) in sub-Sahara Africa.
- Load reduction for WSP enhanced by UASB and micro sieve.
- First application of an UASB reactor before WSP in the sub-Saharan context with peak pCOD removal of 89%.
- First application of a 250 μm micro sieve as pretreatment for municipal wastewater before WSP with a maximum 72% of pCOD reduction.
- Pilot project for many other WSP in Namibia and other countries, especially in Africa.

doi: 10.2166/wrd.2020.021

GRAPHICAL ABSTRACT

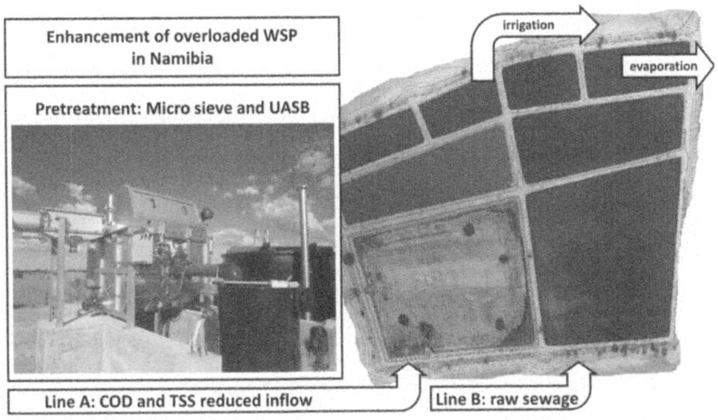

INTRODUCTION

Water scarcity has become an increasing struggle in many regions worldwide. Recent variable weather conditions in sub-Saharan Africa show direct effects of climate change. Most affected by variable rainfall or lack of any precipitation are farmers who need the precipitation for their crops. Therefore, water reuse of treated wastewater is becoming an ever more important mitigation measure. Mara (2009) already postulated the increasing need for reuse of treated municipal wastewater, from e.g. waste stabilization ponds (WSP), in water scarce areas. However, in many towns with sewer systems, the only wastewater treatment systems, if any, are WSP. They require low maintenance and power requirements but are often too small and overloaded due to the very high population growth.

With the combination of extended drought periods due to climate change and increasing water demand, the focus is shifting more and more towards water reuse, e.g. for irrigation. Thereby, WSP face new challenges. They have to fulfil national or international water quality standards, especially for hygienic parameters but also for COD removal. Not only the increasing population and the related need for more irrigation water with valuable nutrients but also the reduction of greenhouse gases impose a growing burden on WSP (Shelef & Azov 2000; Hernandez-Paniagua *et al.* 2014). A typical strategy for capacity enhancement is volume extension (number and or size of the ponds) which is accompanied by growing land requirements. Other options may present themselves through upgrades of existing WSP with more advanced treatment technologies, such as anaerobic biological or mechanical treatment units.

This study compares an upstream anaerobic sludge blanket (UASB) reactor to a micro sieve (MS) as potential pretreatment technologies in order to reduce the load of organic carbon (measured as chemical oxygen demand, COD) and suspended solids entering an existing overloaded WSP in Northern Namibia. The potential of these technologies to reduce COD and solids has been reported by Lazarova & Bahri (2005) with a wide range of values of 40–70% and 20–75% for UASB and MS, respectively. Also, the effect of such measures on the performance of consecutive ponds has not been monitored in depth in any sub-Saharan countries.

Due to high COD and TS loads into primary facultative ponds, sludge accumulation is also an issue and reduces the treatment capacity. A typical WSP set-up would include an anaerobic pond to reduce COD and TSS loads into the facultative pond (Shilton 2005; Sperling 2007). However, this would not reduce potential methane emissions and would not solve the problem of the quickly accumulating sludge. In order to address this, UASB reactors have already been installed as anaerobic biological pretreatment of municipal wastewater in front of WSP in other countries (e.g. in Brazil or India) with similar climate conditions (Frassinetti *et al.* 1996; Khan *et al.* 2011; Dias *et al.* 2017; Bressani-Ribeiro *et al.* 2019; Vassalle *et al.* 2020). Only Müller (2017) already examined UASB reactors before rotating biological contactors (RBC) in Namibia.

At the same time, mechanical pretreatment with micro sieves is becoming increasingly more important for energy recovery (Rusten & Ødegaard 2006; Prösl *et al.* 2013; Paulsrud *et al.* 2014; Walder *et al.* 2015; Hey *et al.* 2016; Jahn *et al.* 2017). In Europe, Jahn *et al.* (2017) and Walder *et al.* (2015) researched an MS within the Austrian context. In Namibia, micro sieves have so far been implemented as pretreatment for an industrial wastewater treatment plant followed by membrane bio-reactors (Prösl *et al.* 2013) and as post-treatment for a municipal wastewater treatment plant (Müller 2017). An MS for the pretreatment before a WSP, however, has been implemented for the first time in this research.

This is the first study which compares a UASB reactor (biological pretreatment) with an MS (mechanical pretreatment) to relieve overloaded WSP, with Namibia as an example. The results are especially important for fast-growing communities in warm climates with the need of water reuse for irrigation and for regions without perennial streams (Butler *et al.* 2017).

MATERIAL AND METHODS

WSP and pretreatment technologies

This project was conducted at an existing WSP system in a regional capital in Northern Namibia. The original WSP system as planned in 2004, consisted of two parallel lines (line A and line B) with four ponds each (Figure 1): one primary facultative pond (1) and three maturation ponds (ponds 2–4) (Supplementary material, Figure S1). According to the design, the feed rhythm alternated between the two lines. The two facultative ponds have the largest volume of 16,000 m^3 each and a surface of 11,000 m^2. Both are followed by three maturation ponds with smaller surfaces and volumes. The total water surface of the WSP is 40,500 m^2 with a total volume of 53,000 m^3. The water level decreases from 1.5 m in the first pond to 1.1 m in the last pond (Supplementary material, Table S1). Due to a high groundwater table and a nearby ephemeral stream, the earth dams constructed above the natural surface are covered with concrete, while the ground of the ponds is lined with clay.

Two pretreatment technologies were tested: an MS and an UASB reactor. Upfront, a coarse screen with 0.01 m bar spacing followed by a 56 m^3 open collection chamber to buffer peak inflows was installed (Figure 1). The pretreatment technologies for the upgrade were installed in line A. The UASB reactor had a volume of 42 m^3 and was continuously fed with 3–4 m^3/h inflow of raw wastewater, resulting in a hydraulic retention time of 10–14 h. In parallel, an MS (Noggerath Rotary Drum Screen RSJ-MG®) with a 250-µm monodur polyamide mesh was installed. Its drum diameter was 1.6 m, with a length of 1.5 m, resulting in a total area of 7.5 m^2. Aside from the standard monitoring, a 2-week intensive testing phase was carried out from day 265 to day 275. During this period, different flows and operational settings were applied. These included an increasing inflow from 28 to 60 m^3/h (maximum capacity of the feed pump: 75 m^3/h), an increase of the impounding depth from 10 to 18 cm and chemical cleaning with a 2% potassium hydroxide solution at 60 °C.

Additionally to the pretreatment, pond A1 was renewed by emptying out the settled sludge and installing two floating

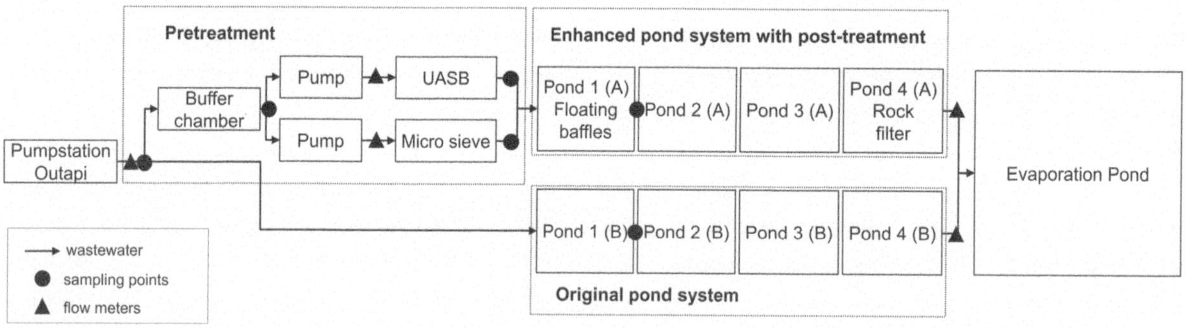

Figure 1 | Schematic drawing of the water reuse project, the black dots represent sampling points.

baffles made of 2-mm thick HDPE (high density polyethylene) sheets down to the full depth over two-thirds of its length. This improvement now avoids short-circuiting and enables better plug flow conditions. The overall project also includes the installation of a rock filter in pond A4 as post-treatment. The focus of this work was, however, on the pretreatment technologies only. These were continuously monitored for 17 months. Due to pump failure after heavy rainfalls with inundations and power failures, the MS had 88 days standstill time from day 110 to day 198.

Sampling and analyses

Electromagnetic flowmeters measured the inflow to the plant, the inflow to the UASB reactor, and the MS, as well as the outflow of ponds A4 and B4. Wastewater quality was monitored with portable probes for pH, dissolved oxygen, and conductivity using a WTW multimeter 3410 (Xylem Analytics Germany) at different sites: untreated wastewater, the effluents of UASB, MS, and A1 and B1 (Figure 1).

Due to high retention times in the sewer system with 15 pump sumps and in the ponds, grab samples were taken at different times of the day. For the untreated wastewater also 2-h mixed samples were analysed for comparison. The samples were analysed regularly with Hach Lange LCK cuvette tests for COD, total nitrogen (TN) ammonium, nitrite, nitrate, total phosphorus (TP), and phosphate with a spectrophotometer DR 2800 (Hach Lange GmbH, Germany). Indicator bacteria were examined with IDEXX Colilert-18 (enumeration of total coliforms and *Escherichia coli*) and Enterolert (*Enterococci*) employing Quanti-Tray/2000 (IDEXX Germany). To obtain total solids (TS) and total suspended solids (TSS), the samples were dried at 105 °C and for TSS also filtered with glass microfiber filters (Whatman 934-AH). Total volatile solids (TVS) and volatile suspended solids (VSS) were measured after ignition at 550 °C. The UASB reactor had five outlets to determine the level of the sludge bed.

RESULTS AND DISCUSSION

Site development in Outapi

Originally about 2,500 of 3,000 inhabitants were connected to the sewer system, and the pond system treated their wastewater. By 2018, the connected population had almost tripled up to nearly 7,000 people. However, due to the high population growth rate of 9.3% per annum, the total population in 2018 was estimated at 12,000 (Mwinga *et al.* 2018). This resulted in only 58% of the total population discharging their wastewater to the sewer and wastewater treatment system, while 69% were connected to the town's water supply system (Mwinga *et al.* 2018).

The WSP were designed with no overflow to the surrounding ephemeral watercourse, so that all water was supposed to evaporate. Already at the early stage of operation, an additional evaporation pond with a surface area of 41,000 m^2 and a volume of 20,500 m^3 (Supplementary material, Figure S1 and Table S1) was built with simple earth dams. However, due to higher flow rates, especially during the rainy season, the evaporation pond was overflowing regularly, posing a potential health risk to humans and grazing animals.

Within this study, enhancement measures at the existing WSP with simple pre- and post-treatment technologies were introduced and tested in order to improve the effluent quality and thereby ensure safe water reuse for irrigation of fodder crops. For comparison, one line (line B) kept its original status of 2004, while the other line (A) was enhanced.

Inflow characteristics

In this study, wastewater characteristics were monitored extensively for a long period of time. Such detailed information has so far not been available for the sub-Saharan context. The collected data included water quality parameter as well as flow patterns.

The mean total inflow to the WSP was 802 (±177) m^3/d during the observation period, with a peak inflow of 1,330 m^3/d on day 267 (Supplementary material, Figure S2). The maximum inflow was reached after rainfall events due to surface water entering the sewer system. Compared to preceding years, there was hardly any rain in the region during summer. This is also an indicator of changing rain patterns. In comparison, previous years recorded the highest hydraulic inflow to the plant with almost 2,000 m^3/d after heavy rainfalls.

Over the past years, the mean daily inflow has increased from 710 m^3/d in 2016 to 811 m^3/d in 2018. This 14%

increase over 2 years very well reflects the growth of the town and more connections to the sewer system. The peak inflow to the WSP was in the morning between 6 and 11 am. Three typical inflow patterns are presented in Supplementary material, Figure S3. After the peak in the morning, there was a second increase towards the early evening followed by reducing values during the night, showing rather typical daily variations in the inflow.

The inflow was divided equally between both lines for the purpose of good comparison between the enhanced line A and the line at its original state (line B). Line A received a mean inflow of 345 (±191) m^3/d pretreated wastewater, the inflow to line B without pretreatment was 472 (±211) m^3/d (Figure 2).

The water quality of the raw wastewater is included in Supplementary material, Table S2. Mean loads and concentrations of 606 (±149) kg/d and 749 (±153) mg/L for tCOD, 61.7 (±15.4) kg/d and 76.0 (±12.5) mg/L for TN, 8.0 (±1.7) kg/d and 9.8 (±1.3) mg/L for TP, and 174 (±67) kg/d and 220 (±81) mg/L for TSS were observed. These values are within a common range for municipal wastewater (Tchobanoglous & Abu-Orf 2014).

Pretreatment technologies

Two different pretreatment technologies, a biological system relying on anaerobic degradation processes (UASB) and a mechanical treatment system (MS) were monitored in parallel to evaluate their potential for reducing the COD and solids load into the WSP. Besides the settling of the pCOD and the TSS in the UASB, anaerobic microorganisms also reduced sCOD, nutrients, and pathogens. In comparison, the MS used a solely mechanical sieving process to reduce pCOD and TSS. The continuous operation of the UASB required a larger buffer volume to compensate for peak inflows, while the MS was very flexible with changing flow patterns.

Over the course of the research period, the MS received on average 357 (±152) m^3/d and the UASB 60 (±33) m^3/d. Changing treated volumes depended on the total inflow fluctuation and peak times as well as on the pump capacities. Peak inflows were 235 m^3/d for the UASB and 873 m^3/d for the MS.

Micro sieve

The different operation phases of the MS are indicated in Figure 3. The commissioning (day 1 to day 43) with unstable COD effluent values was followed by a stable operation phase. Effluent concentrations of the MS from day 77 to day 109 were tCOD 676 (±24) mg/L, pCOD 370 (±26) mg/L, and sCOD 304 (±14) mg/L. After the standstill (day 110 to day 199), the operation was less stable but improved towards the end of the study period with concentrations for tCOD of 708 (±132) mg/L, pCOD = 404 (±133) mg/L, and sCOD = 293 (±68) mg/L. Over the whole study period (day 1 to day 527), the mean sCOD concentrations with 306 (±58) mg/L were more constant than the tCOD concentrations with 740 (±142) mg/L and pCOD = 431 (±128) mg/L.

Over the whole study period, the tCOD removal was 22 (±18)% and pCOD removal was 28 (±24)%. Therefore, the installed MS reached good percentage removal which was within the range of others, e.g. the 15–25% tCOD removal

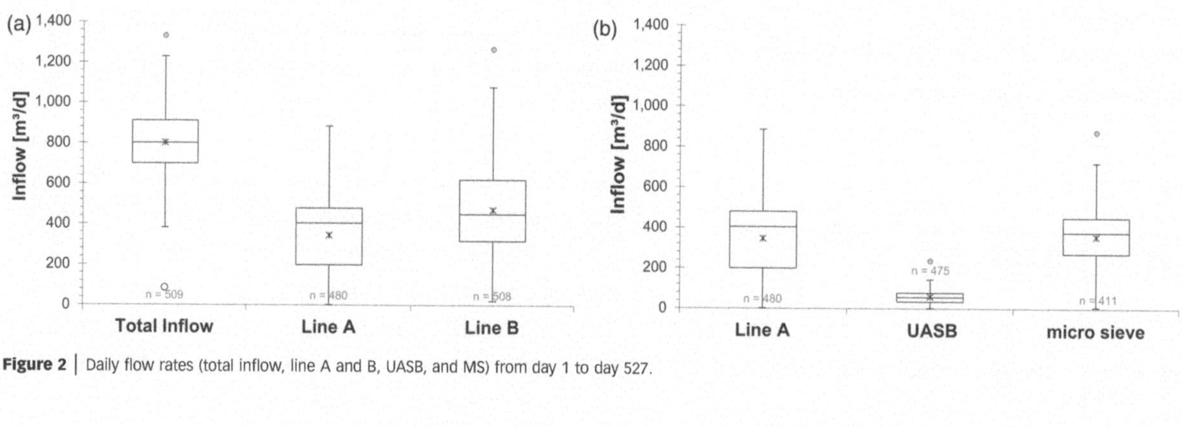

Figure 2 | Daily flow rates (total inflow, line A and B, UASB, and MS) from day 1 to day 527.

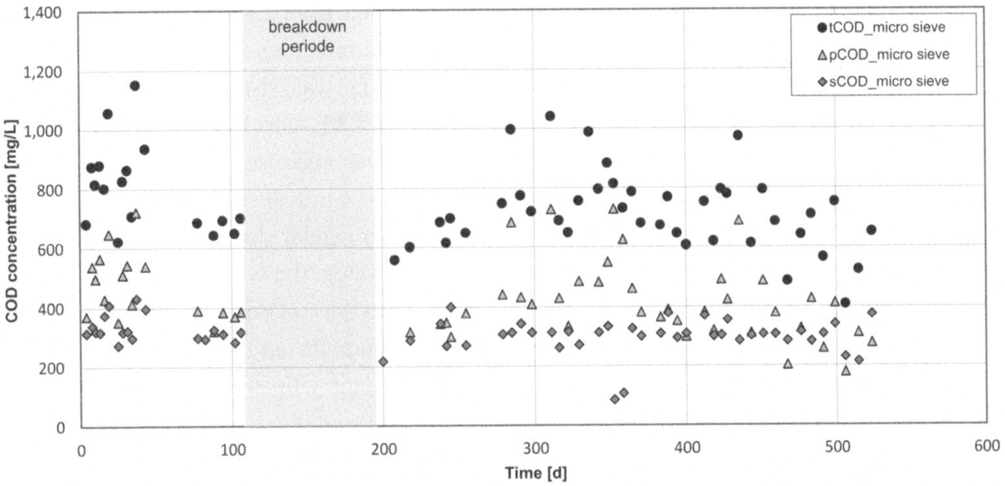

Figure 3 | COD effluent concentrations of the MS (tCOD: total COD, pCOD: particulate COD, and sCOD: soluble COD). The grey area indicates the breakdown period.

reported by Lazarova & Bahri (2005) or the 25% removal reported by Prösl *et al.* (2013). Elimination efficiencies of up to 40% tCOD seem only reachable with flocculation (Walder *et al.* 2015; Jahn *et al.* 2017). Peak tCOD and pCOD removal reached up to 57 and 72%, respectively. At the same time, the sCOD was hardly reduced as expected.

On the contrary, TSS removal reached 45 (±27)% over the whole study period. The removal of VSS was in the same range as 48 (±22)%. These values are better than at an industrial wastewater treatment plant in Windhoek. There, the two MS for the mechanical pretreatment also operate with 250 μm sieves and remove only 30% of TSS (Prösl *et al.* 2013). Sieves in Austria reached a removal of 65% with flocculation (Walder *et al.* 2015; Jahn *et al.* 2017). Our MS reached similar levels as the ones in Austria only during 3 days for its peak performance, which showed a maximum removal of 80% (TSS) and 82% (VSS), respectively. Therefore, the original design value of the manufacturer with an average of 60% TSS removal can only be reached with flocculation.

The intensive 2-week testing phase revealed that with the increasing inflow, the spray time over the cycle time in-between two cleaning events increased linearly (Figure 4(a)). Hence, the maximum theoretical inflow of the tested MS would be about 100 m^3/h. This, however, would require a constant spray operation which is not practical. Therefore, a maximum spray time of 80% of the cycle time was chosen which yielded a maximum inflow of 83 m^3/h. This compares very well with the maximum design value of 82 m^3/h. Based on the total sieve surface of 7.5 m^2, the maximum hydraulic loading rate of 11 $m^3/(m^2\,h)$ can be reached. This value is within the same range of 6–12 $m^3/(m^2\,h)$ as observed in Austria (Jahn *et al.* 2017).

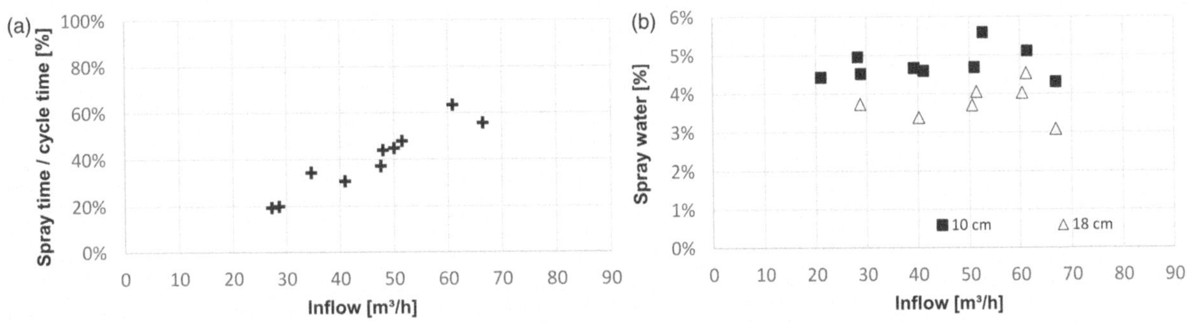

Figure 4 | Micro sieve spray time (a) in % of the cycle time and spray water consumption (b) in % of the treated water.

Another important aspect of the MS was the spray water consumption in relation to the treated wastewater flow. For a sustainable operation process, water from the MS effluent was used to spray the sieve itself instead of using valuable, high-quality tap water. Nevertheless, spray water use should be as low as possible to reduce energy consumption for the process of water pumping. This study shows that the percentage of spray water did not depend on the inflow but on the impounding depth. With an impounding depth of 10 cm, there was an average need of just below 5% while with a depth of 18 cm less than 4% was needed (Figure 4(b)). This is almost double the 2–3% measured by Walder *et al.* (2015) in Austria who have cleaned their sieve with water and a combined air injector.

UASB reactor

For the first 2 months after commissioning of the pretreatment stage, there was only a small COD reduction in the UASB reactor. Total COD was reduced by only 18% from 838 to 685 mg/L, while the sCOD remained constant at around 400 mg/L. After the inoculation of the UASB reactor with anaerobic sludge from the bottom of pond B1 on day 65, the tCOD concentration reduced over 3 months down to 234 mg/L and sCOD to 102 mg/L (Figure 5). This improvement of the performance was partially attributed to this inoculation but also to a rise in water temperature. The temperature increased from 23.5 °C in winter (day 1 to day 30) up to 28.5 °C in summer (day 75 to day 145). The temperature influence was also evident during later periods (day 200 to day 527). During the cold season, the highest tCOD effluent was measured with 720 mg/L (day 370). However, this was also influenced by the accumulating sludge in the UASB. With regular sludge removal every 4–6 weeks from day 380 onwards, the tCOD effluent improved and reduced back down to 274 mg/L. While the tCOD was influenced by temperature and sludge accumulation, the sCOD remained constant with an average concentration of 108 mg/L after the inoculation.

The removal of tCOD was 50 (±21)% during the whole study period. The peak removal was reached on day 311 with 84%. This is lower than the measured average values in Brazil of 63% (Dias *et al.* 2017), but close to the 57% of Vassalle *et al.* (2020). At the same time, the mean sCOD removal was 54 (±28)%, pCOD was removed by 45 (±29)%. Peak removals were 80% for sCOD on day 524 and 89% for pCOD (day 311).

After the inoculation of the UASB reactor with the anaerobic sludge from pond B1 on day 65, the solids in the UASB settled constantly over the following 3 months. In the beginning, the TS concentration at the top level reduced from 11 to 1.0 g/L and TVS from 7 g/L down to 0.4 g/L. At the same time, the TS concentration at the bottom level increased from 43 to 62 g/L, the TVS

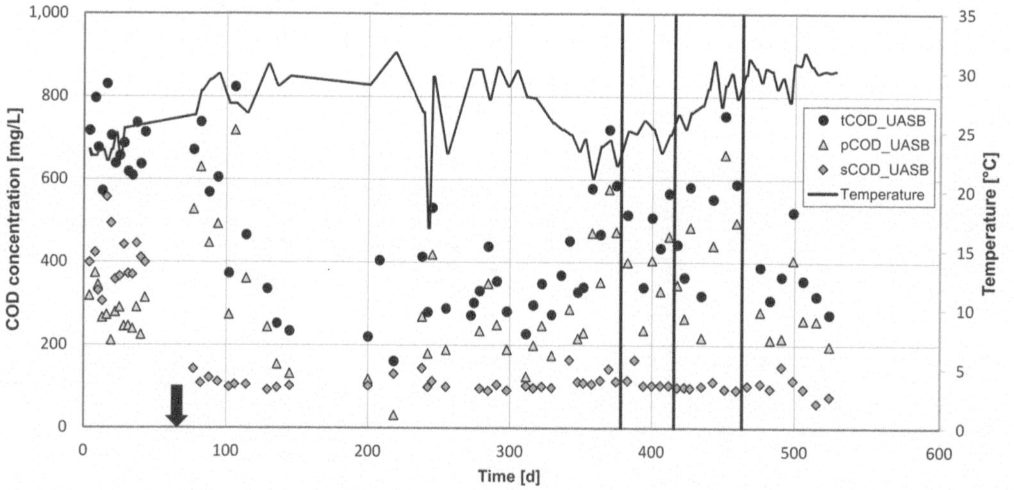

Figure 5 | COD effluent concentrations of the UASB reactor (tCOD: total COD, pCOD: particulate COD, and sCOD: soluble COD) and effluent temperature. The arrow indicates the inoculation (day 65), the black lines represent the sludge removal.

concentration from 30 to 38 g/L. After a continuous feeding period of 1 year, the first sludge was removed from the bottom layer (Figure 6). During the following months, the sludge bed was monitored and when it reached a height of 2.5 m in the reactor, between the second and third sampling level, about 5 m^3 were removed. This was three times during the study period. The mean TS concentration at the bottom level was 58 g/L with a TVS content of 66%, while the top level had an average TS concentration of 3.8 g/L with a TVS content of also 66% (Supplementary material, Figure S4).

Water quality comparison

Both, UASB reactor and MS, are valuable pretreatment technologies that have been implemented in different contexts. Within the local context of this study, the direct comparison of the effluent values showed that the percentage removal of the UASB was often higher than the ones of the MS. With regards to the tCOD removal, the UASB mostly came close or above 50%, while the MS seldom achieved over 50% (Figure 7(a)). This was mainly a result of the sCOD reduction in the UASB that did not exist in the MS. For the pCOD removal, the MS delivered more values above 50% but never above 75% like the UASB did (Figure 7(b)). The average tCOD reduction of the MS was 22 versus 50% of the UASB. TSS were removed by 45% with the MS and by 57% with the UASB reactor.

Even though the main purpose of both technologies was to reduce COD and TSS loads, they also had a visible effect on the microbiology. While the MS removed up to 50% of total coliforms and *E. coli*, the UASB reached even better removal efficiencies of 50–75% for *E. coli* and even above 75% for total coliforms (Figure 7(c) and 7(d)). Negative percentage removal in all cases can be explained by washing out of biofilm that accumulated on the walls of the UASB and the MS. The UASB was less affected by detaching biofilm as this biomass generally settles with the pCOD and is removed via sludge withdrawal. For the MS, detached biofilm was easily washed out with the effluent and was, therefore, more often detected.

On average, the UASB reactor reached tCOD effluent concentrations of 477 (±174) mg/L compared to 740 (±142) mg/L by the MS (Figure 8(a)). This was mainly due to the additional sCOD removal of the UASB. Soluble COD effluent values for UASB and MS were 171 (±126) and 306 (±58) mg/L, respectively (Figure 8(b)). In addition, for the pCOD, the UASB had a better removal of 45% with an outflow of 313 (±136) mg/L compared to 28% and 431 (±128) mg/L for the MS (Supplementary material, Table S2).

With regard to the potential water reuse purpose for irrigation of fodder crops, nutrient removal or conservation is another important aspect to be considered. As shown in Figure 8(c) and 8(d), there was almost no removal of phosphorus and nitrogen. The effluent values for TN were very

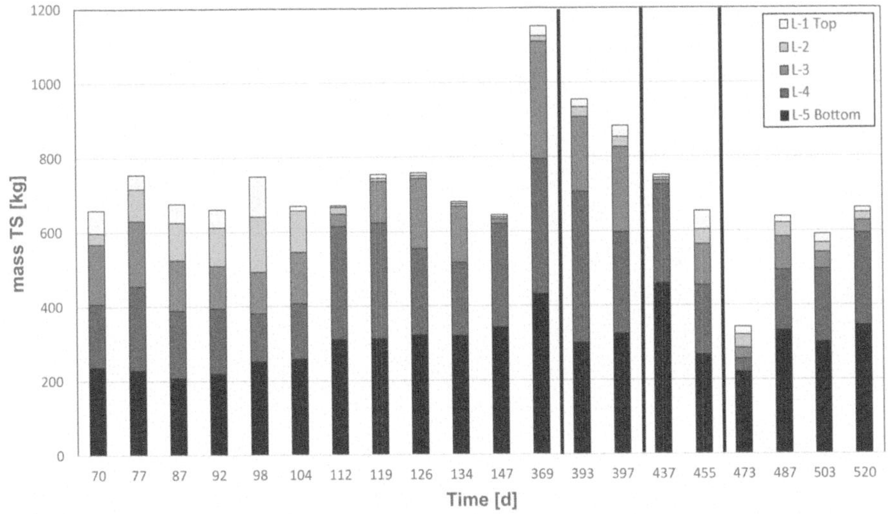

Figure 6 | Mass of total solids (TS) in the UASB in each layer L-1 Top, L-2, L-3, L-4, and L-5 Bottom with the black lines representing sludge removal.

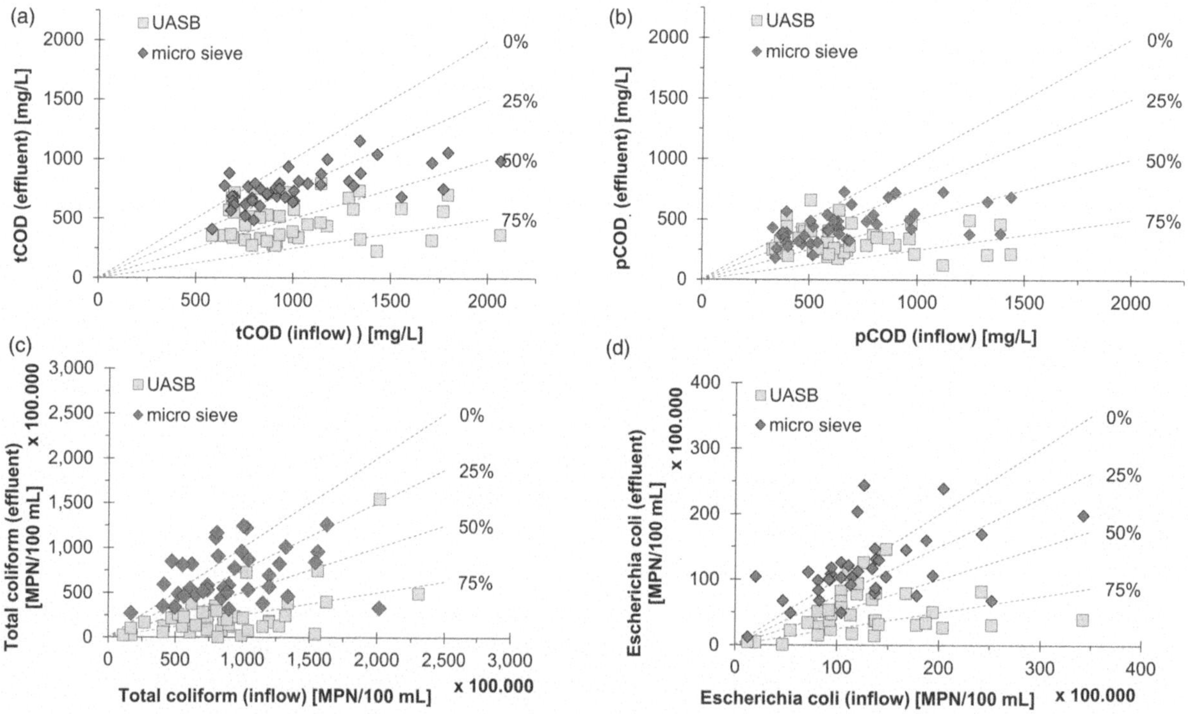

Figure 7 | Comparison of the removal percentages of MS and UASB reactor for tCOD (a), pCOD (b), total coliforms (c), and *E. coli* (d).

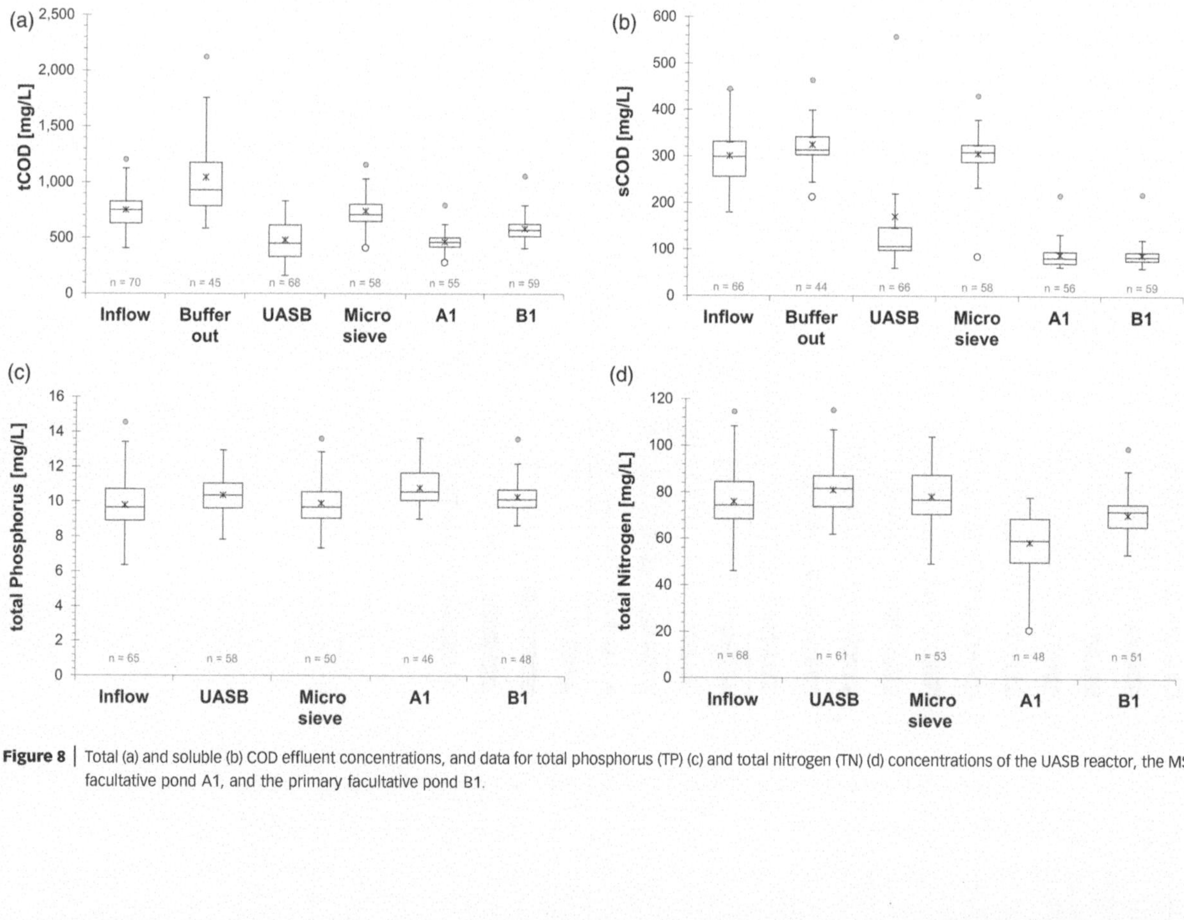

Figure 8 | Total (a) and soluble (b) COD effluent concentrations, and data for total phosphorus (TP) (c) and total nitrogen (TN) (d) concentrations of the UASB reactor, the MS, the secondary facultative pond A1, and the primary facultative pond B1.

close with 81 (±10) mg/L for the UASB and 79 (±12) mg/L for the MS. The same applied for TP with 10.4 (±1.1) and 9.9 (±1.3) mg/L, respectively. This corresponds well with findings of Dias *et al.* (2017) who measured even slight increases in total Kjeldahl nitrogen (TKN) concentrations after an UASB. Jahn *et al.* (2017) and Walder *et al.* (2015) reached 29% of TP removal and 6% of TN with their MS. However, they also implemented a flocculation process which certainly influenced the phosphorus removal.

Facultative ponds

Both first ponds were originally designed as primary facultative ponds, while pond A1 was transferred into a secondary facultative pond with the installation of the pretreatment (Shilton 2005; Sperling 2007). The floating baffles were installed into pond A1 on day 64 as a measure to optimize the flow conditions instead of potential short-circuiting existing in B1, as the last measure of three improvements. The other two main differences of this pond (A1) were the reduced inflow loads of COD and TS (due to the pretreatment) and the full available pond volume due to initial sludge removal. The total inflow into pond A1 was the combined effluent from the UASB reactor (16%) and the MS (84%), while pond B1 received 100% raw sewage.

The tCOD concentration in the effluent of pond A1 was 21% lower than that of B1 with 465 (±93) versus 588 (±123) mg/L, respectively (Figure 8(a)). Also, the pCOD concentrations were lower in A1 with 380 (±92) mg/L than in B1 with 461 (±125) mg/L. However, the sCOD concentrations with 89 (±28) mg/L for A1 and 89 (±22) mg/L for B1 were the same (Figure 8(b)). This clearly showed that the enhancement of the pond system had its main impact on the removal of the pCOD.

At the same time, both ponds (A1 and B1) recorded impacts on the microbial parameters. All measured parameters indicated further but low removal. A1 had maximum one $\log_{10}$ unit better effluent values than pond B1. For *E. coli*, it reached an average of $4.4\times10^{+05}$ ($\pm6.6\times10^{+05}$) MPN/100 mL with B1 of $1.6\times10^{+06}$ ($\pm3.2\times10^{+06}$) MPN/100 mL (Figure 9). Total coliforms were almost at the same level as $1.2\times10^{+07}$ ($\pm1.3\times10^{+07}$) MPN/100 mL and $2.6\times10^{+07}$ ($\pm3.8\times10^{+07}$) MPN/100 mL, respectively. Also for *Enterococci*, there was one $\log_{10}$ unit difference with A1 being $8.6\times10^{+03}$ ($\pm1.4\times10^{+04}$) MPN/100 mL and B1 being at $6.3\times10^{+04}$ ($\pm4.4\times10^{+04}$) MPN/100 mL. Even with those low removal values, it has to be stated that disinfection was not the main task of facultative ponds. This had to be achieved by the following maturation ponds (Shilton 2005; Sperling 2007).

In order to judge the treatment capacity of the ponds, the effluent concentrations were one aspect as they are relevant for the intended irrigation purpose. However, due to high evaporation losses, those concentrations can be misleading in terms of the functionality of the system. Therefore, we also considered loads and their reduction. The mean tCOD loads at the outflow of pond A1 were 158 (±69) kg/d with a removal efficiency of 49% compared

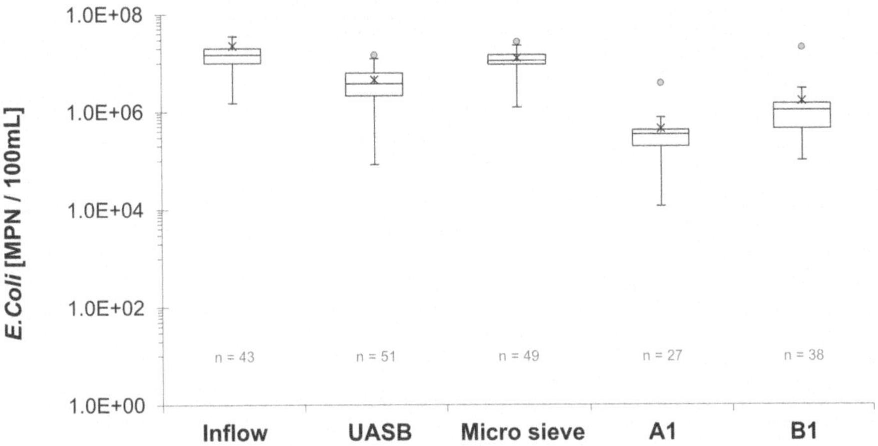

Figure 9 | *E. coli* concentrations of the inflow, the UASB reactor, the MS, and line A and B (MPN: most probable number).

to the inflow into line A. At the same time, the removal after pond B1 was only 36% with an average load of 211 (±117) kg/d (Figure 10). For the sCOD, the removal in both lines was 77% with similar load values of 29 (±15) kg/d for A1 and 32 (±18) kg/d for B1. Also for TP, there was only a slightly reduced load for A1 with 3.4 (±1.5) kg/d compared to 3.8 (±2.0) kg/d at B1. On the contrary, the TN loads were further reduced in line A than in line B with 19 (±9) kg/d and 26 (±14) kg/d, respectively.

Challenges for large-scale implementation

Both technologies proved to be viable options for the pretreatment of raw sewage upfront of WSP in the sub-Saharan context. However, for the large-scale implementation of only one technology, there are certain issues that have to be considered. The UASB had better effluent values but only treated a small fraction of the total inflow. For further improvement of the inflow to pond A1 by treating the entire inflow volume with UASB, they would have to be much larger. Either a much bigger reactor or several reactors in parallel, which would be more complicated to feed and operate, have to be installed. In the local situation, the power supply was needed to operate the feed pump of the UASB. Depending on other situations, it might be possible to feed the UASB with the last sewer pump in town. In this case, proper waste removal has to be ensured at the last pump station. Besides the reactor itself, dry beds are needed for the handling of excess sludge. They have to be designed according to the size of the reactor and the expected sludge accumulation. In order to reduce climate gas emissions, methane has to be transformed into CO_2. This can be either achieved with a flare as in this study or, if higher volumes of CH_4 are produced, with a combined heat and power system. This would also have a positive effect on the energy consumption of the plant. Another consideration for the implementation of a system with UASB reactors is the local temperature. During hot seasons, the removal efficiencies were considerably better, and therefore, it has to be ensured that also during colder periods the effluent values are reached.

In areas with long cold periods, the pretreatment with an MS might be a better option as the temperature has no influence on the MS. However, the constant power supply is necessary for the operation of the MS as well as for the spray water pump. If process water is used for spraying, a buffer tank is required, so that tap water consumption can be reduced. At the same time, the MS was flexible with changing inflow peaks and volumes. The one MS would be able to treat the full inflow as the only pretreatment technology. Nevertheless, regular maintenance such as cleaning of the sieves and replacement of spare parts is necessary, which can be difficult to acquire in the region. Also, further treatment of the sieving residue is required. Ideally, this is done with a fermentation process, so that the produced biogas can be transferred into power. Alternatively, the residue can be composted and after stabilization and drying used as fertilizer.

During the research period, operation and maintenance were conducted jointly by project partners, students, and the local Namibian plant operators. They were trained during the implementation of the project and are afterwards capable of sustaining the required daily works.

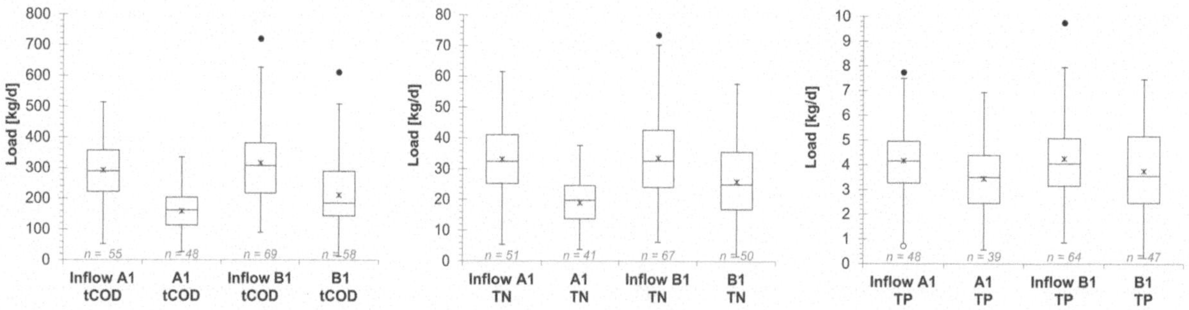

Figure 10 | Reduction of tCOD, TN, and TP loads over the treatment line.

CONCLUSIONS

In this study, two different pretreatment technologies were compared for the enhancement of overloaded WSP in Africa, also providing a detailed dataset of wastewater characteristics and effluent qualities. This full-scale research proved that both UASB reactor as well as MS are capable of reducing COD and TSS from the raw wastewater in the local context of warm climates, especially if there are space restrictions for further extensions. However, with the technical enhancement of existing WSP, the requirements increase. This means that regular power supply is needed for the MS, and it would be a benefit for the operation of the UASB reactor. Secondly, more maintenance for pumps and machinery will be necessary. At the same time, this improves the approach of the operators, who might have interpreted WSP as maintenance-free. Thirdly, spray water is necessary to clean the sieve. Ideally, this is implemented through a recirculation system of process water. Water consumption could be further reduced with a combined air and water cleaning.

The reduction of COD and solid loads into the first pond was achieved by both technologies with better effluent values of the UASB. However, the MS was more flexible with changing inflows and large volumes. The UASB further reduced sCOD, pathogens, and small amounts of nutrients, which is beneficial for further treatment with ponds but not strictly required. In contrast, phosphorus and nitrogen are valuable nutrients for the projected irrigation. Little to no removal of TN and TP was observed with the pretreatment, only later small amounts were consumed by algae and therefore remained in the system and were available for further irrigation purposes.

A positive effect of sludge removal, pretreatment, and baffles in line A was evident by better effluent values from A1 than with the unimproved pond B1. A further benefit of the pretreatment will be lower sludge accumulation rates and therefore longer removal intervals. Further research will focus in detail on the effluent values of ponds A4 and B4 and the suitability of the water quality for irrigation purposes. Such research will also include the maturation ponds and especially the influence of algae on the irrigation water as well as the disinfection and reduction capacity for pathogens. Long-term performance and operation stability data of UASB and MS will also be available for this local context.

ACKNOWLEDGEMENTS

This research was conducted within the joint research project EPoNa (grant number 02WAV1401A-F), which was funded by the German Federal Ministry of Education and Research under the funding measure Future-oriented Technologies and Concepts to Increase Water Availability by Water Reuse and Desalination – WavE. We would like to thank Peter Cornel for his support with micro sieve testing and for valuable discussions of the results. We would also like to thank Emanuel Döser, Andrea Friebe, Mikael Hidulika, Sven Jarczok, Lydia Luvinga, Haikela Nahambelelwe, Jöran Nguyen, Jakob Rauch, and Marcel Reusing for their support with regular sampling and water quality analyses; and the Outapi Town Council (OTC) for their support and the access to their facilities.

DATA AVAILABILITY STATEMENT

All relevant data are included in the paper or its Supplementary Information.

REFERENCES

Bressani-Ribeiro, T., Chamhum-Silva, L. A. & Chernicharo, C. A. L. 2019 Constraints, performance and perspectives of anaerobic sewage treatment: lessons from full-scale sewage treatment plants in Brazil. *Water Science and Technology* **80** (3), 418–425.

Butler, E., Hung, Y.-T., Suleiman Al Ahmad, M., Yeh, R. Y.-L., Liu, R. L.-H. & Fu, Y.-P. 2017 Oxidation pond for municipal wastewater treatment. *Applied Water Science* **7** (1), 31–51.

Dias, D. F. C., Passos, R. G., Rodrigues, V. A. J., de Matos, M. P., Santos, C. R. S. & von Sperling, M. 2017 Performance evaluation of a natural treatment system for small communities, composed of a UASB reactor, maturation ponds (baffled and unbaffled) and a granular rock filter in series. *Environmental Technology* **39** (4), 490–502.

Frassinetti, P., Catunda, C. & van Haandel, A. C. 1996 Improved performance and increased applicability of waste stabilisation ponds by pretreatment in a UASB reactor. *Water Science and Technology* **33** (7), 147–156.

Hernandez-Paniagua, I. Y., Ramirez-Vargas, R., Ramos-Gomez, M. S., Dendooven, L., Avelar-Gonzalez, F. J. & Thalasso, F. 2014 Greenhouse gas emissions from stabilization ponds in subtropical climate. *Environmental Technology* **35** (5–8), 727–734.

Hey, T., Zarebska, A., Bajraktari, N., Vogel, J., Helix-Nielsen, C., la Cour Jansen, J. & Jonsson, K. 2016 Influences of mechanical pretreatment on the non-biological treatment of municipal wastewater by forward osmosis. *Environmental Technology* **38** (18), 2295–2304.

Jahn, L., Haslinger, J., Krampe, J., Walder, C., Prösl, A. & Lahnsteiner, J. 2017 Microsieving as an innovative technology for the energetic use of sewage. *Österreichische Wasser- und Abfallwirtschaft* **69** (9–10), 369–377.

Khan, A. A., Gaur, R. Z., Tyagi, V. K., Khursheed, A., Lew, B., Mehrotra, I. & Kazmi, A. A. 2011 Sustainable options of post treatment of UASB effluent treating sewage: a review. *Resources, Conservation and Recycling* **55** (12), 1232–1251.

Lazarova, V. & Bahri, A. 2005 *Water Reuse for Irrigation – Agriculture, Landscapes, and Turf Grass*. CRC Press, Florida.

Mara, D. 2009 Waste stabilization ponds: past, present and future. *Desalination and Water Treatment* **4** (1–3), 85–88.

Müller, K. 2017 *A Systemic Approach to Implementation of Sanitation and Agricultural Water Reuse*. Verein zur Förderung des Instituts IWAR der TU Darmstadt e.V., Darmstadt, Germany.

Mwinga, M., Siboleka, M. & Kavezuva, C. 2018 *Economy Profile of Outapi*. First Capital Treasury Solutions (Pty) Limited, Windhoek, Namibia.

Paulsrud, B., Rusten, B. & Aas, B. 2014 Increasing the sludge energy potential of wastewater treatment plants by introducing fine mesh sieves for primary treatment. *Water Science and Technology* **69** (3), 560–565.

Prösl, A., Dammmert, R., Klegraf, F., Menge, J., Esterhuizen, J., Lahnsteiner, J., Leopold, P. & du Pisani, P. 2013 Reclamation of blended industrial effluents – pilot tests at Ujams, Windhoek. In *9th IWA International Conference on Water Reclamation & Reuse*, Windhoek, Namibia.

Rusten, B. & Ødegaard, H. 2006 Evaluation and testing of fine mesh sieve technologies for primary treatment of municipal wastewater. *Water Science and Technology* **54** (10), 31–38.

Shelef, G. & Azov, Y. 2000 Meeting stringent environmental and reuse requirements with an integrated pond system for the twenty-first century. *Water Science and Technology* **42** (10–11), 299–305.

Shilton, A. 2005 *Pond Treatment Technology*. IWA Publishing, London.

Sperling, M. 2007 *Waste Stabilisation Ponds*. IWA Publishing, London.

Tchobanoglous, G. & Abu-Orf, M. 2014 *Wastewater Engineering: Treatment and Resource Recovery*. Metcalf & Eddy, Inc., New York.

Vassalle, L., Diez-Montero, R., Machado, A. T. R., Moreira, C., Ferrer, I., Mota, C. R. & Passos, F. 2020 Upflow anaerobic sludge blanket in microalgae-based sewage treatment: co-digestion for improving biogas production. *Bioresource Technology* **300**, Article 122677.

Walder, C., Prösl, A., Klegraf, F., Soerensen, K. & Lahnsteiner, J. 2015 Enhanced primary treatment using microsieving for increased removal rates and energy recovery on WWTPs. In *12th IWA Specialised Conference on Design, Operation and Economics of Large Wastewater Treatment Plants*, Prague.

First received 27 March 2020; accepted in revised form 22 May 2020. Available online 6 July 2020

Modular treatment of arsenic-laden brackish groundwater using solar-powered subsurface arsenic removal (SAR) and membrane capacitive deionization (MCDI) in Vietnam

U. Hellriegel, E. E. Cañas Kurz, T. V. Luong, J. Bundschuh and J. Hoinkis

ABSTRACT

To evaluate energy efficient concepts for the modular treatment of brackish water, pilot trials for groundwater desalination and arsenic (As) removal were carried out in the Mekong Delta, Vietnam. Groundwater here is affected by naturally occurring high iron (Fe^{2+}) and As concentrations, while, in coastal regions, groundwater is additionally contaminated by high salinity mostly due to seawater intrusion. Desalination was conducted by membrane capacitive deionization (MCDI), which shows low specific energy consumption (SEC). Anoxic groundwater with As(III) and Fe^{2+} was treated using a pre-oxidation step called subsurface arsenic removal (SAR) with the main advantage that no As-laden waste is produced. The pilot plant was operated using a photovoltaic system (3 kW_p) and a small wind turbine (2 kW_p). The SEC of drinking water produced was 3.97 kWh/m^3. Total dissolved solids (TDS) of 1,560 mg/L were lowered to 188 mg/L, while Fe^{2+} was reduced from 1.8 mg/L to the below detection limit and As from 2.3 to 0.18 μg/L. The results show that SAR is a feasible remediation technique for Fe^{2+} and As removal in remote areas, and demonstrate the potential of MCDI for brackish water desalination coupled with renewable energies. However, improvements in energy demand of the MCDI module can still be achieved.

Key words | arsenic removal, coastal groundwater, desalination, MCDI, pilot scale, renewable energy

HIGHLIGHTS

- Pilot trials for groundwater treatment were successfully carried out in the Mekong Delta, Vietnam.
- Subsurface arsenic (SAR) and iron removal and membrane capacitive deionization (MCDI) achieved drinking water standard.
- The process was run by renewable energy (photovoltaic and wind turbine).
- Efficiency improvements in energy conversion losses are needed for the MCDI.
- The findings can serve as an example for coastal groundwater treatment.

U. Hellriegel
E. E. Cañas Kurz
J. Hoinkis (corresponding author)
Center of Applied Research,
Karlsruhe University of Applied Sciences,
Moltkestr. 30, 76133 Karlsruhe,
Germany
E-mail: *jan.hoinkis@hs-karlsruhe.de*

U. Hellriegel
E. E. Cañas Kurz
Department of Chemistry and Chemical Technologies,
University of Calabria,
Via Pietro Bucci 12/C, 87036 Arcavacata di Rende, Cosenza,
Italy
and
Institute on Membrane Technology,
National Research Council (ITM-CNR),
Via Pietro Bucci 17/C, 87036 Arcavacata di Rende, Cosenza,
Italy

T. V. Luong
Department of Mechatronics and Sensor Systems Technology,
Vietnamese-German University,
Le Lai Street, 822096 Binh Duong Province,
Vietnam
and
Department of Separation Science, School of Engineering Science,
Lappeenranta-Lahti University of Technology,
Sammonkatu 12, 50130 Mikkeli,
Finland

J. Bundschuh
J. Hoinkis
Faculty of Health, Engineering and Sciences,
University of Southern Queensland,
West Street, Toowoomba, 4350 Queensland,
Australia

J. Bundschuh
UNESCO Chair on Groundwater Arsenic Within the 2030 Agenda for Sustainable Development,
School of Civil Engineering and Surveying,
University of Southern Queensland,
4350 Queensland,
Australia

doi: 10.2166/wrd.2020.031

GRAPHICAL ABSTRACT

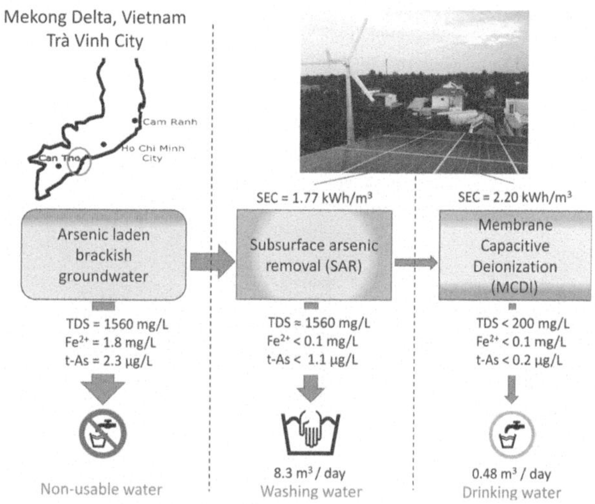

INTRODUCTION

Currently, about 600 million people worldwide live in low-elevation coastal zones affected by progressive salinization (Wheeler 2011; Dasgupta *et al.* 2014a). Groundwater in coastal line aquifers is often affected by seawater intrusion due to overextraction and rising sea levels, which will be aggravated by climate change in the future. The salinity level is subject to seasonal variations during dry and wet periods, and tidal channels are an important potential source of high salinity water (Brecht *et al.* 2012; Dasgupta *et al.* 2014b; Colombani *et al.* 2016). In the Mekong Delta, the salt concentration in the groundwater varies between 1 and >10 g/L (Wassmann *et al.* 2004; Hien *et al.* 2009; Ngo-Duc 2014; Vu *et al.* 2018).

In addition, high arsenic (As) concentrations are widespread in the groundwater of tidal delta plains in South and Southeast Asia such as the Ganges–Brahmaputra–Meghna in Bangladesh, the Irrawaddy delta in Myanmar, and the Mekong and Red river delta in Vietnam (Smedley *et al.* 2002; Buschmann *et al.* 2008; Winkel *et al.* 2011; Ayers *et al.* 2016; Wang *et al.* 2018; Le Luu 2019). As dissolved in groundwater in South Asia is caused by subsurface sediments, where As is adsorbed onto hydrous ferric oxides (HFOs; Hossain *et al.* 2012; Sankar *et al.* 2014; Wang *et al.* 2018). Manganese is also often co-associated with iron. Previous studies confirmed that reductive dissolution of HFOs driven by microbial metabolism of organic matter results in increased concentrations of dissolved As in shallow groundwater in Bangladesh (Nickson *et al.* 2000) and other South Asian countries (Berg *et al.* 2001; Datta *et al.* 2014; Stopelli *et al.* 2020).

Goal and scope

This study was carried out within the German joint research project *WaKap* as part of one of three plants installed around the Mekong Delta for subsurface arsenic removal (SAR) and/or desalination of brackish water (WaKap 2016). This study focused on the implementation of a combined desalination and As remediation technology using membrane capacitive deionization (MCDI) and SAR, respectively.

Combining the two technologies aimed at a robust, sustainable and easy-to-use approach for high removal and water recovery efficiencies as well as low energy demand. Additionally, an important goal was to evaluate the supply of the system with renewable power sources such as solar (photovoltaic, PV) and wind (wind turbine, WT) power. A sustainability assessment and evaluation of the system was carried out taking environmental and socio-economic aspects of the region into consideration (WaKap 2016; Fritz *et al.* 2020).

State of the art

Capacitive deionization (CDI) is a relatively new electrochemical desalination process with the general advantage of low specific energy consumption (SEC) since no high pressure pumps or heat sources are required. In this process, saline water flows between two porous electrodes made of activated carbon, on which a voltage is applied. The positive ions are thereby attracted to the negatively charged electrode and vice versa. After the electrodes are loaded, the salt ions can be removed by reversing the polarity of the electrodes. Through a constant change of charge (desalination) and discharge (regeneration), two water streams with desalinated water or concentrated salt water are produced (AlMarzooqi *et al.* 2014; Wang *et al.* 2019). Mounting ion exchange membranes in front of the electrodes enhances the desalination performance of the CDI module, because co-ion leakage is lowered and the discharge phase can be shortened (Zhao *et al.* 2013). This technology is then defined as MCDI. CDI is ideally suited for decentralized applications since it can be directly powered by solar PV energy (Tan *et al.* 2018).

SAR aims at the removal of As, Fe and Mn, in advance of the MCDI process. For this, part of the extracted groundwater is enriched with oxygen and introduced back into the underground (aquifer). Dissolved iron (Fe^{2+}) and manganese (Mn^{2+}) ions are oxidized and precipitated as sparingly soluble (hydr)oxides forming a zone with treated water around the re-injection zone in the aquifer that is free of dissolved Fe^{2+} and Mn^{2+} (Sarkar & Rahman 2001; Sen Gupta 2008; van Halem 2011; Rahman *et al.* 2014; Kundu *et al.* 2018). Under anoxic conditions, as in South Asia, groundwater principally contains arsenite in the form of H_3AsO_3 which is oxidized through SAR to arsenate (depending on pH conditions to $H_2AsO_4^-$ or $HAsO_4^{2-}$) and is then immobilized in the soil by co-adsorption onto the iron and manganese(hydr)oxides (Rahman *et al.* 2015; Annaduzzaman *et al.* 2018; Luong *et al.* 2018). The salient advantage of the process is that no As-laden (waste)stream is produced, since As is immobilized in the aquifer, where it naturally occurs (Rott & Friedle 2000; van Halem *et al.* 2009; Stopelli *et al.* 2020).

In contrast, every membrane-based technology for removing As from water creates an As-concentrated volume as a co-product, and safe disposal methods need to be evaluated. Examples are reverse osmosis (RO) or nano-, ultra- and microfiltration (Shih 2005; Figoli *et al.* 2016). The RO and dense NF membranes can remove sufficient As to meet drinking water standards (Schmidt *et al.* 2016; Figoli *et al.* 2020). UF and MF treatments need a coagulation/flocculation process upfront to remove As, due to the membranes having a course porosity, which cannot retain dissolved As (Ćurko *et al.* 2011; Elcik *et al.* 2013).

Renewable energy systems, especially PV panels, are mostly needed on the numerous Vietnamese islands without stable grid connections. Due to a generous feed-in tariff of US$93.5/MWh for new projects, together with supporting policies such as tax exemptions, Vietnam experienced a solar PV installation boom, also on the main land in the first half of 2019, with installed capacity increasing to 4,450 MW (Do *et al.* 2020). Underlying drivers include the government's desire to enhance energy self-sufficiency and the public's demand for local environmental quality.

MATERIALS AND METHODS

Pilot plant

The modular pilot plant for the desalination of brackish water including the subsurface removal of Fe^{2+}, Mn^{2+} and As was installed in June 2018 on a household rooftop in Trà Vinh city, in a coastal region in the Mekong Delta, Vietnam. The process scheme of the system consisting of an MCDI (CapDI, Voltea®) plant with upfront SAR (*FERMA-NOX®-Wasseraufbereitung* type BV 104) for iron and As removal is shown in Figure 1. Two different product streams are produced from the modular system: As and Fe^{2+} free treated water from SAR is stored in T1 (e.g. for washing purposes), while drinking water from the MCDI is stored in T2.

(1) *SAR*: A submersible pump (PENTAX® 4ST 14-8) is used to pump the groundwater (water table at 2 m below the surface) through a sand filter (1354 FRP Vessel, sand

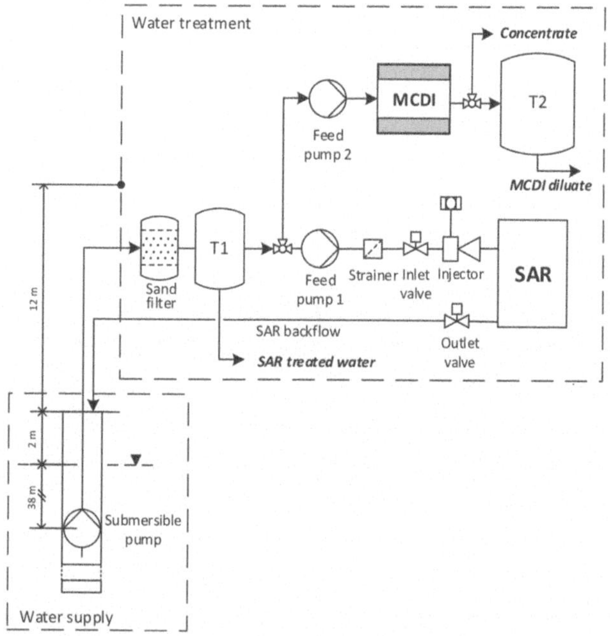

Figure 1 | Scheme of the modular pilot plant including *water supply* with submersible feed pump and *water treatment* consisting of SAR and MCDI with pre-filter (sand filter) and storage tanks T1 (SAR-treated water) and T2 (MCDI diluate).

0.5–1 mm as filtration media) to prevent particles from passing into a storage tank on the rooftop (T1, 12 m above the surface). The treatment process is divided into the aeration and infiltration cycles: firstly, water from T1 is aerated with an air injector using a booster pump (feed pump 1; GRUNDFOS® SCALA2) and stored in the SAR tank (900 L). Then, the oxygen-enriched water is infiltrated back into the aquifer by gravity (SAR backflow). Iron oxidation and As adsorption take place underground, and water with very low, nontoxic concentrations of As is obtained around the well. Subsequently, all extracted water from the well, which is stored in T1, can be used by the household or further treated with MCDI.

Aeration and infiltration cycles must be repeated constantly to maintain the oxygenated zone around the well. When water is extracted from the well, oxidation–adsorption takes place in the oxidation zone and the infiltrated oxygen is gradually consumed. In order to avoid leakage of As or any other contaminant, a daily water extraction limit V_E is calculated as a design parameter based on the previous experience of the manufacturer and fixed for using SAR (Luong *et al.* 2018, 2019; Cañas Kurz *et al.* 2020). With a daily operation of two infiltration cycles (each cycle $V_I = 900$ L), the maximum designed capacity of the SAR plant is $V_E = 8.3\ m^3/d$. This leads to an infiltration/extraction ratio at the pilot site of $Q_E = V_I/V_E = 0.22\ m^3/m^3$. Since the oxygenated water is always infiltrated back into the aquifer, another advantage of the process is that no water is lost and no waste stream is produced.

(2) *MCDI desalination*: Secondly, a home-made MCDI unit with a commercial electrode module (Voltea B.V® Model C3) is used for the desalination of the groundwater to obtain drinking water (see Figure 2). The information provided by the manufacturer is the total surface area of the electrodes at 3.7 m^2 as well as sulfonyl and quaternary ammonium groups as active functionality groups for the cation and anion exchange membranes, respectively. By using a diaphragm pump (feed pump 2; SHURflo® 8000-243-155), SAR-treated water from T1 is passed through the MCDI module in alternate cycles for desalination (charge) and regeneration of the electrodes (discharge). The MCDI module is operated with constant current (CC) during both phases with reversed voltage in the discharge phase using an XP power supply (XP Power HDS 1500 PS12, with a maximum electrical current at

Figure 2 | MCDI pilot plant for desalination including the electrode module (Voltea C3) and power supply (XP Power HDS 1500 PS12).

the outlet $I_{max} = 100$ A). To determine the TDS removal, the electrical conductivity (*EC*) of the water stream was measured at the inlet and the outlet of the MCDI module with conductivity sensors from Supmea (SPE-EC100). The *EC* in μS/cm was then converted to a TDS concentration in mg/L by multiplying the *EC* value with the factor $K = 0.5$ (Walton 1989). This factor was confirmed experimentally to fit to the concentrations used in these studies by comparing the *EC* values of solutions with known TDS concentrations up to 1.65 g/L. The diluate of the MCDI is stored in T2, which is connected to the water supply, whereas the concentrate effluent is discharged directly into the sewer.

The pilot plant is operated by renewable energy (see Figure 3). A 2 kW_p WT (Hummer, 2 kW-Windspot 220 V) is controlled by an on-grid inverter, and a 3 kW_p solar PV system (SUNERGY, SUN72M, monocrystalline) is connected with a hybrid inverter (Goodwee, GW-3648) that includes two lithium batteries (Scared sun, 2,400 Wh 50 Ah) for a total battery storage capacity of 4.8 kWh. The pilot plant is supplied directly by PV power. Excess energy is fed into the regional power grid; if energy is required, for example at night, the plant is powered additionally by the battery storage. If the battery state of charge (SOC) is lower than 20%, the plant can be supplied through the grid. A separate net-metering system controls the in-going and out-going energy of the latter. Additional power generation from the WT is directly fed into the grid.

Furthermore, a home-made weather station including anemometer (Davies, 6410) and pyranometer (Apogee Instr., SP-215) is used for monitoring wind direction, wind speed, ambient temperature and solar radiation at the pilot site.

Laboratory tests

Laboratory tests were carried out in order to compare the performance of two MCDI modules with different setups. The described MCDI module of the pilot plant-treating groundwater in Trà Vinh is hereinafter defined as setup 1. A second MCDI module consisting of the development kit (DK) of Voltea B.V.® (maximum electrical current at the outlet $I_{max} = 60$ A) using the same type of electrode module (C3 model) and the automation and control system is used to desalinate NaCl (>99.8%) diluted with deionized water ($EC = 3.29$ mS/cm) under laboratory conditions. The DK includes conductivity sensors at the inlet and outlet as well as a pressure sensor to constantly record the same parameters as the pilot plant. This installation is subsequently defined as setup 2.

The focus in evaluating the performance of each setup is its *SEC*, which is defined by the used energy divided by the produced volume of the diluate (V_d). Two different types are therefore defined. One SEC_{CDI} for desalinating the water in the MCDI module to separate the salt ions from the water, taking into account the electrical power input measured at the outlet of the according power supply (P_{CDI}) and (t),

Figure 3 | PV system and WT at the pilot plant location in Trà Vinh, Mekong Delta, Vietnam.

and another $SEC_{CDI,sys}$ for the system, which includes the performance of the diaphragm pump (P_p). They are calculated as stated in the following equations:

$$SEC_{CDI} = \frac{P_{CDI} \cdot t}{V_d} \tag{1}$$

$$SEC_{CDI,sys} = \frac{(P_{CDI} + P_p) \cdot t}{V_d} . \tag{2}$$

To compare the performance of the MCDI modules of desalinating water, the charge efficiency η_{ch} can be used. Hereby, the electrical current I_{ch} during charge time is compared to the amount of adsorbed ions in the electrodes as stated in the following equation:

$$\eta_{ch} = \frac{\Delta m / M}{Q / F} \tag{3}$$

where Δm is the removed amount of total dissolved solids (TDS) in g, M the molar mass, Q the electrical charge and F the Faraday constant. For all calculations, the molar mass of NaCl ($M = 58.5$ g/mol) is assumed.

Sample analysis

The raw water samples were taken directly from the pipe behind the submersible pump after water was extracted for some time. Subsequently, the samples were acidified for conservation as prescribed by the standard DIN EN ISO 5667-3.

Values for pH and EC were measured on site using a portable sensor (WTW Multi-Parameter 3430). Total Fe and Mn were measured with the flame technique using atomic adsorption spectrometry (AAS, Analytik Jena, contrAA® 300) and a 50 mm burner with an air/acetylene gas mixture. To measure total As (t-As) concentrations, AAS together with the hydride generation (HS 55 batch system) was used (Analytik Jena AG 2019). Ion chromatography (Metrohm, 883 Basic IC plus) was used to analyze ions. The concentration of cations was determined with a Metrosep C4-150/4.0 column according to DIN EN ISO 14911, and the concentrations of anions with a Metrosep A Supp 5-150/4.0 column. Total organic carbon (TOC) was analyzed with a Shimadzu TOC-L Analyzer. All measurements were repeated at least twice unless specified. Additionally, samples were analyzed by the certified commercial lab QUATEST3 in Ho Chi Minh City, Vietnam.

RESULTS AND DISCUSSION

Raw water quality at the pilot site

The chemical and physical properties of the raw groundwater in comparison to the Vietnamese standard for drinking water quality are summarized in Table 1. Analysis was carried out at different times before commissioning. In addition to initial elevated concentrations of Fe^{2+} ($c_{max} = 3.1$ mg/L) and t-As ($c_{max} = 11$ µg/L), the groundwater at the pilot site has an increased salinization with TDS concentrations of $c > 1.5$ g/L. In order to assess the groundwater quality at the beginning of the pilot trials, relevant water parameters were monitored over a period of 1 month directly before commissioning. During this period, an average of 15.5 m^3 of water was extracted on a daily basis, while no infiltration was carried out. This volume of water is greater than the extraction limit of 8.3 m^3/d. Measurements showed a slow increase in Fe^{2+} and t-As concentrations during this month of sampling until the start of the SAR operation, which indicates that the extraction of large amounts of groundwater without treatment has an influence on the concentrations given hydrogeologically. However, the averaged concentrations of Fe^{2+} and Mn^{2+} were similar compared with the values from March 2018, and the averaged t-As level decreased slightly, remaining below the WHO guideline value of $c = 10$ µg/L, which also indicates that the rise in the contaminant levels during the measurement period was within the naturally occurring fluctuations. The value for the t-As concentration from the pre-commissioning phase differed from the initial measured value of 11 µg/L (see Table 1) 1 year before and can also be attributed to naturally occurring fluctuations.

Iron, As and Mn removal (SAR)

Results for the first 181 days of operation of the SAR plant for t-As, Fe^{2+} and Mn^{2+} removal are shown in Figure 4.

Table 1 | Analysis of raw groundwater at the pilot site in Trà Vinh

Parameter	Unit	Sampling[a] 2 August 2017	Sampling[b] 2 August 2017	Sampling[a] 7 March 2018	Pre-commissioning[a,c]		Vietnamese standard[d]
Ph	–	7.2	NM	6.9	7.1	±0.3	6.5–8.5
EC	mS/cm	3.3	NM	3.3	3.3	±0.3	NA
t-As	μg/L	10	11	2.6	2.3	±1	10
Fe^{2+}	mg/L	1.7	3.1	1.8	1.8	±0.7	0.3
Mn^{2+}	mg/L	0.66	0.69	0.26	0.25	±0.05	0.3
F^-	mg/L	NM	NM	NM	0.92	±0.003	1.5
Cl^-	mg/L	886	827	955	1,035	±7	250
NO_3^-	mg/L	11	NM	NM	8.1	±5	50
SO_4^{2-}	mg/L	95.2	97.8	138	101	±1	250
PO_4^{3-}	mg/L	BDL	BDL	BDL	BDL	–	NA
Na^+	mg/L	343	338	374	379	±22	200
K^+	mg/L	12	12	10	13	±1	NA
Ca^{2+}	mg/L	170	178	NM	204	±21	120[e]
Mg^{2+}	mg/L	111	111	110	120	±8	NA
NH_4^+	mg/L	2.1	2.4	2.1	3.1	±0.3	3
HCO_3^-	mg/L	NM	437	NM	NM	–	NA
TOC	mg/L	16.2	NM	NM	4.2	±2	NA

[a]Analyzed in own laboratories (see the section Materials and methods).
[b]Certified commercial lab: QUATEST3, HCMC, Vietnam.
[c]Averaged value from samples taken before commissioning between 25 May 2018 and 28 June 2018 (number of samples = 8).
[d]QCVN 01 : 2009/BYT.
[e]Calculated from hardness limit $CaCO_3$ = 300 mg/L.
NM: Not measured; BDL: below detection limit; NA: not applicable.

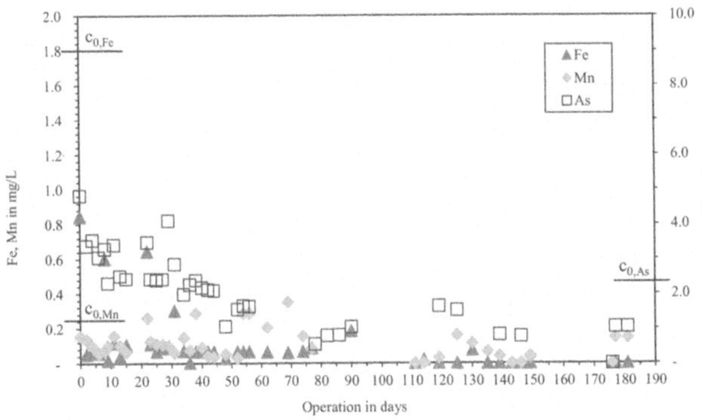

Figure 4 | Iron, As and manganese removal with SAR treatment at the pilot site in Trà Vinh.

The averaged background concentrations measured during 34 days before commissioning (see Table 1) are considered as the initial concentrations for Fe^{2+}, Mn^{2+} and t-As, which are indicated at the *y*-axes as $c_{0,Fe}$, $c_{0,Mn}$ and $c_{0,As}$, respectively. During this time, the t-As and Fe^{2+} concentrations increased to $c_{As} > 4.8$ μg/L and $c_{Fe} > 2.1$ mg/L due to natural fluctuations and the extraction of unusually high amounts of water without infiltration cycles. The Mn concentration remained at $c_{Mn} > 0.24$ mg/L. Thus, the t-As concentration at the first day of SAR operation is $c_{As} > c_{0,As}$, also indicating the slower removal of As over Fe. Measurements in Table 1 also show that the t-As concentrations can be above 10 μg/L on this site.

After the start of operation, the concentration of As was continuously lowered following the removal of Fe^{2+} until the value 0.55 μg/L was reached on day 78, corresponding to elimination rates of 76%. The highest increase in t–As measured from day 91 to 120 from 1.0 to 1.6 μg/L follows the rise in Fe due to the technical failure of the pump outlined above. The t-As mobilization decreases immediately when normal operation continues from day 120. The last measured value of t-As indicated a concentration of 1.1 μg/L and thus a removal of 54%. These two abovementioned removal rates for t-As are acceptable for an already significantly low background concentration of $c_{0,As} = 2.3$ μg/L. In earlier SAR experiments in South Vietnam, Cañas Kurz *et al.* (2020) showed with the same technology for high background concentrations of $c_{As} = 80$ μg/L a removal to $c_{As} < 5$ μg/L.

After initiating the first infiltration cycle, results show an immediate strong decrease in Fe concentrations within the first days of operation. Values as low as $c = 0.1$ mg/L for Fe^{2+} were measured after day 3 until constant values were achieved after day 37. Manganese concentration dropped linearly to below $c = 0.06$ mg/L within the first 7 days of operation, indicating a removal of 75%. However, Fe concentrations spiked irregularly within the first 30 days and increases in Mn^{2+} were observed throughout the 70 days of operation. Subsequently, concentrations do not exceed the initial value of $c_{0,Mn}$. Increases in Fe^{2+} and Mn^{2+} concentrations can be explained by minor technical problems during pilot operation including the incomplete infiltration cycles caused by malfunction of the submersible pump, over-extraction (abstraction above the plant capacity of 8.3 m^3), and the failure of the water meter. These technical complications did not stop the process completely but, by not maintaining a correct extraction/infiltration ratio, interrupted the proper operation of the infiltration and abstraction cycles temporarily, which causes the concentration peaks (Luong *et al.* 2019).

In general, the results show a slow but continuous removal for all three species (Fe^{2+}, As and Mn^{2+}), and after 181 days of operation, no breakthrough above the limits for drinking water standard was observed.

In addition, ammonium (NH_4^+) was lowered from initial concentrations of 3.1 mg/L to an average of 1.1 mg/L corresponding to a removal of 65%.

Desalination results (MCDI)

To determine the ideal operational parameters for the MCDI plant, laboratory tests with synthetic water (setup 2) have been carried out. Subsequently, these parameters were adjusted for the operation at the pilot site with setup 1. The comparison between lab and pilot scale is discussed below.

The optimum process parameters for the different MCDI setups employed are shown in Table 2. The general goal is to achieve a diluate concentration of <450 mg $_{NaCl}$/L or <600 mg $_{TDS}$/L (WHO guidelines for 'good' drinking water; WHO 2017). Different operational parameters are therefore chosen for the different setups, such as the electrical current for CC charge and the adsorption times in addition to the discharge current and discharge phase times, which are adapted accordingly. To obtain a reasonable water recovery rate, the discharge phase times have to be set as short as possible. Thus, the CC in the discharge phase is set always higher than the CC during the charge phase. The TDS concentrations of the feed are set depending on the quality of the groundwater ($EC = 3.3$ mS/cm, converted with the K factor, see the section Materials and methods) and are averaged over the time of the experiment ($t_{tot,av} = 64$ min) as well as the diluate salt concentration. Best results with setup 1 are obtained at a slightly lower feed $EC = 3.12$ mS/cm than the overall seasonal average at the pilot site. The test with setup 2 in the laboratory showed a lower $SEC_{CDI,2} = 0.97$ kWh/m^3 (see Equation

Table 2 | Characteristics of the two MCDI plant setups with optimal operation parameters

Setup number	Units	1	2
Feed EC	mS/cm	3.12	3.29
Feed TDS concentration	mg/L	1,560	1,645
Diluate TDS concentration	mg/L	188	410
TDS reduction	mg/L	1,372	1,235
Energy use in one cycle	kJ	4.7	10.2
Product rate	L/d	477	400
SEC of the desalination module	kWh/m^3	1.75	0.97
Charge efficiency	–	75%	77%
Water recovery	–	33%	65%
TDS removal	–	88%	75%
Constant current (charge phase)	A	50	12.3
Constant current (discharge phase)	A	66	54.1
Voltage limit	V	1.3	1.3
Water flow (charge phase)	L/min	1	0.4
Phase time (charge phase)	s	45	440
Cycle time	s	135	630

(1)). After the transition from the laboratory to the pilot site, a higher $SEC_{CDI,1} = 1.75\ kWh/m^3$ has been achieved, treating a complex salt matrix with setup 1 and thus bivalent cations as well, which can reduce Na^+ and Cl^- ions adsorption (Suss 2017). The corresponding energy values were measured at the outlet of the power supply for the MCDI module, giving the values for the actual desalination performance, without considering the further periphery. The component within the home-made experimental setup with the highest energy consumption is the power supply, which has very poor energy efficiency as low as 20%, and is thus the component with the highest potential for improvement. This is achieved in large commercial modules by a design that allows a more efficient conversion of an alternate current (AC) grid with the voltage of 230 V in a low ampere range to a direct current circuit with a voltage of <2 V and a high ampere range. Therefore, the energy loss of the power supply in the experimental units (setup 1 and setup 2) is not considered in the following (see Equation (2)).

The $SEC_{CDI,sys}$ can be calculated using Equation (2) to $SEC_{CDI,sys,1} = 2.20\ kWh/m^3$ for setup 1 and $SEC_{CDI,sys,2} = 1.08\ kWh/m^3$ for setup 2. The performance of the diaphragm pump ($P_{p,1} = 8.9$ W and $P_{p,2} = 1.8$ W, respectively), which is proportional to the pressure loss over the MCDI module, is then added to the power consumption for the desalination. It should be noted that MCDI can also be operated by gravitational flow in combination with a flow regulating valve, and consequently, the diaphragm pump can be omitted. The comparatively low water recovery in setup 1 of 33% shows that the size of the module should be increased to handle the water under pilot conditions. Thus, a higher water recovery and hence a lower *SEC* value could be achieved.

The desalination with MCDI setup 1 on the pilot site with operational parameters of Table 2 started after the pre-treatment with SAR was running stably on day 112 (see Figure 4). The late application of MCDI also ensures that the Fe^{2+} concentration in groundwater is in compliance with the manufacturer's feed water recommendations of $Fe^{2+} < 0.5$ mg/L (Voltea 2016). Although the desalination efficiency of MCDI can be mainly accessed by the overall salt removal (see Table 2), the adsorption efficiency for selected ions was taken within 12 weeks to evaluate the overall ion removal especially for Fe^{2+} and t-As.

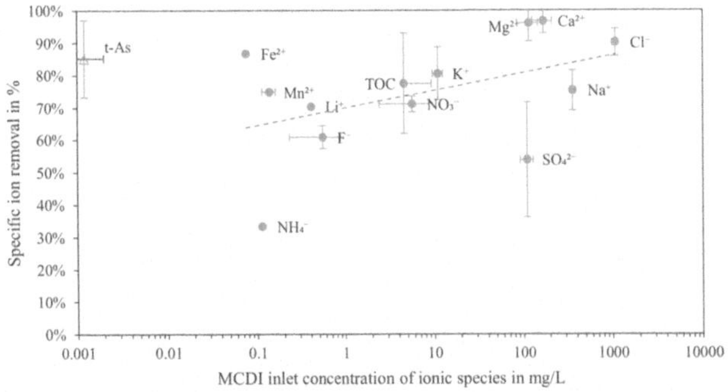

Figure 5 | MCDI averaged specific removal of ionic species and TOC at different initial concentrations. Values averaged from inlet and outlet concentrations. If more than one measurement, horizontal and vertical bars show the standard deviation of the averaged values. *X*-axis: log-scale. Dotted line: logarithmic trend excluding t-As.

Figure 5 shows the elimination rates for ionic species at their given averaged inlet concentrations at the MCDI module (after SAR treatment, Figure 4) in between days 112 and 181. The samples were taken from the inlet and the diluate stream of the MCDI. The electrosorption of the bivalent Fe^{2+} ions was very high, at 87%. The t-As concentration after the SAR treatment of averaged 1.2 μg/L was lowered to below 0.18 μg/L through MCDI, indicating another removal of 85%.

The results also showed specific ion removals of 75% Na^+, 90% Cl^-, and up to 97% for Ca^{2+} and Mg^{2+}, indicating a stronger selective adsorption of bivalent cations as shown in literature results (Suss 2017). However, lower sulfate (SO_4^{2-}) mitigation was observed with only 54% removal, which can be explained by the increasing hydrated radius and decreasing permeability of the bivalent ions Ca^{2+}, Mg^{2+} and SO_4^{2-}, respectively (Tansel *et al.* 2006; Li *et al.* 2016).

Renewable energy supply

Measurements of solar radiation at the pilot plant location and the results of the PV panels show an average daily production of 11.2 ± 1.3 kWh/d (the upper line in gray) for the measured days between May 2018 and December 2019 with an average radiation of 338 ± 53 W/m^2 (the lower line in blue) and maximum radiation values between 990 and 1215 W/m^2 (Figure 6). The average WT power production was 3.4 ± 0.3 kWh/d. The average wind speed measured at

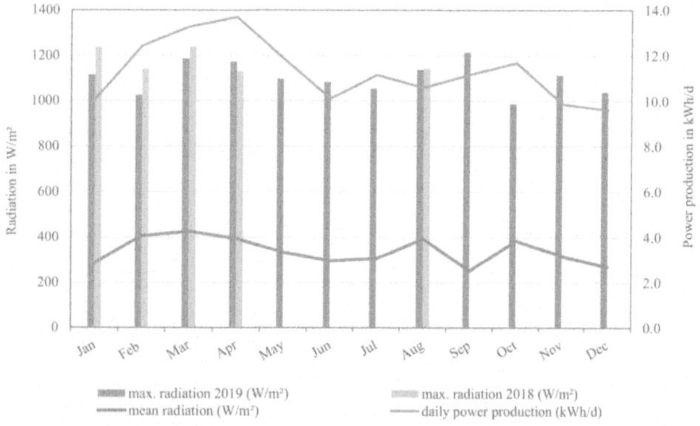

Figure 6 | Monthly solar radiation and solar energy supply at the pilot location averaged from measurements from May 2018 to December 2019.

the pilot location was 4.0 ± 1.3 m/s. However, the WT could not be operated continuously due to noise complaints from neighbors.

The power generation and power consumption at the pilot site for four different scenarios is shown in Figure 7(a)–7(d). Figure 7(a) shows the power production of the PV panels (blue line) on a typical clear-sky, sunny day with no power consumption and a peak performance of $P_{peak} = 2060$ W. The yellow line with the same absolute

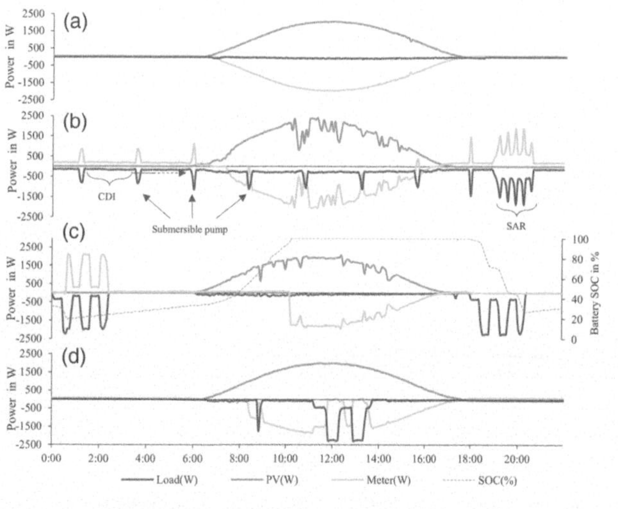

Figure 7 | Solar powered energy supply of the pilot plant for different scenarios: (a) solar power generation (PV, blue line) without consumption (purple line) and feed onto grid (yellow line); (b) typical daily course of load (purple) including MCDI (24 h), SAR (op. at 18:00) and submersible pump (intermittent peaks); (c) semi-autonomous supply using battery (orange line) and grid for SAR operation (two cycles at night 18:00 and 00:00, respectively); (d) close to autonomous supply with PV for SAR (one cycle, 11:30). (Please refer to the online version of this paper to see this figure in colour: doi:10.2166/wrd.2020.031).

values as the power production shows the feeding to the grid (net-metering). Figure 7(b) shows the PV power production and the supply of the pilot plant. The purple line shows the load of the pilot plant at a typical daily operation including the desalination with MCDI (24 h operation) with averaged $P_{CDI,tot} = 176$ W for the MCDI module and its periphery, SAR (abstraction and oxygenation cycle with the SCALA 2 pump, beginning at 18:00) and the submersible pump for feeding T1 during SAR operation with averaged $P_{SAR} = 1{,}732$ W. In this example, the load is supplied by the grid during night time (yellow line: positive). When solar power production starts at sunrise (06:00), the load is supplied by PV. The addition of the power production (blue line, positive) and the load (purple line, negative) equals the amount of energy supplied by or fed into the grid (yellow line, positive or negative, respectively).

Figure 7(c) shows the semi-autonomous supply for the SAR when using both battery (orange line) and grid for two infiltration cycles during the night at 18:00 and 00:00, respectively. Following the discharge of the battery (orange line) after the first SAR cycle (18:00), the second SAR cycle (00:00) is supplied directly from the grid (yellow line). Battery recharge for the next cycle is then powered by PV, and SOC is completed (SOC = 100%) around 10:15. After this, the energy produced is fed into the grid (net-metering: yellow line decreasing) following the absolute values of the power production (blue line). Figure 7(d) shows the autonomous supply with PV for SAR operation without MCDI (one cycle, 11:30).

To use the solar power production more efficiently, the operation of SAR is shifted during day time as shown in Figure 7(d), demonstrating that an autonomous (off-grid) operation is feasible. By doing this, the consumption of energy from the grid can be significantly reduced by 3.65 kWh, which is the average energy of one full infiltration cycle, and thus the degree of autonomy is increased. A high degree of autonomy or even of grid solutions is worthwhile, especially for a possible operation on the numerous small Vietnamese islands.

In total, the average energy consumption for the operation of the SAR and MCDI module was 11.6 kWh/d. A degree of autonomy for the overall pilot plant of 97% could be achieved with a total energy supply of 11.2 kWh with

Table 3 | Summary of all *SEC* values in kWh/m^3 to evaluate the pilot plant

SEC_{SAR}	$SEC_{SAR,supply}$	$SEC_{SAR,tot}$	SEC_{CDI}	$SEC_{CDI,sys}$	$SEC_{pilot,tot}$
0.88	0.89	1.77	1.75	2.20	3.97

Note: Energy values of the following processes were taken into account for *SEC* values with respective indices: *SAR*: SAR treatment; SAR,supply: water supply from subsurface to household; SAR,tot: *SAR* plus SAR.supply; CDI: desalination only and no periphery; CDI,sys: desalination plus pump for the CDI system; pilot,tot: SAR,tot plus CDI,sys.

PV. The energy deficit was supplied by the power grid. If the wind power of 3.4 ± 0.3 kWh/d at an average wind speed of 4.0 ± 1.3 m/s can be used, the degree of autonomy increases to 126% and thus 3.0 kWh/d of electrical energy can be fed into the grid, which provides governmental financial benefit (see the section State of the art), or used for a different purpose.

It should be noted that the *SEC* of the total SAR process consists of an *SEC* for the subsurface treatment (SEC_{SAR}) and an *SEC* for supply of the product water ($SEC_{SAR,supply}$) (see Table 3). On the pilot site, the water is supplied by the PENTAX well pump to the rooftop of the house via an elevation of $H = 14$ m, with an average performance of $P_{wp} = 1{,}332$ W and a maximum water flow of $\dot{V} = 1{,}500$ L/h. Hence, the *SEC* value for supply is $SEC_{SAR,supply} = 0.89$ kWh/m^3. This value is highly dependent on construction and environmental conditions. The advantage of supplying the water on the rooftop is that no additional auxiliary pumps are needed to maintain the necessary pressure to provide treated water to the complete household, which is standard in Vietnamese households due to nonconstant pressure conditions in the public water grid.

The SAR module treated 8.3 m^3 of water, which is potentially available subsurface. However, at the pilot site, on average 3.64 m^3 of water were supplied to the user. The MCDI module produced 0.48 m^3 of drinking water per day. Thus, the *SEC* values equal to $SEC_{SAR} = 0.88$ kWh/m^3 ($SEC_{SAR,supply} = 0.89$ kWh/m^3), $SEC_{CDI} = 1.75$ kWh/m^3 ($SEC_{CDI,sys} = 2.20$ kWh/m^3) and hence the overall $SEC_{pilot,tot} = 3.97$ kWh/m^3.

The two treatment stages produce water in different qualities and amounts, which can be used for different purposes modularly. The water produced from the SAR treatment can be used for washing, for example, and the diluate of the MCDI for drinking.

CONCLUSION AND OUTLOOK

- SAR has proved a feasible *in situ* remediation technique for Fe, Mn and As in brackish anoxic aquifers in combination with MCDI desalination, resulting in a total $SEC_{pilot,tot} = 3.97$ kWh/m^3 of drinking water. Fe, Mn and As concentrations could be lowered below drinking regulation standards. The TDS removal by the MCDI of 1,372 mg/L to a concentration of $c = 188$ mg/L could be achieved.
- Finally, two water streams were produced, which can be modularly adjusted according to the local requirements. One treated by SAR with low Fe, Mn and As but high TDS, for washing purposes and another additionally treated by MCDI, for drinking purposes.
- The very low efficiency of the power supply for the MCDI module shows the improvement potential of the home-made desalination plant on the pilot site.
- Based on the previous experience, it is expected that the efficiency of the system can be increased by upscaling the capacity of the MCDI module.
- The supply with 3 kW$_p$ solar PV energy was feasible and a degree of autonomy of 97% for the water treatment was achieved, producing 0.48 m^3 drinking water and supplying 3.16 m^3 of washing water per day (out of the potential 8.3 m^3). Using a 2 kW$_p$ WT, a degree of autonomy of 126% can be realized. However, the use of the WT was difficult because the high noise level was not accepted by those living in the neighborhood.

ACKNOWLEDGEMENTS

This project received funding from the German Ministry of Education and Research under the grant agreement no. 02WAV1413.

DATA AVAILABILITY STATEMENT

All relevant data are included in the paper or its Supplementary Information.

REFERENCES

AlMarzooqi, F. A., Al Ghaferi, A. A., Saadat, I. & Hilal, N. 2014 Application of capacitive deionisation in water desalination: a review. *Desalination* **342**, 3–15. https://doi.org/10.1016/j.desal.2014.02.031.

Analytik Jena AG 2019 Operating Manual novAA 800 AAS, 05.19.ed. Documentation number: 10-1430-002-23. https://www.analytik-jena.com/fileadmin/content/products/novAA/Manual_novAA_800_en_0519.pdf.

Annaduzzaman, M., Bhattacharya, P., Biswas, A., Hossain, M., Ahmed, K. M. & van Halem, D. 2018 Arsenic and manganese in shallow tubewells: validation of platform color as a screening tool in Bangladesh. *Groundwater Sustainable Dev.* **6**, 181–188. https://doi.org/10.1016/j.gsd.2017.11.008.

Ayers, J. C., Goodbred, S., George, G., Fry, D., Benneyworth, L., Hornberger, G., Roy, K., Karim, M. R. & Akter, F. 2016 Sources of salinity and arsenic in groundwater in southwest Bangladesh. *Geochem. Trans.* **17**, 1–22. https://doi.org/10.1186/s12932-016-0036-6.

Berg, M., Tran, H. C., Nguyen, T. C., Pham, H. V., Schertenleib, R. & Giger, W. 2001 Arsenic contamination of groundwater and drinking water in Vietnam: a human health threat. *Environ. Sci. Technol.* **35**, 2621–2626. https://doi.org/10.1021/es010027y.

Brecht, H., Dasgupta, S., Laplante, B., Murray, S. & Wheeler, D. 2012 Sea-level rise and storm surges. *J. Environ. Dev.* **21**, 120–138. https://doi.org/10.1177/1070496511433601.

Buschmann, J., Berg, M., Stengel, C., Winkel, L., Sampson, M. L., Trang, P. T. K. & Viet, P. H. 2008 Contamination of drinking water resources in the Mekong delta floodplains: arsenic and other trace metals pose serious health risks to population. *Environ. Int.* **34**, 756–764. https://doi.org/10.1016/j.envint.2007.12.025.

Cañas Kurz, E. E., Luong, V. T., Hellriegel, U., Leidinger, F., Luu, T. L., Bundschuh, J. & Hoinkis, J. 2020 Iron-based subsurface arsenic removal (SAR): results of a long-term pilot-scale test in Vietnam. *Water Res.* **181**, 115929. https://doi.org/10.1016/j.watres.2020.115929.

Colombani, N., Osti, A., Volta, G. & Mastrocicco, M. 2016 Impact of climate change on salinization of coastal water resources. *Water Resour. Manage.* **30**, 2483–2496. https://doi.org/10.1007/s11269-016-1292-z.

Ćurko, J., Mijatović, I., Matošić, M., Jakopović, H. K. & Bošnjak, M. U. 2011 As(V) removal from drinking water by coagulation and filtration through immersed membrane. *Desalination* **279**, 404–408. https://doi.org/10.1016/j.desal.2011.06.043.

Dasgupta, S., Akhter, F., Zahirul, K., Khan, Z., Choudhury, S. & Nishat, A. 2014a *River Salinity and Climate Change Evidence From Coastal Bangladesh (No. 6817).*

Dasgupta, S., Hossain, M. M., Huq, M. & Wheeler, D. 2014b *Climate Change, Groundwater Salinization and Road Maintenance Costs in Coastal Bangladesh (No. 7147).*

Datta, S., Johannesson, K., Mladenov, N., Sankar, M., Ford, S., Vega, M., Neal, A., Kibria, M., Krehel, A. & Hettiarachchi, G. 2014 Groundwater-sediment sorption mechanisms and role of organic matter in controlling arsenic release into aquifer sediments of Murshidabad area (Bengal basin), India: Proceedings of the 5th International Congress on Arsenic in the Environment. In: *One Century of the Discovery of Arsenicosis in Latin America (1914–2014) As 2014*, pp. 95–97. https://doi.org/10.1201/b16767-37

Do, T. N., Burke, P. J., Baldwin, K. G. H. & Nguyen, C. T. 2020 Underlying drivers and barriers for solar photovoltaics diffusion: the case of Vietnam. *Energy Policy* **144**, 111561. https://doi.org/10.1016/j.enpol.2020.111561.

Elcik, H., Cakmakci, M., Sahinkaya, E. & Ozkaya, B. 2013 Arsenic removal from drinking water using low pressure membranes. *Ind. Eng. Chem. Res.* **52**, 9958–9964. https://doi.org/10.1021/ie401393p.

Figoli, A., Hoinkis, J. & Bundschuh, J. (Eds.) 2016 *Membrane Technologies for Water Treatment: Removal of Toxic Trace Elements with Emphasis on Arsenic, Fluoride and Uranium, Sustainable Water Developments.* CRC Press.

Figoli, A., Fuoco, I., Apollaro, C., Chabane, M., Mancuso, R., Gabriele, B., De Rosa, R., Vespasiano, G., Barca, D. & Criscuoli, A. 2020 Arsenic-contaminated groundwaters remediation by nanofiltration. *Sep. Purif. Technol.* **238**. https://doi.org/10.1016/j.seppur.2019.116461

Fritz, M., Hohmann, C. & Tettenborn, F. 2020 Framework conditions to design sustainable business models for decentralised water treatment technologies in Viet Nam for international technology providers. *J. Water Reuse Desalin* **10** (4), 317–331. https://doi.org/10.2166/wrd.2020.016.

Hien, L. T., Quy, P. N. & Viet, N. T. 2009 *Assessment of Salinity Intrusion in the Red River Under Effect of Climate Change.* ICEC.

Hossain, M., Williams, P. N., Mestrot, A., Norton, G. J., Deacon, C. M. & Meharg, A. A. 2012 Spatial heterogeneity and kinetic regulation of arsenic dynamics in mangrove sediments: The Sundarbans, Bangladesh. *Environ. Sci. Technol.* **46**, 8645–8652. https://doi.org/10.1021/es301328r.

Kundu, D. K., Gupta, A., Mol, A. P. J., Rahman, M. M. & van Halem, D. 2018 Experimenting with a novel technology for provision of safe drinking water in rural Bangladesh: the case of sub-surface arsenic removal (SAR). *Technol. Soc.* **53**, 161–172. https://doi.org/10.1016/j.techsoc.2018.01.010.

Le Luu, T. 2019 Remarks on the current quality of groundwater in Vietnam. *Environ. Sci. Pollut. Res.* https://doi.org/10.1007/s11356-017-9631-z

Li, Y., Zhang, C., Jiang, Y., Wang, T. J. & Wang, H. 2016 Effects of the hydration ratio on the electrosorption selectivity of ions during capacitive deionization. *Desalination* **399**, 171–177. https://doi.org/10.1016/j.desal.2016.09.011.

Luong, V. T., Cañas Kurz, E. E., Hellriegel, U., Luu, T. L., Hoinkis, J. & Bundschuh, J. 2018 Iron-based subsurface arsenic removal technologies by aeration: a review of the current state and future prospects. *Water Res.* **133**, 110–122. https://doi.org/10.1016/j.watres.2018.01.007.

Luong, V. T., Cañas Kurz, E. E., Hellriegel, U., Luu, T. L., Hoinkis, J. & Bundschuh, J. 2019 Iron-based subsurface arsenic removal by aeration (SAR) – results of a pilot-scale plant in Vietnam. In: *Environmental Arsenic in A Changing World As 2018* (J. Bundschuh & P. Bhattacharya, eds). CRC Press/Balkema, London, UK, pp. 418–419.

Ngo-Duc, T. 2014 Climate change in the coastal regions of Vietnam. In: *Coastal Disasters and Climate Change in Vietnam*. Elsevier, pp. 175–198. https://doi.org/10.1016/B978-0-12-800007-6.00008-3

Nickson, R. T., Mcarthur, J. M., Ravenscroft, P., Burgess, W. G. & Ahmed, K. M. 2000 Mechanism of arsenic release to groundwater, Bangladesh and West Bengal. *Appl. Geochem.* **15**, 403–413. https://doi.org/10.1016/S0883-2927(99)00086-4.

Rahman, M. M., Bakker, M., Freitas, S. C. B., van Halem, D., van Breukelen, B. M., Ahmed, K. M. & Badruzzaman, A. B. M. 2014 Exploratory experiments to determine the effect of alternative operations on the efficiency of subsurface arsenic removal in rural Bangladesh. *Hydrogeol. J.* **23**, 19–34. https://doi.org/10.1007/s10040-014-1179-0.

Rahman, M. M. M., Bakker, M., Patty, C. H. L. H. L., Hassan, Z., Röling, W. F. M. F. M., Ahmed, K. M. M. & van Breukelen, B. M. M. 2015 Reactive transport modeling of subsurface arsenic removal systems in rural Bangladesh. *Sci. Total Environ., Arsenic Environ.* **537**, 277–293. https://doi.org/10.1016/j.scitotenv.2015.07.140.

Rott, U. & Friedle, M. 2000 25 jahre unterirdische Wasseraufbereitung in Deutschland. *Wasser Spec.* **141**, 99–107.

Sankar, M. S., Vega, M. A., Defoe, P. P., Kibria, M. G., Ford, S., Telfeyan, K., Neal, A., Mohajerin, T. J., Hettiarachchi, G. M., Barua, S., Hobson, C., Johannesson, K. & Datta, S. 2014 Elevated arsenic and manganese in groundwaters of Murshidabad, West Bengal, India. *Sci. Total Environ.* **488–489**, 570–579. https://doi.org/10.1016/J.SCITOTENV.2014.02.077.

Sarkar, A. R. & Rahman, O. T. 2001 *In-situ Removal of Arsenic – Experiences of DPHE-Danida Pilot Project*. Available from: http://archive.unu.edu/env/Arsenic/Sarkar.pdf (accessed 14 January 2020).

Schmidt, S.-A., Gukelberger, E., Hermann, M., Fiedler, F., Großmann, B., Hoinkis, J., Ghosh, A., Chatterjee, D. & Bundschuh, J. 2016 Pilot study on arsenic removal from groundwater using a small-scale reverse osmosis system towards sustainable drinking water production. *J. Hazard. Mater.* **318**, 671–678. https://doi.org/10.1016/j.jhazmat.2016.06.005.

Sen Gupta, B. 2008 *Subterranean Arsenic Removal Technology for In Situ Groundwater Arsenic and Iron Treatment [WWW Document]*.

Shih, M.-C. 2005 An overview of arsenic removal by pressure-driven membrane processes. *Desalination* **172**, 85–97. https://doi.org/10.1016/j.desal.2004.07.031.

Smedley, P. L., Nicolli, H. B., Macdonald, D. M. J., Barros, A. J. & Tullio, J. O. 2002 Hydrogeochemistry of arsenic and other inorganic constituents in groundwaters from La Pampa, Argentina. *Appl. Geochem.* **17**, 259–284. https://doi.org/10.1016/S0883-2927(01)00082-8.

Stopelli, E., Duyen, V. T., Mai, T. T., Trang, P. T. K., Viet, P. H., Lightfoot, A., Kipfer, R., Schneider, M., Eiche, E., Kontny, A., Neumann, T., Glodowska, M., Patzner, M., Kappler, A., Kleindienst, S., Rathi, B., Cirpka, O., Bostick, B., Prommer, H., Winkel, L. H. E. & Berg, M. 2020 Spatial and temporal evolution of groundwater arsenic contamination in the Red River delta, Vietnam: interplay of mobilisation and retardation processes. *Sci. Total Environ.* **717**, 137143. https://doi.org/10.1016/j.scitotenv.2020.137143.

Suss, M. E. 2017 Size-based ion selectivity of micropore electric double layers in capacitive deionization electrodes. *J. Electrochem. Soc.* **164**, E270–E275. https://doi.org/10.1149/2.1201709jes.

Tan, C., He, C., Tang, W., Kovalsky, P., Fletcher, J. & Waite, T. D. 2018 Integration of photovoltaic energy supply with membrane capacitive deionization (MCDI) for salt removal from brackish waters. *Water Res.* **147**, 276–286. https://doi.org/10.1016/j.watres.2018.09.056.

Tansel, B., Sager, J., Rector, T., Garland, J., Strayer, R. F., Levine, L., Roberts, M., Hummerick, M. & Bauer, J. 2006 Significance of hydrated radius and hydration shells on ionic permeability during nanofiltration in dead end and cross flow modes. *Sep. Purif. Technol.* **51**, 40–47. https://doi.org/10.1016/j.seppur.2005.12.020.

van Halem, D. 2011 *Subsurface Iron and Arsenic Removal for Drinking Water Treatment in Bangladesh*. Water Management Academic Press.

van Halem, D., Heijman, S. G. J., Amy, G. L. & van Dijk, J. C. 2009 Subsurface arsenic removal for small-scale application in developing countries. *Desalination* **248**, 241–248. https://doi.org/10.1016/j.desal.2008.05.061.

Voltea 2016 *Technical Bulletin: Feedwater Guidelines*. Sassenheim. Available from: https://voltea.com/wp-content/uploads/2016/03/402D021_Rev02-Technical-Bulletin_Water-Quality-Criteria.pdf (accessed 08 August 2020).

Vu, D. T., Yamada, T. & Ishidaira, H. 2018 Assessing the impact of sea level rise due to climate change on seawater intrusion in Mekong Delta, Vietnam. *Water Sci. Technol.* **77**, 1632–1639. https://doi.org/10.2166/wst.2018.038.

WaKap 2016 *WaKap – Ein Projekt zur Wasseraufbereitung mittels Kapazitiver Deionisierung [WWW Document]*. Available from: www.wakap.de (accessed 30 August 2019).

Walton, N. R. G. 1989 Electrical conductivity and total dissolved solids – what is their precise relationship? *Desalination* **72**, 275–292. https://doi.org/10.1016/0011-9164(89)80012-8.

Wang, Y., Le Pape, P., Morin, G., Asta, M. P., King, G., Bártová, B., Suvorova, E., Frutschi, M., Ikogou, M., Pham, V. H. C., Vo, P. L., Herman, F., Charlet, L. & Bernier-Latmani, R. 2018 Arsenic speciation in Mekong delta sediments depends on their depositional environment. *Environ. Sci. Technol.* **52**, 3431–3439. https://doi.org/10.1021/acs.est.7b05177.

Wang, L., Dykstra, J. E. & Lin, S. 2019 Energy efficiency of capacitive deionization. *Environ. Sci. Technol.* **53**, 3366–3378. https://doi.org/10.1021/acs.est.8b04858.

Wassmann, R., Hien, N. X., Hoanh, C. T. & Tuong, T. P. 2004 Sea level rise affecting the Vietnamese Mekong Delta: water elevation in the flood season and implications for rice production. *Clim. Change* **66**, 89–107. https://doi.org/10.1023/B:CLIM.0000043144.69736.b7.

Wheeler, D. 2011 *Quantifying Vulnerability to Climate Change: Implications for Adaptation Assistance (No. 240).* Washington, DC.

WHO 2017 *Guidelines for Drinking-Water Quality: Fourth Edition Incorporating the First Addendum*, 4th edn. World Health Organization, Geneva, Switzerland.

Winkel, L. H. E., Trang, P. T. K., Lan, V. M., Stengel, C., Amini, M., Ha, N. T., Viet, P. H. & Berg, M. 2011 Arsenic pollution of groundwater in Vietnam exacerbated by deep aquifer exploitation for more than a century. *Proc. Natl. Acad. Sci.* **108**, 1246–1251. https://doi.org/10.1073/pnas.1011915108.

Zhao, Y., Wang, Y., Wang, R., Wu, Y., Xu, S. & Wang, J. 2013 Performance comparison and energy consumption analysis of capacitive deionization and membrane capacitive deionization processes. *Desalination* **324**, 127–133. https://doi.org/10.1016/j.desal.2013.06.009.

First received 3 April 2020; accepted in revised form 2 September 2020. Available online 5 October 2020

Waterminer – a regional spatio-temporal approach to water reuse management in mining areas in Vietnam

S. Greassidis, V. Trinh Quoc, K. Brömme and H. Stolpe

ABSTRACT

Mining is usually associated with draining of large quantities of water. On the other hand, mining operations themselves are water users for purposes like coal or ore washing, dust mitigation, recultivation, etc. The normally existing water surplus can be seen as a resource for public purposes instead of discharging the water to rivers and the sea. It is typical for the mining operation that these water amounts vary spatio-temporally according to the mining process. The R&D project WaterMiner investigates the example of the Hon Gai hard coal mining area in Ha Long, Vietnam, as to how and to what extent under the variable spatio-temporal conditions the mine water can cover the water demand in the mines themselves and contribute to the public water supply in the surrounding settlement area of Hon Gai. The spatio-temporal change of mine water drainage, mine water treatment and mine-internal water use and the potential mine-external water use volumes in the project region are investigated by a material flow model and visualized by Sankey diagrams, maps, and tables. Several options for delivery of mine water for mine-external water uses as a contribution to the public water supply are shown.

Key words | material flow analysis, mine water management, mine water treatment, mine water use, spatio-temporal, Vietnam

S. Greassidis (corresponding author)
V. Trinh Quoc
K. Brömme
H. Stolpe
Environmental Engineering and Ecology,
Ruhr University of Bochum,
Universitaetsstr. 150, 44801 Bochum,
Germany
E-mail: *sandra.greassidis@rub.de*

HIGHLIGHTS

- The combination of material flow analysis and a geoinformation system is a strong instrument for water reuse management, taking into account the spatial and temporal changes in mining.
- Treated mine water can be regarded as a complementary water resource for public purposes instead of discharging the water into rivers and the sea.

INTRODUCTION

Mining is usually associated with draining of large quantities of water, which up to now, have mostly been discharged unused. On the other hand, mining operations themselves often represent water users for purposes such as coal washing, dust mitigation, etc. Treated mine water can be regarded as a complementary water resource for public purposes instead of discharging the water into rivers and the sea. It is typical for the mining operation that these water amounts vary spatio-temporally depending on the mining process.

The joint R&D project WaterMiner investigates under different spatio-temporal situations how and to what extent the existing water demand in the mines and parts of the water demand in the surrounding urban area can be met using treated mine water. Over the course of stricter implementation of environmental legislation as well as transition from open pit to underground mining, there is a need for a comprehensive approach in mine water management.

doi: 10.2166/wrd.2020.045

The joint R&D project WaterMiner 'Spatio-temporally coordinated recycling and reuse of mining waste water exemplified by an urban mining area' (established in 2016, project coordination by EE + E Environmental Engineering + Ecology, Ruhr University, Bochum) was sponsored by the German Federal Ministry of Education and Research (BMBF).

The WaterMiner project consists of the following work packages: project coordination (EE + E); basic and system analysis, material flow analysis, water management system (EE + E); GIS for water infrastructure (Disy Informationssysteme Ltd, Karlsruhe); monitoring information system (ribeka Ltd, Bornheim); technical concept for water and sediment management (DGFZ, Dresden); economic concept (environmental economics, University of Koblenz-Landau); exemplary technical implementation of water and sediment management (DGFZ, Dreseden and LUG Engineering Ltd, Cottbus); project host and cooperation (VINACOMIN Vietnam National Coal and Mineral Industries Holding Corporation Ltd, Hanoi).

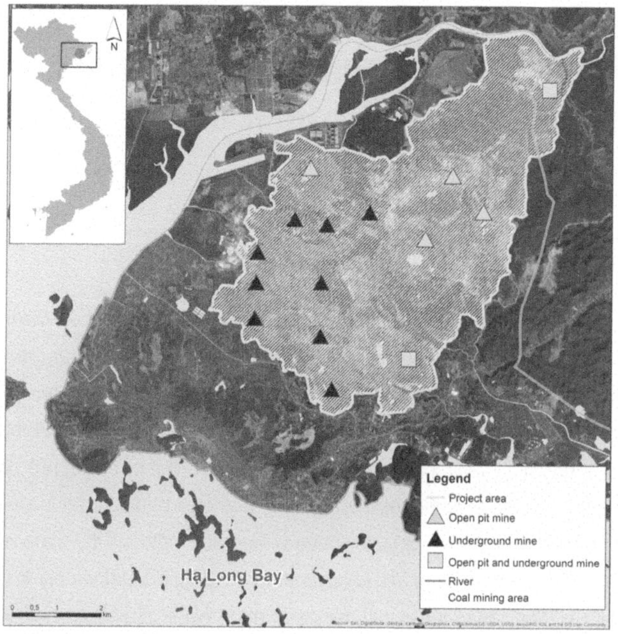

Figure 1 | Project area in Hon Gai.

PROJECT AREA: HON GAI

The Hon Gai hard coal mining area is located in Quang Ninh province in the northeast of Vietnam. The project area covers the eastern part of the city of Ha Long, the Hon Gai peninsula, with an area of 125 km^2 (see Figure 1).

The coal is mined in open pit mines and increasingly also in underground mines (VN PC Ha Long City 2014). The open pit mining will be continued until 2025, the underground mining beyond the year 2030 (GoV 2016). The spatio-temporal changes in the mining process and also increasing improvements in the treatment of mine water, lead to continuous changes in the location, quality, and quantity of mine water (VINACOMIN 2013).

Currently, 13 mines (open pit and underground mines) are operating, all belonging to the state-owned parent company Vietnam National Coal and Mineral Industries Holding Corporation Ltd (VINACOMIN). Within the mining area of Hon Gai, also mining infrastructure facilities exist such as coal processing plants, a coal-fired power plant, coal stores, mine water treatment plants, facilities for the mine workers like canteens, washing rooms, etc.

Basic investigations and the development of environmental protection concepts for the mining area including mine water treatment have been performed by the R&D project RAME (Research Association Mining and Environment) from 2005 to 2015 (Broemme *et al.* 2014; Broemme & Stolpe 2015).

The average annual rainfall in Hon Gai area is 1,800 mm/year, but the rainfall is very unevenly distributed in two seasons. The rainy season starts from May and ends at the end of October. The rainfall in the rainy season accounts for 85% of the annual rainfall. The dry season starts from November and ends in April. The average rainfall is very low (mostly of less than 100 mm/month). The rainfall in the dry season accounts for 15% of the annual rainfall.

The uneven distribution of the annual rainfall usually means large drainage rates from the open pit mines in the rainy season and low drainage rates in the dry season. The drainage from the underground mines is partly independent of the annual rainfall distribution. Surpluses after deduction of mine-internal water uses occur mainly in the rainy season.

The water supply in Quang Ninh province, and particularly in Hon Gai, is operated by the Quang Ninh Clean Water Joint Stock Company (QUAWACO). The project area Hon Gai is supplied with approximately 20,000 m^3 per day.

The water is taken from the Cao Van reservoir. The water resource of the reservoir is limited and the supply is difficult in the dry season. The current water resources plan of Quang Ninh province prognoses a water deficit for Ha Long City including Hon Gai during the dry season (DONRE Quang Ninh 2016).

Against the background of the growing water demand in Hon Gai due to the increasing tourism in the form of numerous newly built apartments, the use of additional water resources will be essential for the further development of Hon Gai.

METHODOLOGY

The investigation for mine water management using the example of the Hon Gai hard coal mining area under the existing continuously changing spatio-temporal conditions consists of the following work steps:

1. **System analysis**: identification of the system boundaries and the relevant system elements of mine waters in the project area (mine drainage water, mine water treatment, mine-internal water uses, potential mine-external water uses), defining of interconnections of the mine water infrastructure and GIS based visualization.
2. **Spatio-temporal investigation:** investigation of the mining and the water amounts in the project area for the existing situation, the situation 2021–2026, and the final situation from 2030 onwards. The spatio-temporal investigation is performed by using Geoinformation System (GIS).
3. **Material flow analysis (MFA):** based on the system analysis and the spatio-temporal investigation, material flows (mine drainage water, mine water treatment, auxiliary material, sludge, etc.) for the existing situation in 2016, the situation 2021–2026, and 2030 onwards are estimated. The MFA is done by using the software Umberto® NXT Efficiency (ifu 2016).
4. **Recommendations:** the recommendations for the future water management are derived. The recommendations refer to general options for development of the mine water management and water delivery for mine-external water uses as a basis for general decisions. Based on such general decisions, further technical and economic studies should be carried out.

The results form the basis for developing strategies in the utilization of treated mine water in the Hon Gai peninsula, Vietnam. The results of this study can be transferred to other coal mining areas in Vietnam or areas with similar climatic and socio-economic conditions.

RESULTS AND DISCUSSION

The results of the single above-mentioned work steps are reported and discussed in detail.

System analysis

A detailed system analysis was performed to define the system boundaries, the relevant system components, and the system interconnections. In the area of the individual mines, this system differs in detail; the generalized result of the system analysis is shown in Figure 2.

Figure 2 shows the following main mine water-related system components and interactions.

The surface runoff and (possible) treatment is the objective of another WaterMiner part project (Ulbricht *et al.* 2018; Ulbricht & Bilek 2020).

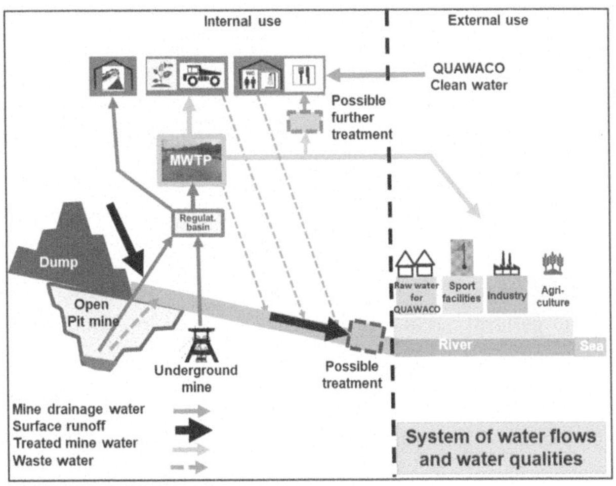

Figure 2 | Generalized mine water system.

The mine water sources (mine drainage water), mine water treatment plants (MWTP), mine-internal water use and the potential mine-external water use are objectives of this study. The mine water sources include also clean water from the water supply system (QUAWACO, drinking water). Another component of the mine water system is also wastewater, e.g., from coal screening, car washing, domestic wastewater, etc.

The mine drainage water flows via regulating basins to the mine water treatment plants (MWTPs) and is treated there mainly by removing iron and manganese and, by compensation, the acidity. In the rainy season, and especially during heavy rainfall events, mine drainage water is only partially treated or discharged into the surface waters without treatment.

Existing mine-internal and potential mine-external uses of mine water are also included in the system description. Mine water from the regulating basin without treatment is used for coal screening. Mine water after treatment is used mine-internal for irrigation of recultivation areas, truck washing, and dust mitigation. With further treatment, mine water also can be used for additional mine-internal services like drinking water or in canteens and can substitute the water from QUAWACO.

Potential mine-external water uses are shown in the right-hand part of Figure 2: raw water for QUAWACO for further treatment, water for irrigation of agriculture and sport facilities, water for industry (e.g., cooling water for the thermal coal power plant in Hon Gai).

Spatio-temporal investigation

The mining process causes continuous spatio-temporal changes by closing mines (open pit mines), opening mines (underground mines), by dumping mine waste material, by refilling open pit mines, by revegetation and reuse of abandoned mining areas. This study focuses on the resulting effects on the sources and quantities of mine water.

The general spatial mine water situation, shown in Figure 2, varies in the different drainage units (DUs) of the mining area. The DUs, shown in Figure 3, are five small water basins within the mining area belonging to different mines or groups of mines: Dien Vong, Lo Phong, Ha Tu, Cai Da, and Suoi Lai.

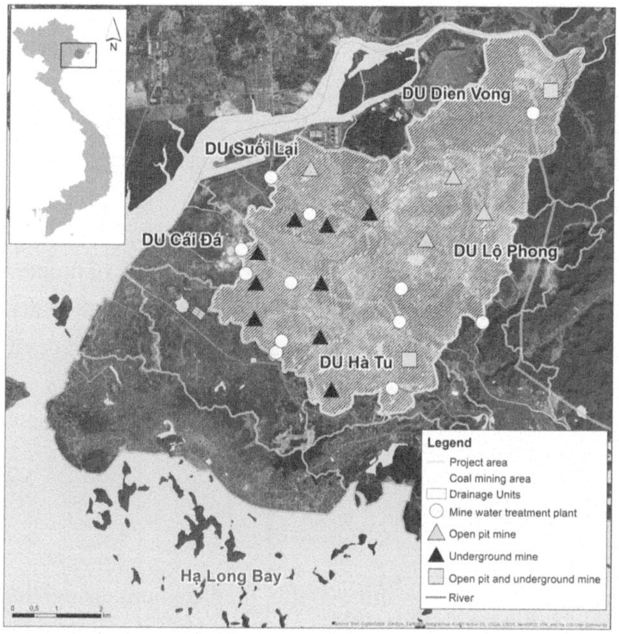

Figure 3 | Drainage units (DUs), mine water sources (open pit and underground mines), and mine water treatment plants (MWTPs).

The relevant temporal mine water related situations are defined as follows:

- Existing situation at the project start (2016)
- Intermediate situations (2021, 2026)
- Final situation (2030 onwards).

The existing situation characterizes the situation when the project was established. The intermediate situations were selected due to significant changes in mine drainage water volumes and the enforcement of environmental legislation. The final situation is the year in which the current National Coal Mining Master Plan ends (GoV 2016).

Within the five DUs, 12 mine water treatment plants (MWTPs) are located (see Figure 3). Table 1 shows the number of MWTPs in the related DUs according to the defined temporal situations.

In the existing situation, 12 MWTPs are operating in the mining area. The number will be reduced to six MWTPs by the year 2030. The proportion of treated mine water averages 74% (VINACOMIN 2014). Especially during heavy rainfall events, mine water is only partially treated. This is improved particularly by extending MWTPs and modifying the operation management.

Table 1 | Number of mine water treatment plants within the drainage units

Drainage unit	2016 (MWTPs)	2021 (MWTPs)	2026 (MWTPs)	2030 (MWTPs)
Cai Da	5	5	3	2
Suoi Lai	2	2	2	2
Ha Tu	1	1	1	1
Lo Phong	3	3	0	0
Dien Vong	1	1	1	1
Total	**12**	**12**	**7**	**6**

As mine operations close down gradually, the mine drainage water will decrease, the MWTPs will reduce or end their operation and the water consumption for mining purposes will reduce. At the same time, along with the expansion of industry, urban areas, and tourism, the water demand of external users will increase.

Material flow analysis (MFA)

The Umberto® NXT Efficiency software (ifu 2016) is applied to identify, quantify, and assess the potentials in recycling and reuse of mine water in the individual DUs and summarize for the entire mining area. The investigated materials are water, auxiliary substances, and sludge (Broemme *et al.* 2018). As well, energy and costs are considered by the material flow analysis and by a separate study (Do *et al.* 2018).

Figure 4 shows the general structure of the material flow models, as they were built in a modified way for each DU. The material flow (mainly water flow) takes place from the left side from mine water sources, via mine water treatment to mine-internal and mine-external water use.

According to the respective structures of the water flows within the individual DUs, the water flows were recorded and modeled by the Umberto software. The basis was the quantity data provided by the mines. The quantities were combined into a total flow of the year 2016 in order to determine plausible proportions of partial flows: starting with the mine drainage water quantities, through the mine water treatment quantities, to the internal water uses. On this basis, simulations for future changes in mine drainage water quantities and thus also for the determination of water surpluses for external water uses were possible.

In addition to this quantity consideration, cost sharing is also possible by linking the costs of energy, auxiliary substances, maintenance costs, etc.

For all five DUs the relevant water-related components and interconnections were identified and the material balances of the components were estimated and implemented in the Umberto model (see Figure 4):

- Mine water sources (mine drainage water)
- Mine water treatment
- Mine-internal water uses
- Mine-external water uses.

Based on the existing situation in 2016, the defined future situations (2021, 2026, 2030) are simulated. The results were compiled in the form of water flow diagrams

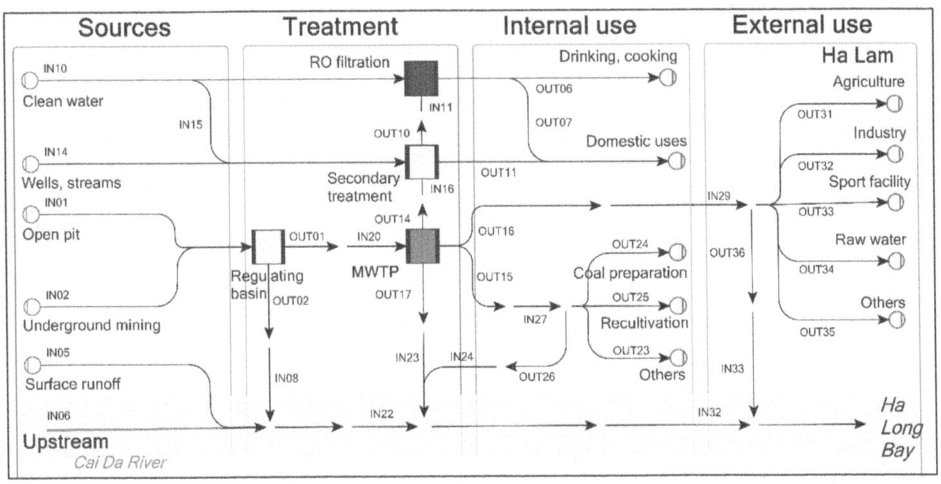

Figure 4 | Generalized material flow model (IN = input; OUT = output (section of the complete model, therefore numbering not continuous)).

(Sankey diagrams) and water balance tables for the individual DUs. Here, the results are presented in summarized form.

Within the Umberto model, the treated mine water volume is seen as the water source to cover the mine-internal water uses and as a possible complementary source for mine-external water uses.

Table 2 shows the summarized average amounts of treated mine water for the investigated spatio-temporal situations. From 2016 to 2021, there is an increase of the treated mine water volumes as all the large open pit mines are still active and the MWTPs further increase their operation to fulfill the stricter environmental regulations. After 2021, most of the open pit mines end their operation and the total mine water volume decreases, as the drainage volumes of underground mines are smaller.

Figure 5 illustrates the amounts of treated mine water within the DUs. The treated amounts within the individual DUs are each part of the total amount of treated mine water within the Hon Gai mining area.

In the following, the treated mine water amounts are compared with the mine-internal and the mine-external water demands. The amounts for potential mine-external water uses are calculated after deduction of mine-internal uses, if there is a surplus, mine-external water demands can be covered.

The annual mine-internal water demands (see Table 3) take into account all water demands for mining activities such as coal selecting, dust control, truck washing, revegetation, as well as domestic uses for mine staff (e.g., canteens, laundry, bathrooms). From the perspective of mining, the mine-internal water uses have priority.

Table 2 | Results of the material flow model: average amounts of treated mine water

Drainage unit	2016 (m³/d)	2021 (m³/d)	2026 (m³/d)	2030 (m³/d)
Cai Da	27,000	20,000	19,000	22,000
Suoi Lai	6,000	21,000	14,000	14,000
Ha Tu	12,000	34,000	27,000	27,000
Lo Phong	13,000	7,000	0	0
Dien Vong	3,000	7,000	11,000	11,000
Total	**61,000**	**89,000**	**71,000**	**74,000**

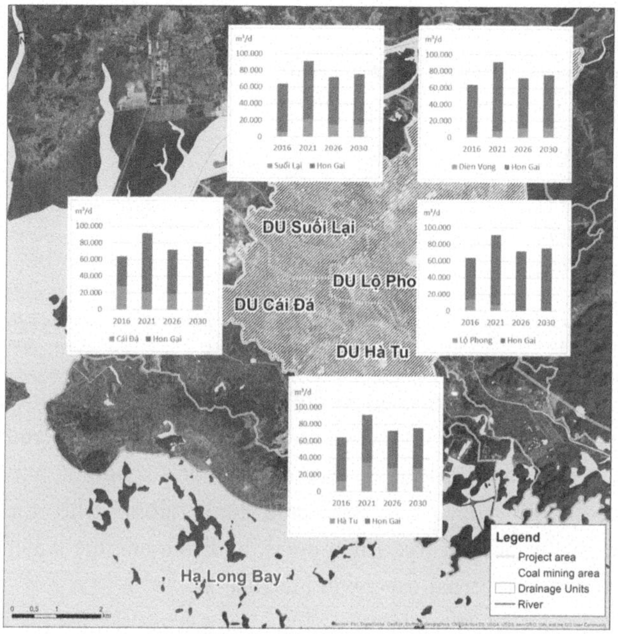

Figure 5 | Treated mine water within the drainage units as part of the summarized treated mine water within the Hon Gai mining area.

Table 3 | Results of the material flow simulation: average amount of mine-internal water uses

Drainage unit	2016 (m³/d)	2021 (m³/d)	2026 (m³/d)	2030 (m³/d)
Cai Da	5,000	3,000	3,000	3,000
Suoi Lai	1,000	1,200	1,000	1,000
Ha Tu	4,500	4,000	3,000	3,000
Lo Phong	3,000	3,000	0	0
Dien Vong	1,000	1,000	1,000	500
Total	**14,500**	**12,200**	**8,000**	**7,500**

Mine-external water demands (see Table 4) are demands for domestic water supply, tourism and public services, irrigation of urban green and golf courses, road cleaning, etc. With the rapid development of Ha Long City, according to the urban planning of housing areas (2010: 230,000 inhabitants; 2030: 350,000 inhabitants), of recreation areas, tourism (2010: 2 million international tourists, 3 million national tourists), and of services and industries, the water demand of Ha Long City, respectively Hon Gai, will apparently increase, requiring substantial additional water sources in the future.

Table 4 shows the expected development of the annual mine-external water demands of the settlement area of

Table 4 | Expected mine-external water demand in Hon Gai (VIUP 2013), amount for 2026 extrapolated by the author

Mine-external water demand	2016 (m^3/d)	2021 (m^3/d)	2026 (m^3/d)	2030 (m^3/d)
Domestic	16,000	23,000		37,000
Irrigation urban green, golf courses	1,300	2,000		3,000
Tourism, public services	1,300	2,000		4,000
Industry	2,000	2,000		2,000
Leakage	3,000	5,000		7,000
Water supply plant (QUAWACO)	1,000	2,000		2,500
Total	**24,600**	**36,000**	**46,000**	**55,000**

Hon Gai area according to the 'Review of the construction plan of Ha Long City' carried out by the Vietnam Institute for Urban and Rural Planning (VIUP 2013). The sum for 2026 was extrapolated by the authors from 2021 and 2030.

Table 5 compares the surplus amounts of treated mine water (treated mine water minus mine-internal water uses) with the mine-external water demands and shows that there is normally a surplus of treated mine water.

The comparison shows, that the mine water can generally cover parts of the mine-external water demand, particularly for the later spatio-temporal situations (2021, 2030). However, seasonal variations of the water amounts have to be considered, e.g., less available water during the dry season and water storage in the rainy season.

This general result needs to be viewed critically and further specified with regard to the question of water quality for the different applications and with regard to the seasonal variation of precipitation.

Table 5 | Surplus amounts of treated mine water, compared with mine-external water demand

Water amounts	2016 (m^3/d)	2021 (m^3/d)	2026 (m^3/d)	2030 (m^3/d)
1. Treated mine water	61,000	89,000	71,000	74,000
2. Mine-internal water demand	14,500	12,200	8,000	7,500
3. Surplus water amount[a]	**46,500**	**76,800**	**63,000**	**66,500**
Mine-external water demand	24,600	36,000	46,000	55,000

[a]The surplus water amount is the difference between 1. Treated mine water and 2. Mine-internal water demand.

It should be considered that the treated mine water complies with the Vietnamese water quality standard QCVN 40 (2011) and is usable, e.g., for irrigation of urban green or some industrial purposes. For drinking water purposes, a further treatment is needed according to the Vietnamese Quality standard for drinking water (QCVN 01 2009).

The water amounts in Tables 3–5 are averages. The seasonal variability of the water amounts must be taken into account by establishing water reservoirs, e.g., in the form of small reservoirs to store the surplus water in the rainy season and to compensate for the water deficit in the dry season.

A potential additional water resource can be the residual lake in the area of Lo Phong DU if it is established.

Recommendations

The WaterMiner project results are conceptual as a basis for general considerations and decisions. The following detailed studies show the feasibility of selected alternatives. Treated mine water for mine-external uses can be delivered at the boundary of the mining area to the public water supply (e.g., QUAWACO).

The selected alternatives are in the form of:

- Decentralized delivery of treated mine water (quality standard according to QCVN 40 2011) from the different Dus.
- Centralized or semi-centralized delivery of treated mine water (quality standard according to QCVN 40 2011) using collection pipes.
- Decentralized delivery of further treated mine water for irrigation (according to QCVN 39 2011).
- Decentralized delivery of further treated mine water for drinking water purposes (according to QCVN 01 2009).

The selected alternatives of delivery have to be investigated by the mine owner with the question: what is possible from the point of view of mining and water occurrence and, on the other hand, by the potential water user (e.g., QUAWACO) with the question: what is possible from the view of the technical infrastructure and distribution of water demands in Hon Gai.

It is recommended the two parties have a detailed discussion for finding suitable alternatives, to start further investigations of feasibility, and to set up a road map for further investigation and planning steps.

CONCLUSIONS

The R&D project WaterMiner shows the usability of the implemented research steps: system analysis, spatio-temporal investigation, material flow analysis, and deriving of recommendations of optional concepts of mine water reuse under the spatio-temporally changing conditions of mining.

The developed procedure enables a flexible prediction and management of mine water flows. In particular, the material flow modeling is an approach, which can be adapted to the sometimes quickly changing situations of mining.

DATA AVAILABILITY STATEMENT

All relevant data are included in the paper or its Supplementary Information.

REFERENCES

Broemme, K. & Stolpe, H. 2015 *Handbook Mining and Environment in Vietnam.* Handbook on the results of the project 'Mining and environment in Vietnam 2005-2015', Bochum, Germany, p. 535. doi:10.2314/GBV:868016799.

Broemme, K., Stolpe, H., Jolk, C., Greassidis, S., Borgmann, A. & Zindler, B. 2014 Development of a methodology for post-mining land use planning in Quang Ninh, Vietnam. In *Hanoi University of Mining and Geology: Proceedings of the Third International Conference on Advances in Mining and Tunneling*, 21–22 October 2014, Vung Tau. Publishing House for Science and Technology, Hanoi, pp. 526–534.

Broemme, K., Trinh, V. Q., Greassidis, S. & Stolpe, H. 2018 Material flow analysis for spatiotemporal mine water management in Hon Gai, Vietnam. In *Proceedings of the International Mine Water Association Congress IMWA*, 10–14 September 2018, Pretoria, South Africa.

Do, H. H., Trinh, V. Q. & Froer, O. 2018 The Economic Benefit Of Mine Water In Halong Peninsula, Vietnam. In: *Proceedings of the International Mine Water Association Congress IMWA*, 10–14 September 2018, Pretoria, South Africa.

DONRE Department of Natural Resources and Environment Quang Ninh 2016 *Quy Hoạch tài Nguyên Nước Tỉnh Quảng Ninh đến năm 2020, Định Hướng đến năm 2030 (Water Resources Plan of the Province Quang Ninh Until 2030, Strategy Until 2030)*, Quang Ninh, Vietnam.

GoV 2016 *Decision No. 403/QĐ-TTg Dated 16 March 2016, on Adjusted Master Plan for Vietnam's Coal Industry Development by 2020, with Perspective to 2030*, Hanoi, Vietnam.

ifu 2016 *Umberto® NXT Efficiency (v7.1.13)* User Manual. Hamburg, Germany. www.ifu.com; www.umberto.de

QCVN 01 2009 *National Technical Regulation on Drinking Water Quality.* QCVN 01:2009/BYT, Hanoi, Vietnam.

QCVN 39 2011 *National Technical Regulation for Water Quality for Irrigation.* QCVN 39:2011/BTNMT, Hanoi, Vietnam.

QCVN 40 2011 *National Technical Regulation on Industrial Waste Water*, QCVN 40:2011/BTNMT, Hanoi, Vietnam.

Ulbricht, A. & Bilek, F. 2020 Planning and assessment of water and sediment retention measures within a mining area based on a rainfall-runoff and sediment transport model. *Journal of Water Reuse and Desalination.* (this issue).

Ulbricht, A., Bilek, F. & Broemme, K. 2018 Development of a technical concept of spatial and temporal coordinated mine water recycling exemplified by a mining area with urban influence. In *Proceedings of the International Mine Water Association Congress IMWA*, 10–14 September 2018, Pretoria, South Africa.

VINACOMIN 2013 Đề án bảo vệ môi trường vùng than Quảng Ninh đến 2020, định hướng 2030 (Project on environmental protection in Quang Ninh coal area till 2020, orientation of 2030), Ha Long City, Vietnam.

VINACOMIN 2014 *Evaluation of the Effectiveness of Technology, Techniques and Management for the Mine Water Treatment Stations*, Ha Long City, Vietnam.

VIUP Vietnam Institute for Urban and Rural Planning 2013 Điều chỉnh quy hoạch chung xây dựng Thành phố Hạ Long – tỉnh Quảng Ninh đến năm 2030, tầm nhìn ngoài năm 2050 (Review of the construction plan of Ha Long City – Quang Ninh province until 2030, vision after 2050), Hanoi, Vietnam.

VN PC Ha Long City 2014 Environmental protection planning for Ha Long City until 2020, vision to 2030, Ha Long City, Vietnam.

First received 6 May 2020; accepted in revised form 10 November 2020. Available online 7 December 2020

IWA Publishing's authorised EU representative for General Product Safety Regulations is Diane D'Arras, 15 rue Duret, 75116 Paris, France, e-mail: safety@iwap.co.uk.

Printed and bound by CPI Group (UK) Ltd, Croydon, CR0 4YY
12/05/2026
02108868-0003